BREAKUP OF LIQUID SHEETS AND JETS

This book is an exposition of what we know about the physics underlying the onset of instability in liquid sheets and jets. Wave motion and breakup phenomena subsequent to the onset of instability are also carefully explained. Physical concepts are established through rigorous mathematics, accurate numerical analyses, and comparisons of theory with experiment. Exercises provided for students new to the subject help familiarize the reader with the required mathematical tools.

This book further provides a rational basis for designing equipment and processes involving the phenomena of sheet and jet breakup. Researchers interested in the transition to turbulence, hydrodynamic stability, or combustion will find this book a highly useful resource, whether their backgrounds lie in engineering, physics, chemistry, biology, medicine, or applied mathematics.

S. P. Lin is Professor of Mechanical and Aeronautical Engineering at Clarkson University. He is the author of more than 100 significant scientific papers.

BREAKUP OF LIQUID SHEETS
AND JETS

S. P. LIN

Clarkson University,
Department of Mechanical and Aeronautical Engineering

CAMBRIDGE
UNIVERSITY PRESS

CAMBRIDGE UNIVERSITY PRESS
Cambridge, New York, Melbourne, Madrid, Cape Town, Singapore,
São Paulo, Delhi, Dubai, Tokyo, Mexico City

Cambridge University Press
The Edinburgh Building, Cambridge CB2 8RU, UK

Published in the United States of America by Cambridge University Press, New York

www.cambridge.org
Information on this title: www.cambridge.org/9780521806947

First published 2003

A catalogue record for this publication is available from the British Library

ISBN 978-0-521-80694-7 Hardback

To my wife, Charlotte
daughter, Anna
and son, Martin

Contents

Notation List

Symbol	Usual meaning	Any exception
a	nozzle inner radius	initial disturbance amplitude
c	speed of sound	
d_0	nozzle diameter	
d	displacement from free surface	
f	dimensionless free surface displacement	
$\mathbf{g}$	gravitational acceleration	
h	dimensionless half sheet thickness	enthalpy
$(\mathbf{i}, \mathbf{j}, \mathbf{k})$	unit vectors in (x, y, z) direction	
k	complex wave number in flow direction	
$\mathbf{k}$	complex wave vector	
l	radius ratio, thickness ratio	characteristic length
n	wave number in θ-direction	
$\mathbf{n}$	unit normal vector	
o	magnitude as small as	
p	dimensionless pressure	
$\mathbf{r}$	dimensionless position vector	
(r, θ, z)	dimensionless cylindrical coordinates	

s	distance along surface	
t	time	
$\mathbf{v} = (u, v, w)$	dimensionless velocity vector in (x, y, z) direction	
(x, y, z)	dimensionless Cartesian coordinates	
$A = Re/(WeQ)^{1/2}$	Taylor parameter	
B_0	Bonds number $= We/Fr$	
C_a	capillary number $= We/Re$	
C_p	constant pressure specific heats	
C_v	constant volume specific heats	
$\mathbf{D}$	characteristic matrix	
D	characteristic determinant	
$Fr = W_0^2/gH_0$	Froude number	
$\mathbf{G}$	inverse Fourier transform of disturbance	
$\mathbf{H}$	position vector of solid surface	amplitude vector
H_i	dimensional half ith layer thickness	
H_0	characteristic length	
$\mathbf{I}$	identity matrix	Fourier integral
$J = \left[SH_0/(\rho_1\nu^2)\right]^{1/2}$	Ohnesorge number	
K	adiabatic module of elasticity	initial rate of axial stretching
L	characteristic length	intact length
M	kinematic viscosity ratio	Mach number W_1/c_2
$Ma = S_{,T} T_0/\rho_1 U_0^2 H_0$	Marangoni number	
M_i	Mach number in fluid i	
$N = \bar{N}_2 = \mu_r$	dynamic viscosity ratio	
$N_j = \nu_j/\nu_1$	kinematic viscosity ratio	

$\bar{N}_\alpha = \mu_\alpha/\mu_1$	dynamic viscosity ratio	
O	magnitude as large as	
$Oh = \left[SH_0/\left(\rho_1 v_1^2\right)\right]^{1/2} = J$	Ohnesorge number	
P	pressure	
$Q = Q_2 = \rho_r$	gas to liquid density ratio	
$\dot{Q}$	volumetric flow rate	
(R, θ, Z)	dimensional cylindrical coordinate	
$Re = \rho_1 U_0 H_0/\mu_1$	Reynolds number	
$St_0 = Re/Fr$	Stokes number	
S	surface tension	
T	temperature	period of oscillation
U_0	characteristic velocity	
$\mathbf{V} = (U, V, W)$	dimensional velocity vector	
$We = \rho_1 U_0^2 H_0/S$	Weber number	
α	wave number in flow direction	
β	wave number in direction perpendicular to flow	
$\gamma = C_p/C_v$	specific heat ratio	swirl number $= \Gamma/R_0 W_1$
δ	Dirac delta function	
ε	small parameter	
τ	dimensionless time	
$\mathcal{T}$	deviatory stress	
ς	dimensionless sheet or jet thickness	
η	dimensionless free surface displacement	
θ	azimuthal angle	phase angle, spray angle
κ	mean curvature	
λ	wavelength	
μ	dynamic viscosity	
ν	kinematic viscosity	

ρ	density	
ψ	stream function	
ϕ	velocity potential	amplitude of ψ
σ	stress tensor	
$\omega = \omega_r + i\omega_i$	complex wave frequency	dimensional frequency
Ω	dimensional frequency	
∇	gradient operator	
Γ	circulation	

Superscripts

$\bullet$	time rate of change	
T	transpose	
$'$	perturbation	differentiation
$\hat{}$	amplitude	

Subscripts

adj	adjoint	
$,$	partial differentiation	
i	inner surface	ith interface
o	reference quantity	outer surface
A, B	fluids A, B	
1, 2, 3	fluids 1, 2, 3	
α	αth layer	
l	liquid	
g	gas	

Preface

The phenomena of the breakup of liquid sheets and jets are encountered in nature as well as in various industrial applications. A good understanding of these phenomena requires a sound basic scientific knowledge of the dynamics of flows involving interfaces between different fluids. This book is the outcome of the author's inquiry into this fundamental knowledge. My understanding of the subject matter has been consolidated gradually through direct and indirect collaborations with my students and colleagues. The objective and scope of this book in the context of related existing works are explained in Chapter 1. Chapters 2 to 5 are devoted to exposition of the onset of sheet breakup. Chapters 6 to 10 discuss jet breakup. A perspective of the challenging aspects of the subject, including the nonlinear evolution subsequent to the onset of instability and nanojets, is sketched in Chapter 11. Some additional topics related to the breakup of a liquid body into smaller parts are discussed in the epilogue. Readers are expected to have the equivalent of at least an undergraduate background in science or engineering. In the theoretical development I have strived for mathematical rigor, numerical accuracy, and rational approximation. However, mathematics has not been used just for the sake of mathematics. I have depended on comparisons between different theories and experiments to establish physical concepts. Practical applications of the concepts are pointed out in appropriate places. The references relevant to each chapter are listed at the end of the chapter. Exercises given at the end of each section or chapter facilitate the exposition. This book has been written with the intention of providing a sound scientific basis for rationally applying the principles to many industrial processes involving jet and sheet breakups. It is hoped that readers who are interested mainly in scientific understanding as well as those interested in creative applications of principles will find this book useful.

Many illustrative figures in Chapters 2 and 3 were obtained by Wei-Yuan Jiang. Financial sources of support for my work presented in this volume can be found in the referenced publications and thus will not be acknowledged here.

1

Introduction

1.1. Overview

When a dense fluid is ejected into a less dense fluid from a narrow slit whose thickness is much smaller than its width, a sheet of fluid can form. When the fluid is ejected not from a slit but from a hole, a jet forms. The linear scale of a sheet or jet can range from light years in astrophysical phenomena (Hughes, 1991) to nanometers in biological applications (Benita, 1996). The fluids involved range from a complex charged plasma under strong electromagnetic and gravitational forces to a small group of simple molecules moving freely with little external force. The fluid sheet and jet are inherently unstable and breakup easily. The dynamics of liquid sheets was first investigated systematically by Savart (1833). Platou (1873) sought the nature of surface tension through his inquiry of jet instability. Rayleigh (1879) illuminated his jet stability analysis results with acoustic excitation of the jet. In some modern applications of the instability of sheets and jets, it is advantageous to hasten the breakup, but in other applications suppression of the breakup is essential. Hence knowledge of the physical mechanism of breakup, aside from its intrinsic scientific value, is very useful when one needs to exploit the phenomenon to the fullest extent. Recent applications include film coating, nuclear safety curtain formation, spray combustion, agricultural sprays, ink jet printing, fiber and sheet drawing, powdered milk processing, powder metallurgy, toxic material removal, and encapsulation of biomedical materials. Current applications can be found in the annual or biannual conference proceedings of several professional organizations, such as the International Conference on Liquid Atomization and Spray Systems (ICLASS) and the Institute for Liquid Atomization and Spray Systems (ILASS) organizations in the Americas, Europe, and Asia, and European and American Coatings Conferences.

Because of the diverse applications, books on the subject tend to focus on specific applications. For example, the book by Lefebvre (1989) centers

1

around internal combustion, and that of Masters (1985) focuses on powdered milk formation. Intended for immediate practical applications, these books rely heavily on phenomenological correlations. The book by Yarin (1993) provides a mathematical treatment of recent applications involving non-Newtonian fluids. In contrast, this book deals exclusively with Newtonian fluids, which are encountered in most of the known applications. It does not cover such topics as atomization and emulsification of liquid in liquid (Kitamura and Takahashi, 1986; Grandzol and Tallmadge, 1973; Villermaux, 1998; Richards, Beris, and Lenhoff, 1993). Electromagnetic effects on the jet breakups (Balachandran and Bailey, 1981), or the electromagnetic effects on atomization and drop formation (Bailey, 1998; Fenn et al., 1989).

We address first the issue of the origin of the breakup or the physical reasons for the breakup. Therefore the mathematical tool used is linear stability analysis, which predicts the onset of jet and sheet instability. The disturbance consisting of all Fourier components is allowed to grow both spatially and temporally in the sheet or jet flows. If only the classical temporally growing disturbance is considered, one arrives at a paradoxical situation as illustrated in the first section of the next chapter. The onset of instability appears to largely dictate the ultimate outcome of the breakup, as exemplified by Rayleigh's linear stability analysis of a liquid jet. However, the detailed process leading to the eventual breakup requires nonlinear theories to describe. Nonlinear descriptions are given in Chapter 11. The results related to the last stage of breakup and topics that still need further development will be addressed in the Epilogue.

1.2. Governing Equations

The governing equations and the corresponding boundary conditions listed below will be referred to in subsequent chapters. Their derivation can be found in standard text books, some of which are given at the end of the chapter. The same notation will be used to denote the same physical variable throughout the book, with few exceptions. When such exceptions on notation take place they will be pointed out; otherwise the same symbol will not be redefined after its first appearance. A list of notations is provided at the front of the book.

Newton's second law of motion applied to a fluid particle gives

$$\rho \frac{D\mathbf{V}}{Dt} = \mathbf{g} + \nabla \cdot \boldsymbol{\sigma}, \tag{1.1}$$

$$\frac{D\mathbf{V}}{Dt} \equiv \mathbf{V}_{,t} + \mathbf{V} \cdot \nabla \mathbf{V},$$

where ρ is the fluid density, $\mathbf{V}$ is the velocity vector, and t is the time. The subscript variable following a comma signifies partial differentiation with that variable, D/Dt is the substantial derivative as defined, ∇ is the gradient operator, $\mathbf{g}$ is the gravitational acceleration, and $\boldsymbol{\sigma}$ is the stress tensor. For an incompressible Newtonian fluid

$$\boldsymbol{\sigma} = -P\mathbf{I} + \mu[\nabla\mathbf{V} + (\nabla\mathbf{V})^T], \tag{1.2}$$

where $\mathbf{I}$ is the identity matrix, μ is the dynamic viscosity, P is the pressure, and the superscript T denotes transpose.

The conservation of mass requires

$$\frac{D\rho}{Dt} + \rho\nabla\cdot\mathbf{V} = 0. \tag{1.3}$$

For an incompressible fluid $D\rho/Dt = 0$, and (1.3) is reduced to

$$\nabla\cdot\mathbf{V} = 0. \tag{1.4}$$

Equations (1.1) to (1.4) are valid for each fluid involved in a flow. The i-th interface between two adjacent fluids is infinitesimally thin and is mathematically defined by a function $F_i(\mathbf{r}, t) = 0$, $\mathbf{r}$ being the position vector. The balance of forces exerted on a unit area of interface gives

$$S_i\nabla\cdot\mathbf{n} + [\mathbf{n}\cdot\boldsymbol{\sigma}\cdot\mathbf{n}]_{B_i}^{A_i} + \nabla_{\prime\prime}S_i = 0, \tag{1.5}$$

where S is the interfacial tension, $\mathbf{n}$ is the surface unit normal vector positive if pointed from fluid B_i to fluid A_i on the opposite side, $\nabla_{\prime\prime}$ is the surface gradient operator, and

$$[\mathbf{n}\cdot\boldsymbol{\sigma}\cdot\mathbf{n}]_{B_i}^{A_i} \equiv \mathbf{n}_i\cdot\boldsymbol{\sigma}_{A_i}\cdot\mathbf{n}_i - \mathbf{n}_i\cdot\boldsymbol{\sigma}_{B_i}\cdot\mathbf{n}_i,$$
$$\mathbf{n}_i = \nabla F_i/|\nabla F_i|.$$

For viscous fluids, the kinematic condition at the interface is

$$[\mathbf{V}]_{B_i}^{A_i} = 0, \tag{1.6}$$

$$W_i = \frac{DF_i}{Dt}, \tag{1.7}$$

where W_i is the component of the i-th interfacial velocity in the direction in which the distance F_i from a reference position to the interface is measured. If a fluid is inviscid, then (1.6) does not hold, and (1.7) must be applied for each fluid separately. A viscous fluid sticks to a nonpermeable solid surface, and thus $\mathbf{V} = 0$ at the solid-viscous fluid interface. If the fluid is inviscid, then it is allowed to slide along the solid surface, but is not allowed to penetrate it. Derivations of Equations (1.1) to (1.7) can be found in the books on fundamental fluid mechanics cited in the references section at the end of the chapter. Note that non-Newtonian fluids as well as more general interfacial conditions allowing phase changes to take place are not treated in this work.

1.3. Dimensionless Parameters

Even for simple Newtonian fluids, the number of dimensionless groups involved in interfacial fluid dynamics is relatively large. To bring out the relevant dimensionless parameters, we nondimensionalize the governing differential system. Identifying the characteristic velocity and length with U_0, length with H_0, time with H_0/U_0, and stress with $\rho_1 U_0^2$, where ρ_1 is the density of the fluid designated by subscript 1, we have the following dimensionless governing equations for incompressible Newtonian fluids:

$$Q_i \frac{\mathbf{D}\mathbf{v}_i}{\mathbf{D}\tau} = \frac{Q_i}{Fr} - \nabla p_i + \frac{N_i}{Re}\nabla^2 \mathbf{v}_i, \tag{1.8}$$

$$\nabla \cdot \mathbf{v}_i = 0, \tag{1.9}$$

kinematic interfacial condition,

$$w_i = h_{i,\tau} + \mathbf{v}_i \cdot \nabla h_i, \qquad h_i = F_i/H_0,$$

dynamic interfacial condition,

$$We_i^{-1}\nabla \cdot \mathbf{n}_i = [\mathbf{n} \cdot \boldsymbol{\tau} \cdot \mathbf{n}]_{B_i}^{A_i}, \tag{1.10}$$

and the no-slip condition at the solid wall at $\mathbf{H}/H_0$, where $\mathbf{H}$ is the position vector defining the solid wall. The lower case letters are used to denote dimensionless variables corresponding to their dimensional counterparts expressed in capital letters, except for τ and $\boldsymbol{\tau}$, which are dimensionless time and stress respectively. The dimensionless groups revealed in these equations are

$$\begin{aligned} &\text{density ratio } Q_i = \rho_i/\rho_1, \\ &\text{viscosity ratio } N_i = \mu_i/\mu_1, \\ &\text{Reynolds number } Re = \rho_1 U_0 H_0/\mu_1, \\ &\text{Froude number } Fr = \rho_1 U_0^2/g H_0, \\ &\text{Weber number } We = \rho_1 U_0^2 H_0/S, \\ &\text{geometric parameters } \mathbf{H}/H_0, H_i/H_0. \end{aligned} \tag{1.12}$$

The interface is considered to be homogeneous, otherwise Marangoni numbers associated with $\nabla_{\!s} S$ in (1.5) will arise. The interface is also assumed to be isotropic. The quantitative sensitivity of the dynamics of the flow to the variation of these dimensionless groups will be used to reveal the relative importance of shear, inertial, body, and surface forces in various modes of interfacial instabilities.

Exercises

1.1. Show that if temperature varies along an interface, the surface gradient term in (1.5) leads to the temperature Marangoni number $Ma = S_{,T} T_0/\rho U_0^2 H_0$, where $S_{,T}$ is the change of surface tension per unit change of temperature, T_0 is a reference temperature, and U_0 is a characteristic velocity. If the fluids on both sides of the interface are stationary, what is the relevant expression for U_0?

1.2. If the solute concentration varies along an interface, find the expression of the solute Marangoni number.

1.3. Show that the Bond number $Bo = We/Fr$, the capillary number $C_a = We/Re$, and the Stokes number $St_0 = Re/Fr$ represent respectively the ratios of body force to surface force, viscous force to surface force, and body force to viscous force.

1.4. Show that if $U_0 = 0$, the Ohnesorge number $\equiv [SH_0/(\rho \nu^2)]^{1/2}$ is a parameter representing the ratio of the surface force to the viscous force.

1.5. Show that the mean curvature $\nabla \cdot \mathbf{n}$ in (1.10) at a point on a surface $z = h(x, y, \tau)$ in the Cartesian coordinate (x, y, z) is given by

$$\nabla \cdot \mathbf{n} = -\frac{h_{,xx} + h_{,yy}}{\left(1 + h_{,x}^2 + h_{,y}^2\right)^{3/2}}. \tag{1.11}$$

1.6. Show that the mean curvature of a surface $r = h(z, \theta, t)$ is given by

$$\nabla \cdot \mathbf{n} = \frac{1}{q^2}\left(h_{,z}\, q_{,z} + h_{,\theta}\, q_{,\theta}\,/h^2\right)$$

$$+ \frac{1}{q}\left(\frac{1}{h} + h_{,\theta}^2\,/h^3 - h_{,zz} - h_{,\theta\theta}\,/h^2\right), \tag{1.12}$$

where $q = [1 + (h_{,\theta}/h)^2 + h_{,z}^2]^{1/2}$.

References

Bailey, A. G. 1998. *Electrostatic Spraying of Liquids*. Wiley, New York.
Balachandran, W., and Bailey A. G. 1981. *J. Electrostatics* **10**, 189–196.
Batchelor, G. K. 1967. *An Introduction to Fluid Dynamics*. Cambridge University Press.
Benita, S. 1996. *Microencapsulation*. Marcel Dekker.
Chandrasekhar, S. 1961. *Hydrodynamic and Hydromagnetic stability*. Oxford University Press.

Fenn, J. B., Mann, M., Meng, C. K., Wong, S. F., and Whitehouse, C. M. 1989. *Science* **246**, 64.

Grandzol, R. J., and Tallmadge, J. A. 1973. Water jet atomization of molten steel. *AIChE Journal*, **19**, 1149–1158.

Hughes, P. A. 1991. *Beams and Jets in Astrophysics*. Cambridge University Press.

Joseph, D. D., and Renardy, Y. Y. 1992. *Fundamentals of Two-Fluid Dynamics*. Springer-Verlag.

Kitamura, Y., and Takahashi, T. 1986. Stability of liquid-liquid jet systems. *Encyclopedia of Fluid Mechanics*. **3**, 474–510.

Landau, L. D., and Lifshitz, E. M. 1982. *Fluid Mechanics*. Pergamon Press.

Lefebvre, A. H. 1989. *Atomization and Sprays*, Hemisphere Publishing.

Masters, K. 1985. *Spray Drying Hand Book*. John Wiley & Sons.

Panton, R. L. 1996. *Incompressible Flow*. John Wiley & Sons.

Platou, J. 1873. *Satique Experimentale Et Theoretique Des Liquid Soumie Aux Seuls Forces Molecularies*. Canthier.

Rayleigh, L. 1879. *Proc. Lond. Math Soc.* **10**, 4.

Richards, J. R., Beris, A. N., and Lenhoff, A. M. 1993. Steady laminar flow of liquid-liquid jet at high Reynolds numbers. *Phys. Fluids.* A **5**, 1703–1717.

Savart, F. 1833. *Ann. Chim. Phys.* **54**, 55–87, 113–145.

Savart, F. 1833. *Ann. de Chim. Phys.* **55**, 257–310.

Villermaux, E. 1998. Mixing and spray formation in coaxial jet. *J. Propul. Power.* **14**, 807–817.

Yarin, A. L. *Free Liquid Jets and Films: 1993 Hydrodynamics and Rheology*. John Wiley & Sons.

2

Uniform Inviscid Liquid Sheets

2.1. Temporal Instability

The dynamics of the formation of liquid sheets was studied by Savart (1833). His work was extended by Taylor (1959) who also studied the wave motion on the sheet. The instability of a plane liquid sheet of a constant thickness in the presence of ambient gas was analyzed by Squire (1953), Hagerty and Shea (1955), and Fraser et al. (1962). They considered the instability with respect to the classical temporal growing single Fourier component of the disturbance. Their analysis is outlined here to illustrate how the classical approach might lead to a paradoxical situation and thus to pave the way to the more recent approach adopted in this book.

2.1a. Perturbed Flow

Consider the onset of instability of a uniform flow of constant velocity U_0 in a liquid sheet of a constant thickness $2H_0$. The liquid as well as the ambient gas are assumed to be inviscid and incompressible. Moreover U_0 is so large that the gravitational force in the flow is negligible compared to the inertial force. Hence the terms associated with the Reynolds number and the Froude number in (1.8) can be omitted, and the governing equation is reduced to

$$Q_i \left(\mathbf{v}_{i,\tau} + \mathbf{v}_i \cdot \nabla \mathbf{v}_i \right) = -\nabla p_i, \tag{2.1}$$

$$\nabla \cdot \mathbf{v}_i = 0 \qquad (i = 1, 2), \tag{2.2}$$

where $\mathbf{v}$ in the Cartesian coordinates (x, y, z) has the components (u, v, w). Here $i = 1$ is used to designate the liquid, and 2 the gas. The x-axis is oriented along the flow direction and placed at the mid-depth of the liquid sheet whose thickness is measured along the z-axis (cf. Fig. 2.1). The corresponding kinematic and dynamic boundary conditions at the interface $z = h(x, y, \tau)$ are

Uniform Inviscid Liquid Sheets

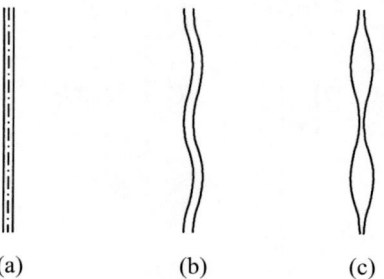

(a)   (b)   (c)

Figure 2.1. Definition sketch. (a) Uniform liquid sheet. (b) Sinuous wave. (c) Varicose wave.

respectively (cf. (1.7), (1.5), and Exercise 1.5)

$$w_i = h,_\tau + u_i\, h,_x + v_i\, h,_y, \tag{2.3}$$

and

$$We^{-1}\, \nabla \cdot \mathbf{n} = p_2 - p_1, \tag{2.4}$$

$$\nabla \cdot \mathbf{n} = -\frac{h,_{xx} + h,_{yy}}{\left(1 + h,_x^2 + h,_y^2\right)^{3/2}}.$$

It is easily verified that the basic flow velocity field $\bar{\mathbf{v}}_1 = (1, 0, 0)$ and $\bar{\mathbf{v}}_2 = (0, 0, 0)$ satisfy the governing Equations (2.1) and (2.2) and the boundary conditions (2.3) and (2.4) if $\bar{p}_2 = \bar{p}_1$ and $h = \bar{h} = 1$. Note that the upper bar will be used henceforth to signify the basic flow whose instability is to be investigated.

To investigate the onset of instability, we perturb the basic flow with disturbances denoted with primes

$$\mathbf{v}_i = \bar{\mathbf{v}}_i + \mathbf{v}'_i = (1 + u', v', w'), \tag{2.5}$$

$$p = \bar{p} + p'. \tag{2.6}$$

Substituting the perturbed flow (2.5) and (2.6) into the governing differential equations, subtracting the basic flow which has already satisfied the differential system, and neglecting the nonlinear terms, we have

$$Q_i\, (\mathbf{v}'_{i},_\tau + \delta_i\, \mathbf{v}'_{i},_x) = -\nabla p'_i, \tag{2.7}$$

$$\nabla \cdot \mathbf{v}'_i = 0. \tag{2.8}$$

2.1b. Two-dimensional Disturbance

For simplicity consider only two-dimensional disturbances for which $v' = 0$. Equation (2.8) is the necessary and sufficient condition for the existence of a stream function such that

$$(u'_i, w'_i) = (\psi_{i,z} - \psi_{i,x}).\qquad(2.9)$$

If we further assume that the disturbance is irrotational, then

$$\mathbf{j} \cdot (\nabla \times \mathbf{v}_i) = \psi_{i,xx} + \psi_{i,zz} = 0.\qquad(2.10)$$

In terms of ψ_i the perturbed pressure according to (2.7) is given by

$$Q_i (\psi_{i,z\tau} + \psi_{i,zx}) = -p_{i,x},\qquad(2.11)$$

$$Q_i \psi_{i,x\tau} = p'_{i,z}.\qquad(2.12)$$

The boundary conditions in terms of ψ_i at the perturbed interface $z = 1 + f'$ are

$$\psi_{1,x} + f'_{1,\tau} + f'_{1,x} = 0,\qquad(2.13)$$

$$\psi_{2,\tau} + f'_{2,x} = 0,\qquad(2.14)$$

$$\mp We^{-1} f'_{,xxx} + \psi_{1,z\tau} + \psi_{1,zx} - Q_2 \psi_{2,z\tau} = 0.\qquad(2.15)$$

To arrive at (2.15), we first differentiate (2.4) with respect to x, substituting (2.11) and (2.12) into the resulting equation. Expanding the variable about $y = 1$ by use of the Taylor series, we then linearize it. The upper and lower signs in (2.15) are respectively for the upper and lower interfaces. To be consistent with the linearization, the boundary conditions (2.13) to (2.15) are to be applied at $z = \pm 1$.

2.1c. Normal Mode Solution

The two independent solutions of (2.10) for any single Fourier component of disturbance in the liquid are $h_s \cosh(kz) e^{i(kx-\omega t)}$ and $h_v \sinh(kz) e^{i(kx-\omega t)}$, where k is the number of waves per unit length, in this case measured in the unit of h_0, and $\omega = \omega_r + i\omega_i$ is the complex wave frequency, the imaginary and real parts of which have respectively the temporal rates of exponential amplification and oscillation both measured in the unit of h_0/U_0. The even solution in z with amplitude h_s leads to the in-phase motion of the upper and lower interfaces and has been termed the sinuous mode. The odd solution displaces the two interfaces antisymmetrically out of phase and has been termed the varicose mode or dilatational mode (cf. Fig. 2.1). The bounded normal Fourier mode solution for the gas phase is $h_2 e^{\mp kz + i(kx - \omega \tau)}$, where the upper and lower signs apply respectively in the regions $z > 0$ and $z < 0$. The corresponding solution for interfacial displacement is $f' = f e^{i(kx - \omega \tau)}$.

First consider the sinuous mode. Because of the symmetry with respect to z of (2.13) and (2.14) and the antisymmetry with respect to z of (2.15), we need to consider only the upper half of the sheet. Substituting the sinuous solution into (2.13)-(2.15), we have

$$\begin{vmatrix} -i(\omega - k) & ik\cosh(k) & 0 \\ -i\omega & 0 & ike^{ik} \\ iWe^{-1}k^3 & ik(\omega - k)\sinh(k) & i\omega Q_2 k e^{-k} \end{vmatrix} \begin{vmatrix} f \\ h_s \\ h_2 \end{vmatrix} = 0, \qquad (2.16)$$

A nontrivial solution of (2.16) for the amplitude vector $[f, h_s, h_2]^T$ exists if the determinant of the coefficient matrix vanishes (Squire, 1953), i.e., if

$$(\omega - k)^2 \tanh(k) + Q_2\omega^2 - We^{-1}k^3 = 0. \qquad (2.17)$$

The derivation of characteristic equations for the varicose mode is left as an exercise. For given values of the parameters We and Q_2, (2.17) relates the characteristic frequency ω of two linearly independent disturbances to the wave number k, which is considered real in this part of analysis. The solution of (2.17) yields for the sinuous mode

$$\omega = \frac{k}{Q_2 + \tanh(k)} \left\{ \tanh(k) \right.$$
$$\left. \pm \left[kWe^{-1}(Q_2 + \tanh(k)) - Q_2 \tanh(k) \right]^{1/2} \right\}. \qquad (2.18)$$

The imaginary part of ω is zero for all k when $Q_2 = 0$. Thus the sheet can sustain an undamped oscillation of disturbances of all wavelengths in an inviscid sheet in the absence of the ambient gas. When $Q_2 \neq 0$, the quantity inside the square root is positive if

$$We < \frac{k}{\tanh(k)} + \frac{k}{Q_2}. \qquad (2.19)$$

The right side of (2.19) is always greater than one. This led Squire (1953) to conclude that the sheet in the presence of the gas cannot be unstable if $We < 1$ because $\omega_i = 0$. This conclusion contradicts the experimental observation of Brown (1961) who observed that when $We < 1$ in a liquid sheet, it tends to rupture. On the other hand if

$$We > \frac{k}{\tanh(k)} + \frac{k}{Q_2}, \qquad (2.20)$$

the disturbance grows exponentially with time everywhere along the liquid sheet at the same rate because $\omega_i > 0$. As $k \to 0$, the right side of (2.20) approaches 1. This led to Squire's instability criterion that the sheet is unstable with respect to long waves when $We > 1$. This conclusion appears to contradict the observation of Crapper, Dombrowski, and Pyott (1975) that a temporally growing disturbance is actually convected downstream in a liquid sheet with $We > 1$, leaving the upstream region relatively unruffled. This paradoxical situation is resolved in the next section.

Exercises

2.1. Show that the characteristic equation for the varicose mode is given by

$$Q_2\omega^2 + (k - \omega)^2 \coth(k) - k^3 We^{-1} = 0, \qquad (2.17a)$$

and the characteristic frequency is given by

$$\omega = \frac{k}{Q_2 + \coth(k)} \left\{ \coth(k) \right.$$
$$\left. \pm \left[kWe^{-1}(Q_2 + \coth(k)) - Q_2 \coth(k) \right]^{1/2} \right\}.$$

2.2. Show that when $Q_2 \ll 1$, the disturbance frequencies for the sinuous and varicose modes are given respectively by

$$\omega = k \left\{ 1 \pm \left[\frac{kWe^{-1} - Q_2}{\tanh(k)} \right]^{1/2} \right\},$$

and

$$\omega = k \left\{ 1 \pm \left[\frac{kWe^{-1} - Q_2}{\coth(k)} \right]^{1/2} \right\}.$$

Show that if Squire's instability criterion is satisfied, the temporal growth rate of the sinuous mode is larger than that of the varicose mode for long wave disturbances but is of the same order of magnitude for sufficiently short waves. Show that $\omega_i = 0$ at $k = 0$ and $k = WeQ_2$, and that ω_i possesses a positive maximum value between these two values of k.

2.2. Convective Instability

The methodology of convective and absolute stability analyses can be found in the works of Bers (1975), Lin and Lian (1989), Huerre and Monkewitz (1990) and references cited therein. Here we apply the method and develop the physical concept through a generic example (de Luca and Costa, 1997; Chomaz and Costa, 1998). Neither of these instabilities can be predicted by considering only temporally growing disturbances as was done in the previous section. However, some information can still be extracted from the temporal analysis, as will be discussed after we perform the stability analysis of the same uniform liquid sheet with respect to spatially and temporally evolving disturbances.

2.2a. Superposition of Disturbances

The double Fourier transform with respect to x and τ of (2.10) and (2.13)-(2.15) gives

$$H_{i,zz} - k^2 H_i = 0, \tag{2.21}$$

$$-(i\omega - ik)F + ikH_1 = 1, \tag{2.22}$$

$$-i\omega F + ikH_2 = 1, \tag{2.23}$$

$$\pm iWe^{-1}k^3 F + i(\omega - k)H_{1,z} - i\omega Q_2 H_{2,z} = 0, \tag{2.24}$$

where

$$H_i = \iint \psi_i e^{-ikx+i\omega\tau} dx d\tau, \tag{2.25}$$

$$F = \iint f' e^{-ikx+i\omega\tau} dx d\tau.$$

The nonhomogeneous terms on the right sides of (2.22) and (2.23) arise from the unit Dirac impulse displacement introduced at $\tau = 0$ and $x = x_0$. Hence downstream or upstream refers to $x_0 > 0$ or $x_0 < 0$. Note that x_0 is located at an arbitrarily assigned distance downstream of the sheet nozzle exit. The integrals in (2.25) extend from negative to positive infinity. The two independent solutions of (2.21) for the liquid phase (i.e., H_1) are the sinuous mode solution $h_s \cosh(kz)$ and the varicose mode $h_v \sinh(kz)$. The bounded solution for the gas phase is again $h_2 e^{\mp kz}$.

2.2b. Response to Impulse

Consider the sinuous mode first. Substituting the sinuous solution into the boundary conditions (2.22) to (2.24), we find a system of three algebraic equations

$$[\mathbf{D}]\mathbf{A} = [1, 1, 0]^T,$$

where [**D**] is the coefficient matrix and **A** is the amplitude vector in (2.16). It follows from (2.26) that

$$
\mathbf{A} = \begin{vmatrix} f \\ h_s \\ h_2 \end{vmatrix} = \frac{[\mathbf{D}]_{adj}}{D} \begin{vmatrix} 1 \\ 1 \\ 0 \end{vmatrix},
$$

where $[\mathbf{D}]_{adj}$ is the adjoined of [**D**], and D is its determinant. It follows that

$$
\mathbf{H} = \begin{vmatrix} F \\ H_1 \\ H_2 \end{vmatrix} = \frac{[\mathbf{D}]_{adj}}{D} \begin{vmatrix} 1 \\ 1 \\ 0 \end{vmatrix} \cdot \begin{vmatrix} 1 \\ cosh\,(k) \\ e^{-k} \end{vmatrix}. \tag{2.27}
$$

The inverse Fourier transformation of **H** then constructs the perturbation displacement and stream functions from superposition of disturbances of all wavelengths and frequencies

$$
\mathbf{G}\,(x,\tau) = \begin{vmatrix} f' \\ \psi_1 \\ \psi_2 \end{vmatrix} = \frac{1}{(2\pi)^2} \iint \mathbf{H} e^{i(kx-\omega\tau)} dk d\omega. \tag{2.28}
$$

Note that $D^{-1} \to 0$ faster than k^{-1}, as $k \to \infty$. It should be pointed out that ω as well as $k = k_r + i k_i$ are now complex, so that spatial and temporal growth are allowed to interact in a disturbance. D appears in the denominator of **H**, and thus the dynamic behavior of the disturbance depends crucially on the type of singularity associated with the characteristic equation

$$
D\,(k,\omega) = Q_2\omega^2 + (k-\omega)^2\, tanh\,(k) - k^3 We^{-1} = 0. \tag{2.29}
$$

Equation (2.29) is identical to (2.17).

   The analysis for the sinuous mode applies equally to the varicose mode except $tanh\,(k)$ is changed to $coth\,(k)$. It will be seen that the characteristic equation (2.29) has simple roots or repeated roots, depending on the parameter values. The appearance of simple poles or higher order singularities in (2.28) leads to a uniquely different dynamic behavior. The spatial amplification rates k_i are obtained as isolated roots of $D = 0$ for various wave numbers k_r, at $We = 10$, $Q_2 = 0.0013$ and $\omega_i = 0$. Two branches of k_i as functions of k_r are plotted in Figure 2.2(a). Both branches of the amplification curve $\omega_i = 0$ can be approached from the upper half plane where $\omega_i > 0$ and $k_i > 0$, as indicated with dotted and dashed lines with arrows in the figure. Thus both branches are relevant to the downstream region because it satisfies the causality condition that the disturbance is nonexistent when $\tau \to -\infty$ for all $0 \le x \le \infty$. When the lower branch of the dispersion curve slightly overshoots $\omega_i = 0$ from $\omega_i > 0$ (i.e., $\omega_i = -\varepsilon \to 0$), it gives the spatial growth rate at the onset of

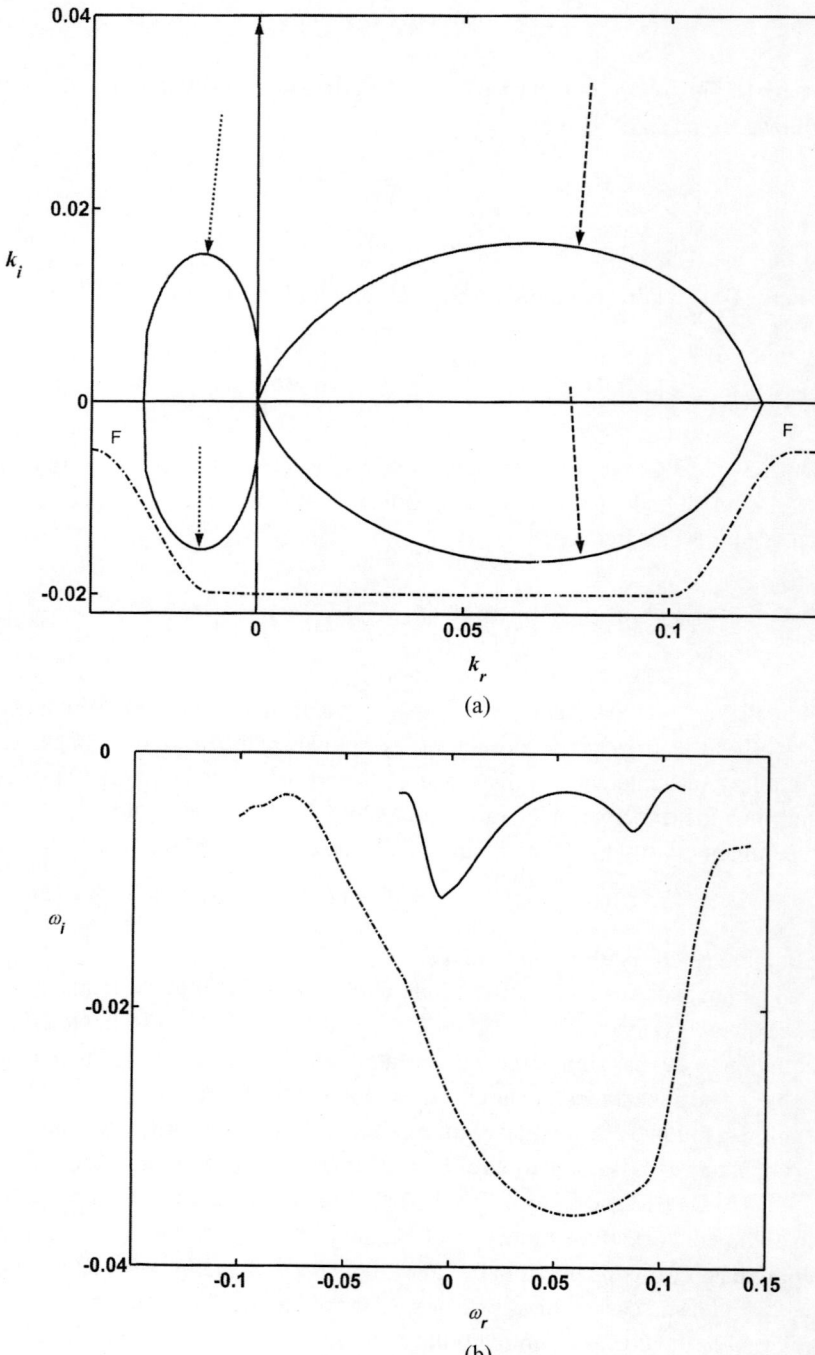

Figure 2.2. Sinuous mode convective instability. $We = 10$, $Q_2 = 0.0013$. (a) Spatial amplification curve. $\omega_i = 0$, and absolute convergence in wave number space. _____, $\omega_i = 0$; - - > $\omega_i \geq 0$, $\omega_r = 0.0675$; ·····> $\omega_i \geq 0$, $\omega_r = -0.045$. (b) Absolute convergence in frequency space. _ _ _, map of F.

temporal instability. However, the precise spatial-temporal evolution of the disturbance at the onset of instability can be determined only after integration over the entire range of wave numbers and wave frequencies in (2.27) is evaluated. Fourier inverse transformation can be accomplished by deforming the integration path F below both branches and completing the counter integral in the upper plane for absolute convergence as shown in Figure 2.2(a). The integration over the complex frequency can be carried out along the ω_r-axis and completes the counter integral in the domain of absolute convergence, that is, in the upper half plane as shown in Figure 2.2(b). Note that the maps of the deformed Fourier counter through (2.29) are all in the lower half ω-plane. The explicit expression of the double integral in terms of x, τ, We, and Q then provides the detailed space-time evolution of the collective disturbance. In practice the detailed initial evolution is frequently less important than large time asymptotic behavior.

2.2c. Time Asymptotics

Before actually determining the time asymptotic behavior we note first that $d\omega_r/dk_r > 0$ for the present case. Hence the group velocity of all Fourier components propagates in the downstream direction. Therefore some time after the disturbance is introduced at $x_0 = 0$, the upper curve in Figure 2.2(a) represents disturbances whose amplitudes decay spatially in the downstream direction and the lower curve represents disturbances whose amplitudes grow spatially in the downstream direction. Because both groups of disturbances are swept downstream relative to a given point in space, it is possible that after a sufficiently long time the disturbance appears to die out at that point. That this is indeed the case can be seen by evaluating the integral (2.28) at a given point x_1. The integration over the wave number in (2.28) can be performed along the integration path shown in Figure 2.2(a) for absolute convergence. Since $D^{-1} \to 0$ at least as fast as k^{-1}, and **H** contains only simple poles, the integral can be evaluated by summing up the residues of all poles,

$$\mathbf{I}(x, \omega) = \int \mathbf{H} e^{ikx} dk$$

$$= i \sum_{k_d} \bar{\mathbf{H}} [k_d(\omega)] \frac{exp\,[i k_d(\omega)\,x]}{(\partial D/\partial k)_{k_d}},$$

where k_d denotes the complex wave numbers along the two downstream propagating branches, and $\bar{\mathbf{H}}$ is the vector appearing in the numerator of (2.27). For determining the large time asymptotic behavior of the disturbance, it is not necessary to know the explicit expression of $\mathbf{I}(x, \omega)$, which is a regular function. The integration over ω can be carried out along the ω_r-axis indicated in Figure 2.2(b) to give, at any finite $x = x_1$

$$\mathbf{G}(x_1, \tau) = \frac{1}{2\pi} \int \mathbf{I}(\omega, x_1) e^{i\omega\tau} d\omega. \tag{2.30}$$

Because the counter integral in the upper half ω-plane in (2.30) contains no singularity,

$$\lim_{\tau \to \infty} \mathbf{G}(x_1, \tau) = 0, \tag{2.31}$$

according to the Cauchy theorem.

2.2d. Follow Disturbance with Group Velocity

However, because the group velocity of the spatially growing disturbances is in the downstream direction (i.e., $d\omega_r/dk_r > 0$), the disturbance as a whole must also grow in time as it travels downstream. To capture this phenomenon mathematically, we travel with the wave packet consisting of all the Fourier components at a yet to be determined velocity U_g. Since the integrand in (2.28) has only simple poles in the k-plane, the Fourier integral can be evaluated with the help of the theory of residue, such that

$$\mathbf{I}(\omega, x = U_g\tau) = \sum \frac{i\hat{\mathbf{H}}[k_d(\omega_r)]}{\partial D/\partial k} \exp\{[-i(\omega_r - U_g k(\omega_r))]\tau\}, \tag{2.32}$$

where we have absorbed $e^{-i\omega\tau}$ into $\mathbf{I}(x, \omega)$ and put $\omega_i = 0$ because we are going to evaluate the integral over ω along $\omega_i = 0$, and the summation is over all downstream branches. In evaluating the integral over ω, most of the Fourier components in (2.32) will destructively interfere with each other except around a particular component which makes the exponent in (2.32) stationary. Putting the derivative with respect to ω_r of the exponent to zero, we have

$$1 - U_g\left(\frac{dk_r}{d\omega_r}\right) - iU_g\left(\frac{dk_i}{d\omega_r}\right) = 0.$$

It follows that

$$U_g = \left(\frac{d\omega_r}{dk_r}\right)_{\omega=\omega_s},$$

where ω_s is determined from the imaginary part,

$$\frac{dk_i}{d\omega_r} = 0.$$

As expected, U_g is the group velocity at which the energy of the wave packet travels (Lamb, 1932; Lighthill, 1987; Whitham, 1974). Expanding the exponent in (2.32) about the stationary phase frequency ω_s, we have

$$\lim_{\tau \to \infty} \mathbf{G}(\tau, x = U_g \tau)$$

$$= \int \sum \frac{i \bar{\mathbf{H}} d\omega_r}{(\partial D/\partial k)_{k_d}} \, exp \left\{ -i \left[\omega_s - U_g k(\omega_s) \right. \right.$$

$$\left. \left. -\frac{1}{2} U_g \left(d^2 k/d\omega_r^2 \right)_{\omega_r = \omega_s} (\omega_r - \omega_s)^2 + O\,(\omega_r - \omega_s)^3 \right] \tau \right\}$$

$$= \sum \frac{\bar{\mathbf{H}}\,[k_d\,(\omega_s)]\,exp\,[i(U_g k(\omega_s) - \omega_s)\tau]}{(\partial D/\partial k)_{k_d} \left[\frac{1}{2} U_g \left(d^2 k/d\omega_r^2 \right)_{\omega_s} \right]^{1/2}} \frac{1}{\sqrt{\tau}} \int_{-\infty}^{\infty} e^{-s^2} ds$$

$$\sim \tau^{-1/2} exp\,[-U_g k_i(\omega_s)\tau]. \tag{2.33}$$

The above method of evaluation of the integral is frequently called the methods of stationary phase, steepest descent, or saddle point. According to (2.33) the upper curve having $k_i > 0$ in Figure 2.2(a) represents an evanescent wave, and the lower curve having $k_i < 0$ represents a convectively unstable disturbance that becomes unbounded when one follows the wave packet, as $\tau \to \infty$, that is,

$$\lim_{\tau \to \infty} \mathbf{G}(\tau, x = U_g \tau) \to \infty. \tag{2.34}$$

Equation (2.31) and (2.34) together constitute the definition of convective instability.

The convective instability of varicose and sinuous modes are compared in Figures 2.3 and 2.4. The upper branches represent downstream propagating evanescent waves and the lower branches represent downstream propagating convectively unstable disturbances. Both branches can be approached from the upper k-plane with $\omega_i > 0$. Notice that as the density ratio is increased by ten times, the amplification rates as well as the unstable wave bandwidth are increased by approximately ten times. Moreover the amplification rate of the varicose mode is smaller than that of the sinuous mode for the case of $Q_2 = 0.0013$, but the two rates are almost equal over the entire unstable wave spectrum for the case of $Q_2 = 0.013$. This fact can be exploited to enhance

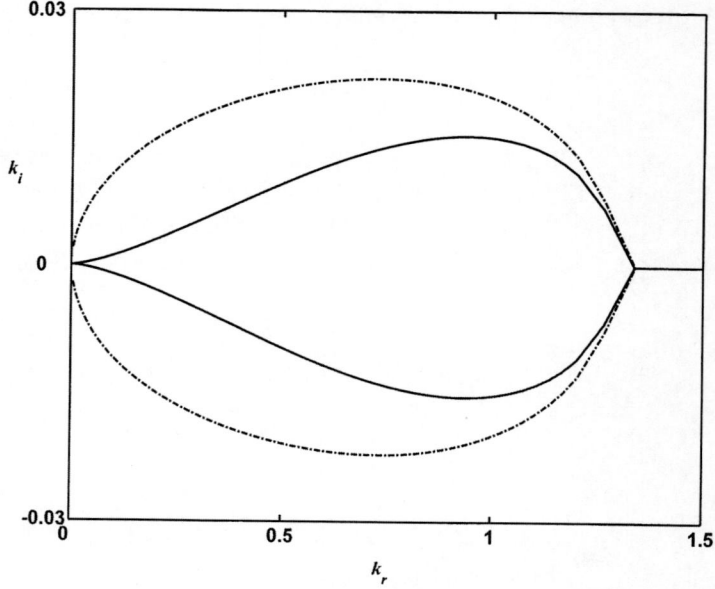

Figure 2.3. Spatial amplification curves. $\omega_i = 0$, $Q_2 = 0.0013$, $We = 1000$. _____, varicose mode; $_ - _ - _$, sinuous mode.

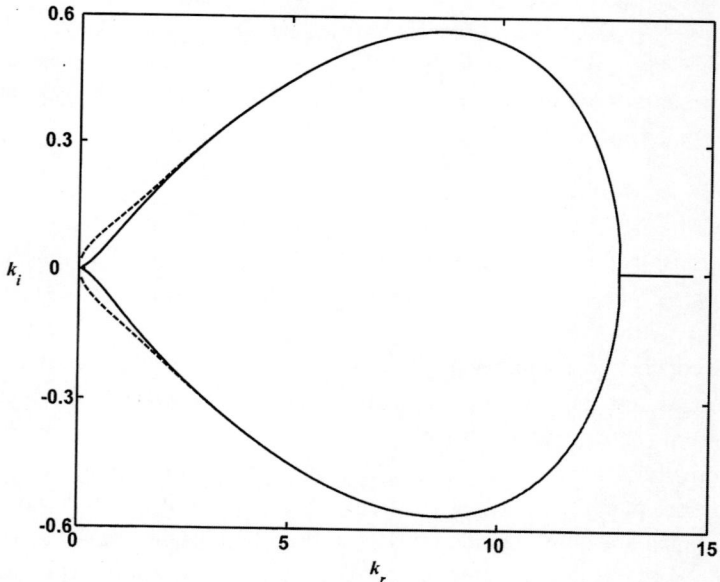

Figure 2.4. Spatial amplification curves. $\omega_i = 0$, $Q_2 = 0.013$, $We = 1000$. _____, varicose mode; $_ - _ -$, sinuous mode.

sheet atomization processes to produce smaller droplets by increasing the ambient gas density (by increasing the gas pressure, for example). For $We = O(10)$, the amplification rate of the varicose mode is much smaller than that of the sinuous mode.

2.3. Absolute Instability

In the previous section we considered the case of $We > 1$ and found that the uniform sheet in the presence of ambient gas is convectively unstable. In this section we show that when the Weber number is reduced to less than one, the sheet becomes absolutely unstable. The disturbance then grows in time and spreads in both the upstream and downstream directions.

Figure 2.5 gives representative spatial amplification curves $\omega_i = 0$ obtained from (2.17). The curves denoted by d_1 and d_2 are both obtained by reducing ω_i from a positive value in the upper k-plane. Curve d_1 crosses over to the lower half plane while curve d_2 remains in the upper half plane. As explained in the previous section, these two curves are relevant to the downstream region. Similarly the other two curves u_1 and u_2 are relevant to the upstream region. Again one branch u_1 crosses over from the lower k-plane and curve u_2 remains in the lower half plane. Along d_1 and d_2, $d\omega_r/dk_r > 0$. Hence the branches d_1 and d_2 represent respectively the growing and decaying

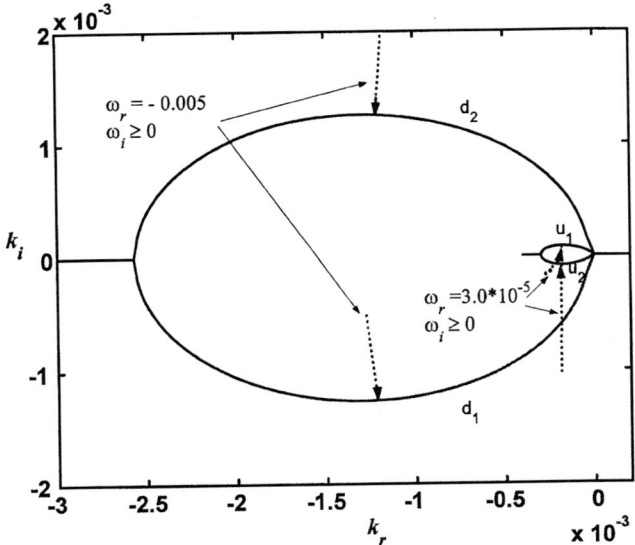

Figure 2.5. Absolute instability. Singularity in wave number space. $Q = 0.0013$, $We = 0.8$ (From Lin and Jiang, 2003)

wave packets as they propagate downstream. Similarly the other two curves u_1 and u_2 are relevant to the upstream region. The branch u_1 crosses over from the lower half k-plane and the other branch u_2 remains in the lower half plane. Along u_1 and u_2, $d\omega_r/dk_r < 0$. Hence the branches u_1 and u_2 represented respectively the growing and decaying wave packets as they propagate upstream. These four curves form an inflectional saddle point at $k = 0$ on the D-surface as $\omega \to 0$. Hence the Laplace integration over ω_r encounters a higher order singular point and its evaluation must be carried out properly. The Fourier integration over k can be achieved by deforming the integration counter as we did in the previous section, but separately for the upstream and downstream branches, and then adding them together. Using the theory of residue we have

$$\mathbf{I}(\omega, z) = i H(x) \sum_{k_d} \frac{exp\left[i k_d(\omega) x\right]}{(\partial D/\partial k)_d} \bar{\mathbf{H}}\left[k_d(\omega)\right]$$

$$- i H(-x) \sum_{k_u} \frac{exp\left[i k_u(\omega) x\right]}{(\partial D/\partial k)_u} \bar{\mathbf{H}}\left[k_u(\omega)\right], \qquad (2.35)$$

where $H(x)$ is the heavy side unit step function. To evaluate the residue, we first note that at $k = \omega = 0$

$$\partial D/\partial k = \partial D/\partial \omega = \partial^2 D/\partial k^2 = (\partial^2 D/\partial k \partial \omega) = 0.$$

Thus the Taylor series expansion of D about $(k_0, \omega_0) = (0, 0)$ gives

$$D = \frac{1}{3!}\left(\frac{\partial^3 D}{\partial k^3}\right)_0 (k - k_0)^3 + \frac{1}{2}\left(\frac{\partial^2 D}{\partial \omega^2}\right)(\omega - \omega_0)^2 + \cdots = 0.$$

Hence near (k_0, ω_0), we have

$$k - k_0 = \left[-3\frac{(\partial^2 D/\partial \omega^2)_0}{(\partial^3 D/\partial k^3)_0}\right]^{1/3}(\omega - \omega_0)^{2/3}. \qquad (2.36)$$

The cubic roots in (2.36) will be designated with C_1, C_2, and C_3. C_3 is the complex conjugate of C_2, that is, $C_3 = C_2^*$. C_1 can be associated with one of the curves in Figure 2.4(b) if the derivatives in the cubic root sign are evaluated. C_2 and C_3 then correspond to the other two curves, which are complex conjugates to each other. The inflectional saddle point of D is formed by the coalescence of these curves at k_0. Without loss of generality D can be written as

$$D(k, \omega) = [k - k_{d1}(\omega)] [k - k_{u1}(\omega)] [k - k_{u2}(\omega)] R(k, \omega), \qquad (2.37)$$

where $R(k, \omega)$ is regular, and

$$k_{dl} = k_0 + C_1 (\omega - \omega_0)^{2/3},$$

$$k_{u1} = k_0 + C_2 (\omega - \omega_0)^{2/3}, \qquad (2.38)$$

$$k_{u2} = k_0 + C_2^* (\omega - \omega_0)^{2/3}.$$

Substituting (2.37) and (2.38) into (2.35) we have

$$
\mathbf{I}(\omega, x) = \left\{ i H(x) \frac{\bar{\mathbf{H}}[k_{d1}(\omega)] \exp(i k_{dl} x)}{(C_1 - C_2)(C_1 - C_2^*)} - i H(-x) \right.
$$

$$
\times \left[\frac{\bar{\mathbf{H}}[k_{u1}(\omega)] \exp(i k_{u1} x)}{(C_2 - C_1)(C_2 - C_2^*)} + \frac{\bar{\mathbf{H}}[k_{u2}(\omega)] \exp(i k_{u2} x)}{(C_2^* - C_1)(C_2^* - C_2)} \right] \right\}
$$

$$
\times R(k_0, \omega_0)^{-1} (\omega - \omega_0)^{-4/3} + \mathbf{I}_R(\omega, x), \qquad (2.39)
$$

where $\mathbf{I}_R(\omega, x)$ is the remaining summation over values of k_d and k_u that are not near k_0. It follows that as $k \to k_0$, $\omega \to \omega_0$

$$
\lim_{\tau \to \infty} |\mathbf{G}(x, \tau)| \sim \int_{BP} \frac{d\omega}{2\pi} \frac{\exp(-i\omega\tau)}{(\omega - \omega_0)^{4/3}}
$$

$$
\sim \tau^{1/3} \exp[i(k_0 x - \omega_0 \tau)] \sim \tau^{1/3}. \qquad (2.40)
$$

It follows from (2.39) and (2.40) that the disturbance for this case of $We < 1$, grows algebraically as $\tau^{1/3}$ for all x as $\tau \to \infty$. Hence the disturbance once introduced must spread upstream and downstream as it propagates. The definition of absolute instability is then

$$
\lim_{\tau \to \infty} \mathbf{G}(x, \tau) \to \infty \text{ for all } x, \qquad (2.41)
$$

regardless of whether the time growth is algebraic or exponential. It should be pointed out that the occurrence of the pinch point at $k = 0$, $\omega = 0$ in this example does not imply that the disturbance wavelength observed is infinitely long. It merely means that the large time asymptotic behavior of the wave packet amplitude is dominated by the infinitely long waves. The disturbance is actually constructed by Fourier superposition of all wavelengths.

No absolute instability is found for the varicose mode. Regardless of whether We is smaller or greater than one, it remains convectively unstable if $Q_2 \neq 0$. If $Q_2 = 0$, the varicose mode is stable.

Exercises

2.3. Obtain the spatial amplification curves similar to Figure 2.2(a) for the case of varicose mode with $Q_2 = 0.0013$, $We = 0.5$, and $We = 2$.

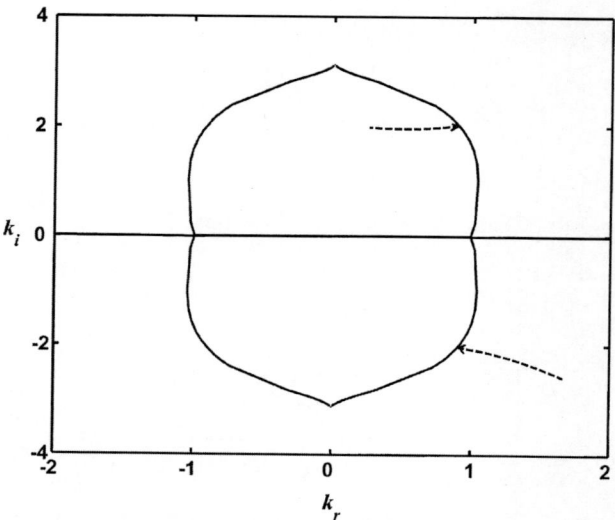

Figure 2.6. Spatial amplification curve. $Q_2 = 0$, $We = 2$, sinuous mode. _____, $\omega_i = 0$; - - >, $\omega_i > 0$.

2.4. It is found from the lowest Taylor series expansion of $D(k, \omega) = 0$ that a generalized pinch point exists at (k_0, ω_0) near which

$$(k - k_0)^{p+2} \sim (\omega - \omega_0)^n \tag{2.42}$$

where $p \geq 0$ and $n \geq 1$. Using the approach similar to that used to arrive at (2.40), show that for a generalized pinch point

$$\mathbf{G}\,(\tau \to \infty) \to \tau^{\nu-1} exp\,(-i\omega_0\tau), \tag{2.43}$$

where $\nu = n(p + 1)/(p + 2)$.

2.5. Given that $Q_2 = 0$ and $We > 1$ ($We = 2$, for example), plot from (2.29) the spatial amplification curve $\omega_i = 0$ in the (k_r, k_i) space (Fig. 2.6). Show that the curves in the upper and lower planes are approached from $\omega_i > 0$ in each of their own planes, but never cross over to the opposite planes as indicated by arrows in Figure 2.6. Therefore the supercritical ($We > 1$) sinuous disturbances are stable (evanescent waves) in the absence of ambient gas.

2.6. For the branch of the amplification curve $k_i = \omega_i = 0$ in the previous problem, $D(k, \omega) = 0$ and $\partial D/\partial k = 0$ are plotted in Figure 2.7. Near the intersection of these two curves, $(k - k_0)^2 \sim (\omega - \omega_0)$, where (k_0, ω_0) is any one of the intersection points. By use of (2.43), show that the disturbance decays as fast as $\tau^{-1/2}$ at any finite spatial position as $\tau \to \infty$ (de Luca and Costa, 1997).

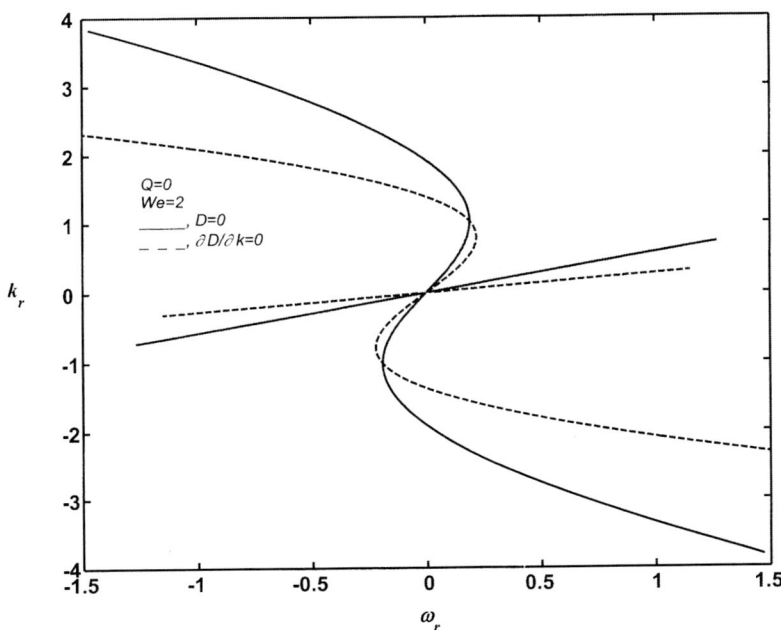

Figure 2.7. Singular points at finite real k and ω. $Q_2 = 0$, $We = 2$, sinuous mode. ———, $\omega_i = 0$; - - -, $\partial D/\partial \omega = 0$.

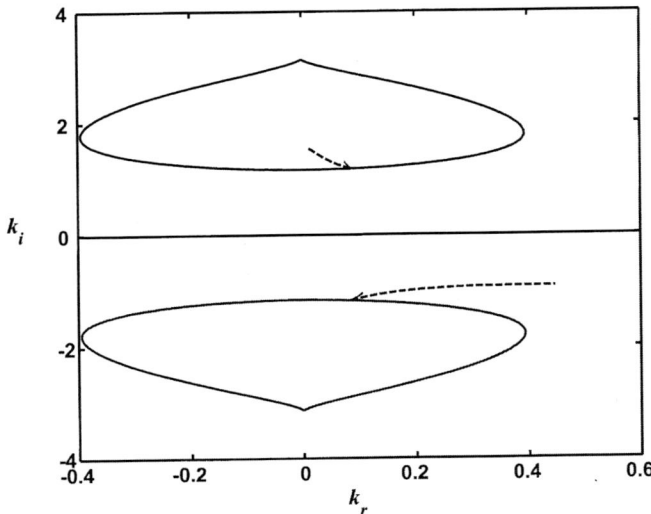

Figure 2.8. Spatial amplification curve. $Q_2 = 0$, $We = 0.5$, ———, $\omega_i = 0$; - - >, $\omega_i > 0$.

2.7. Consider the case of $Q_2 = 0$ and $We < 1$. Show that near the singularity $(k_0, \omega_0) = (0, 0)$ of (2.29)

$$(k - k_0)^2 \sim (\omega - \omega_0)^2,$$

and thus

$$G(\tau \to \infty) \to \tau^0 \exp(ik_0 x + \omega_0 \tau).$$

Therefore in the absence of ambient gas, the subcritical $(We < 1)$ sinuous disturbance amplitude remains nonvanishing for all time and at any finite x (the pseudo-absolute instability, Lin et al., 1990).

2.8. Obtain the spatial amplification curve for $Q_2 = 0$, $We = 0.5$ and 10 for the case of varicose mode, and show that the sheet is stable with respect to varicose mode in the absence of ambient gas. (See Fig. 2.8.)

2.4. Summary

In Section 2.1 the instability of a uniform inviscid liquid sheet was analyzed by considering only an isolated Fourier component of disturbance allowed to grow only temporally. The analysis led to the finding that no instability can occur if $We < 1$. This finding contradicts Brown's observation that when $We < 1$ the disturbance can propagate upstream and tends to rupture the liquid sheet. Although the fluids used in Brown's experiments were viscous, we will show later that fluid viscosity does not make the sheet stable when $We < 1$. The source of this paradox is the unrealistic assumption made in the temporal analysis that any one Fourier component of the disturbance grows exponentially at the same time rate everywhere in space. However, in reality, all the Fourier components grow in space as well as in time. In the particular case of a uniform inviscid sheet with $We < 1$, as discussed in Section 2.3, the disturbance grows algebraically as it propagates upstream and downstream of a point where the disturbance is impulsively introduced. When $We > 1$, temporal theory predicts time-instability but cannot describe the space-time relation of disturbance, which is illustrated in Section 2.2. The convective instability analysis results are consistent with the observations of Taylor. He observed that when $We > 1$ in a uniform sheet, both sinuous and varicose waves are convected downstream and leave the sheet surface relatively unperturbed. He also demonstrated the vestiges of these waves by inserting an obstacle in the sheet. The wave motion in a liquid sheet will be discussed in Chapter 5.

Crapper et al. (1975) also observed convectively unstable sinuous disturbances in a radially expanding and thinning liquid sheet issued from a fan-spray nozzle. The instability of a radially expanding sheet is discussed in Chapter 3. Practical implications of the findings have been discussed in the works of Squire (1953), Hagerty and Shea (1955), and Fraser et al. (1962). Some applications will be discussed after viscosity and other effects are considered.

We have seen that the solution of the differential system for linear instability problems by use of Fourier transformation and by use of normal mode solutions leads to the same characteristic equation. Moreover, if we are not concerned with the initial transience but only interested in the large time asymptotic behavior of the disturbance, then we need not carry out the inverse Fourier transformation but must only know the nature of the singularities of the characteristic equation. For this reason we will use normal mode analysis in the subsequent chapters, but the stability characteristic will be discussed on the basis of the nature of the singularities of the characteristic equations with complex wave numbers and frequencies.

References

Bers, A. 1975. Linear waves and instabilities. In *Physique des Plasmas* (Ed. C. De Witt and J. Peyraud), pp. 452–516. Gordon & Breach.

Brown, D. R. 1961. *J. Fluid Mech.* **10**, 297–305.

Chomaz, J. M., and Costa, M. 1998. In *Free Surface Flows*. Ed. by C. Heindrick and H. J. Roth, p. 45. Springer, New York.

Crapper, G. D., Dombrowski, N., Jepson, W. P., and Pyott, 1973. *J. Fluid Mech.* **57**, 671.

Crapper, G. D., Dombrowski, N., and Pyott, G. A. D. 1975. *Proc. R. Soc. Lond.* A **342**, 209–224.

de Luca, L., and Costa, M. 1997. *J. Fluid Mech.* **331**, 127.

Fraser, R. P., Eisenklam, P., Dombrowski, N., and Hasson, D. 1962. *AIChE J.* **8**, 672.

Hagerty, W. W., and Shea, J. F. 1955. *J. Appl. Mech.* **22**, 509–514.

Huerre, P., and Monkewitz, P. A. 1990. *Ann. Rev. Fluid Mech.* **22**, 473–537.

Lamb, H. 1945. *Hydrodynamics*. p. 395. Dover, New York.

Lighthills, J. 1978. *Waves in Fluids*. p. 254. Cambridge University Press.

Lin, S. P. and Lian, Z. W. 1989. *Phys. Fluids* A**1**, 490–493.

Lin, S. P. and Jiang, W. Y. 2003. *Phys. Fluids* **15** (in print).

Savart, F. 1833a. *Ann. Chim. Phys.* **LIX**, 55–87.

Savart, F. 1833b. *Ann. Chim. Phys.* **LIX**, 257–310.

Squire, H. B. 1953. *Brit. J. Appl. Phys.* **4**, 167–169.

Taylor, G. I. 1959. *Proc. R. Soc. Long.* A **253**, 296–321.

Whitham, G. B. 1974. *Linear and Nonlinear Waves.* p. 374. John Wiley & Sons. New York.

3

Nonuniform Inviscid Liquid Sheets

Nonuniform liquid sheets are encountered in various industrial applications including radiation cooling in space (Chubb et al., 1994), and paper making (Soderberg and Alfredson, 1998). Many of the works cited below were motivated by applications in surface coating, fuel spray formation, nuclear safety, and other industrial processes. The spatial variation of sheet thicknesses in these applications is necessitated by the conservation of mass flow across the cross section perpendicular to the flow direction. For example, the thickness of a planar sheet of constant width must decrease in the flow direction due to the gravitational acceleration. Consequently the local Weber number, based on the local thickness and velocity, changes spatially. In particular *We* may be greater than one in part of the sheet and smaller than one in the rest. If one locally applies the concept of absolute and convective instability in a uniform sheet, then part of the sheet may experience convective instability, while the remaining part may experience absolute instability. Depending on the relative location of the regions of *We* > 1 and *We* < 1, one would expect different physical consequences to the entire flow. The objective of this chapter is to elucidate the effect of the spatial variation of *We* on the dynamics of sheet breakup by properly applying the concept developed in the previous chapter for a sheet of uniform thickness.

3.1. Expanding Liquid Sheet

3.1a. Basic Flow

A radially expanding circular liquid sheet can be created by impinging two identical circular liquid jets vertically against each other (Savart, 1833; Huang, 1970) or impinging a jet against a disk (Taylor, 1959a; Clanet and Villermaux, 2002; Villermaux and Clanet, 2002). The known experiments showed that the effect of liquid viscosity is negligible in axisymmetric sheets. The skin friction

26

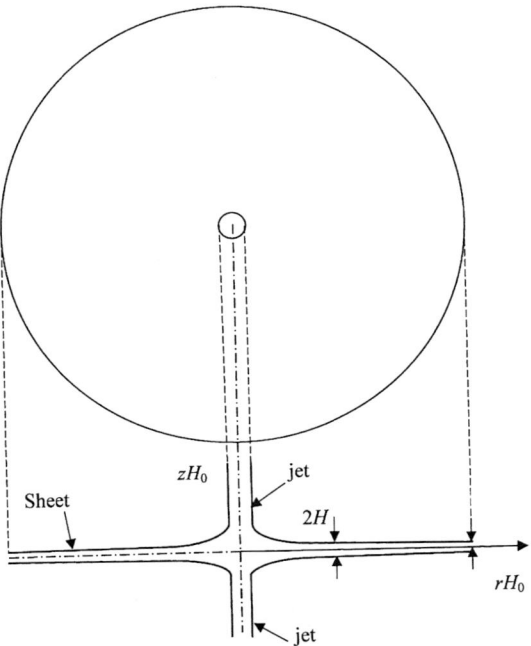

Figure 3.1. Definition sketch for basic flow.

caused by air is also quite small (Taylor, 1959a). Therefore viscous effects may be neglected. The governing equations for the axisymmetric flow of an inviscid and incompressible fluid in the cylindrical coordinates (r, θ, z) are (cf. Fig. 3.1)

$$u_{,\tau} + uu_{,r} + wu_{,z} = -p_{,r}, \tag{3.1}$$

$$w_{,\tau} + uw_{,r} + ww_{,z} = -p_{,z}, \tag{3.2}$$

$$(ru)_{,r} + (rw)_{,z} = 0. \tag{3.3}$$

where r and z are respectively the radial and axial distance nondimensionalized with the half sheet thickness H_0 at a given $R = R_0$, the subscripts denote partial differentiations, (u, w) are velocity components in the (r, z)-directions nondimensionalized with the velocity U_0 at $(R_0, \theta, 0)$, τ is time divided by H_0/U_0, and p is the pressure normalized with ρU_0^2, where ρ is the fluid density. Equations (3.1)-(3.3) apply equally to the liquid and the ambient gas, except that the pressure terms must be multiplied by the gas to liquid density ratio $Q = \rho_g/\rho_l$ for the gas domain.

The boundary conditions at the liquid-gas interface $z = h(r)$ include the kinematic condition

$$h,_\tau + uh,_r = w \tag{3.4}$$

for each fluid, and the dynamic balance of the surface tension force and the pressure force

$$We_0^{-1} \nabla \cdot \mathbf{n} = -p_g + p_l\,[r, h(r)]\,, \tag{3.5}$$

where $\mathbf{n}$ is the unit normal vector pointing from the liquid to the gas and thus $\nabla \cdot \mathbf{n}$ is the mean curvature, p_g and p_l are respectively the gas and the liquid pressure, and We_0 is the Weber number defined by

$$We_0 = \frac{\rho U_0^2 H_0}{S},$$

S being the surface tension. In the basic flow the ambient gas is quiescent, and thus $(u, w) = 0$ and $p_g = $ constant. The steady basic flow in the liquid sheet is irrotational and axisymmetric. Substituting p_l from the Bernoulli equation into (3.5), we have

$$We_0^{-1}\left[\frac{h,_r}{r\left(1 + h,_r^2\right)^{1/2}} + \frac{h,_{rr}}{\left(1 + h,_r^2\right)^{3/2}}\right] + \frac{1}{2} - \frac{1}{2}(u^2 + w^2) = 0. \tag{3.6}$$

Equation (3.3) is the necessary and sufficient condition for the existence of the stream function ψ such that

$$ru = \psi,_z, \qquad rw = -\psi,_r\,.$$

In terms of ψ, the kinematic condition (3.4) can be written as

$$h,_\tau + \frac{1}{r}\psi,_z\, h,_r = -\frac{1}{r}\psi,_r\,. \tag{3.7}$$

The vanishing of vorticity requires

$$\frac{1}{r}\psi,_{zz} + \left(\frac{1}{r}\psi,_r\right),_r = 0.$$

In the experiments of Taylor (1959a,b), Huang (1970), Matsumoto and Takashima (1971), and Crapper, Dombrowski, and Jepson (1975), the thickness $2H$ of the radially expanding sheets is inversely proportional to the radial distance R after some distance R_0 from the origin; that is, $H = K/2R, R > R_0$. R_0 can be taken to be the radial distance where Clanet and Villermaux were able to detect the first optical fringe beyond which the sheet thickness was measured to decrease inversely with the radial distance. In this region, the value of K range from 10^{-4} to 10^{-2} cm^2 if thickness is measured in cm.

Hence the slope of the liquid gas interface is of order $K/2R^2 = H/R$ in $R > R_0$. In this regime $H/R < \varepsilon = H_0/R_0$ is therefore very much smaller than one beyond a radial distance R_0 of order cm, H_0 being the half thickness at $R = R_0$. We rescale the radial distance with

$$\varepsilon r = \xi.$$

Thus we have two length scales (Nafeh and Mook, 1979; Bender and Orszag, 1978). Because $r = R/H_0$ and $\varepsilon = H_0/R_0$ in $R \geq R_0$, then $\xi \geq 1$. The irrotationality condition can be written in the slow variable ξ as

$$\psi_{,zz} - \frac{\varepsilon^2}{\xi}\psi_{,\xi} + \varepsilon^2\psi_{,\xi\xi} = 0. \tag{3.8}$$

The radial distance and the radial derivatives in (3.6) and (3.7) can be written in the rescaled variable by the following substitutions:

$$\partial_r = \varepsilon\,\partial_\xi, \qquad \frac{1}{r} = \varepsilon\frac{1}{\xi}.$$

The solution of the differential system for the steady basic flow will be expanded as

$$\psi = \psi_0 + \varepsilon\psi_1 + \varepsilon^2\psi_2 + \cdots \tag{3.9}$$

Substituting (3.9) into (3.8) and (3.7) and collecting $O(\varepsilon^0)$ terms, we have

$$\psi_{0,zz} = 0,$$

$$\psi_{0,z}\,h_{,\xi} = -\psi_{0,\xi} \qquad \text{at } z = h(\xi).$$

The solution for ψ_0 is

$$\psi_0 = Cz/h(\xi),$$

where C is a constant to be determined from the dynamic boundary condition. Similarly, collecting $O(\varepsilon^1)$ solution terms and solving for ψ_1, we have

$$\psi_1 = C_1 z/h(\xi),$$

where C_1 is to be determined together with C. The irrotational and the free surface kinematic conditions for the $O(\varepsilon^2)$ solution are respectively

$$\psi_{2,zz} = \frac{\psi_{0,\xi}}{\xi} - \psi_{0,\xi\xi}$$

and

$$\psi_{2,z}\,h_{,\xi} = -\psi_{2,\xi}\,.$$

The solution for ψ_2 that satisfies the above two equations is

$$\psi_2 = C\frac{z^3}{6h^2}\left[h_{,\xi\xi} - 2\frac{h_{,\xi}^2}{h} - \frac{h_{,\xi}}{\xi}\right] + C_2(\xi)z,$$

where the arbitrary function $C_2(\xi)$ is so chosen that the free surface $z = h(\xi)$ remains a stream surface. This requirement leads to

$$\psi_2 = \frac{C}{3}\left(\frac{z}{h}\right)\left[\left(\frac{z}{h}\right)^2 - 1\right]\left[h\,h_{,\xi\xi} - \frac{h\,h_{,\xi\xi}}{\xi} - 2h_{,\xi}^2\right] + B\frac{z}{h},$$

where B is a constant to be determined together with C. Substituting this expression into the kinematic boundary condition, we have

$$\frac{C}{3}\frac{h_{,\xi}}{h^2}(h - 1)\left[h\,h_{,\xi\xi} - \frac{h\,h_{,\xi}}{\xi} - 2h_{,\xi}^2\right] = 0.$$

Thus the kinematic condition requires

$$h_{,\xi\xi} - \frac{h_{,\xi}}{\xi} - \frac{h_{,\xi}^2}{h} = 0. \tag{3.10}$$

To $O(\varepsilon^2)$, the stream function for the liquid sheet is therefore given by

$$\psi_l = C\frac{z}{h} + \varepsilon^2\left\{C\frac{z}{6h}\left[\left(\frac{z}{h}\right)^2 - 1\right]h\left[h_{,\xi\xi} - 2\frac{h_{,\xi}^2}{h} - \frac{h_{,\xi}}{\xi}\right] + B\frac{z}{h}\right\}. \tag{3.11}$$

The stream function for the quiescent gas is

$$\psi_g = 0.$$

Substitution of (3.11) with (3.10) into (3.6) yields

$$\varepsilon^2 We_0^{-1}\left[\frac{h_{,\xi}}{\xi\left(1 + \varepsilon^2 h_{,\xi}^2\right)^{1/2}} + \frac{h_{,\xi\xi}}{\left(1 + \varepsilon^2 h_{,\xi}^2\right)^{3/2}}\right]$$
$$+ \frac{1}{2} - \frac{\varepsilon^2}{2\xi^2 h^2}\left[C^2 + \varepsilon(2CC_1) + \varepsilon^2(2CB)\right].$$

Substituting $h = h_0 + \varepsilon h_1 + \varepsilon^2 h_2$ and $C = b_0 + \varepsilon b_1 + \varepsilon^2 b_2$ into (3.12) and solving the system order by order, we obtain

$$O(1):\qquad b_0 = \frac{1}{\varepsilon} = \frac{R_0}{H_0},\qquad h_0 = \frac{1}{\xi}, \tag{3.12}$$

$$O(\varepsilon):\qquad C_1 = 0,\qquad h_1 = 0,$$

$$O(\varepsilon^2):\qquad h_2 = We_0^{-1}\xi^{-4} + b_1\xi^{-1},$$

where $b_1 = -We_0^{-1}$, so that $h = 1$ at $\xi = 1$. In (3.11) $B = We_0^{-1}$ so that the initial condition $u = 1$ at $\xi = 1$ is satisfied.

To the order ε^2 approximation, the variation of the local sheet thickness and velocity are thus given respectively by

$$h = \xi^{-1} + \varepsilon^2 We_0^{-1} [\xi^{-4} - \xi^{-1}]$$

and

$$u(\xi) = \left\{ 1 + We_0^{-1} \varepsilon^2 [\xi^{-3} - 1] \right\}^{-1}.$$

The larger the inertial force imparted to the sheet for a given surface tension, the smaller the rate of decrease of h in the direction of the flow, as it should be. Note that the fluid velocity increases in the radial direction, but its rate is much smaller than the rate of decrease of h. The variation of the local Weber number is given by

$$We = \frac{\rho(U_0 u)^2 (H_0 h)}{S} = We_0 \frac{(ruh)^2}{r^2 h}. \tag{3.13}$$

Since $2\pi ruh$ is the constant discharge and $rh = 1$ to the first order of approximation, the local Weber number decreases inversely with the distance in the downstream condition. This is contrary to the falling liquid sheet in which the local Weber number increases in the flow direction due to gravitational acceleration. This difference brings about a fundamental difference in instability characteristics, although the sheet thickness decreases in the flow direction in both cases, as will be shown in the next section.

3.1b. Stability Analysis

To investigate the onset of instability of the slowly varying basic flow given in the previous section, we rescale the short distance with local sheet thickness $h(\xi) H_0$ instead of H_0, and the time with $h H_0 / U_0 u(\xi)$ instead of H_0 / U_0. However, the notations τ, z, and r will be retained for simplicity, which should not cause confusion. The governing equation (3.8) and its boundary conditions (3.6) and (3.7) remain the same in form, except ψ is now a function of ξ as well as r. ξ and r will be treated as independent variables. We perturb the basic flow $[\psi(\xi), h(\xi)]$ with infinitesimal disturbances $[\psi'(\xi, r, z, t) \, h'(\xi, r, z, t)]$ in both fluids. By substituting the sum of the basic flow and the perturbation into (3.8) and (3.7), subtracting out the basic flow quantities, and neglecting nonlinear terms in perturbations, we have

$$\frac{\varepsilon}{\xi}\psi'_{i,zz} + \frac{\varepsilon}{\xi}\psi'_{i,rr} + \frac{\varepsilon^2}{\xi^2}\psi'_{i,r} = 0 \qquad (i = l, g), \tag{3.14}$$

$$h'_{,\tau} + \frac{\varepsilon}{\xi}\psi_{l,z}\,h'_{,r} + \frac{\varepsilon}{\xi}\psi'_{l,r} + \frac{\varepsilon^2}{\xi}\psi'_{l,z}\,h_{,\xi} = 0, \tag{3.15}$$

$$h'_{,\tau} + \frac{\varepsilon}{\xi}\psi'_{g,r} + \frac{\varepsilon^2}{\xi}\psi'_{g,z}\,h_{,\xi} = 0, \tag{3.16}$$

where the subscripts $i = l$ and $i = g$ denote the liquid and gas phases respectively. Differentiating (3.5) with respect to r, evaluating the pressure derivative terms by use of (3.1) and (3.2), substituting the perturbed flow, and subtracting out the basic flow quantities reveals the dynamic boundary condition

$$\mp\frac{1}{2}We^{-1}\left[h'_{,rrr} + \frac{\varepsilon}{\xi}h'_{,rr} + O(\varepsilon^2)\right] + \frac{\varepsilon}{\xi}\psi'_{l,z\tau} - Q\frac{\varepsilon}{\xi}\psi'_{g,z\tau} + O(\varepsilon^2) = 0, \tag{3.17}$$

where Q is the density ratio ρ_g/ρ_l. The upper and lower signs in front of We refer to the upper and lower interfaces respectively. The solution of (3.14)-(3.17) will be expanded as

$$\frac{\varepsilon}{\xi}\psi'_i = \psi'_{i0} + \varepsilon\psi'_{i1} + \varepsilon^2\psi'_{i2} + \cdots,$$

$$h' = h'_0 + \varepsilon h'_1 + \varepsilon^2 h'_2 + \cdots.$$

Substituting the above regular perturbation series solution into (3.14)-(3.17), collecting the first-order terms, and taking the double Fourier transform in τ and r results in

$$H_{i,zz} - k^2 H_i = 0, \tag{3.18}$$

$$(i\omega - ik)F - ikH_l = 1, \tag{3.19}$$

$$i\omega F - ikH_g = 1, \tag{3.20}$$

$$\mp\frac{i}{2}We^{-1}k^3 F + i\omega H_{l,z} - Qi\omega H_{g,z} = 0, \tag{3.21}$$

where

$$H_i = \iint \psi'_{i0}e^{-ikr+i\omega\tau}\,dr\,d\tau, \tag{3.22}$$

$$F = \iint h'_0 e^{-ikr+i\omega\tau}\,dr\,d\tau.$$

The nonhomogeneous terms on the right sides of (3.19) and (3.20) arise from the unit Dirac impulse introduced at $\tau = 0$ and $r = (R_o/H_o)\xi_0 + x$. Hence downstream or upstream refers to $x > 0$ or $x < 0$, where ξ_0 is a fixed distance in ξ. The integrals in (3.22) formally extend to infinity. This is probably reasonable in an asymptotic sense when $\varepsilon \to 0$ and if there is no upstream influence when local absolute instability occurs, as will be seen shortly to be the case in the present problem. The system of equations (3.18) to (3.22) is formally the same as the system (2.21) to (2.24). However, all variables in the present system are functions of ξ. Since $\varepsilon r = \xi$, the results of Chapter 2 can be applied locally in the limit of $\varepsilon \to 0$.

3.1c. Sinuous Modes

Typical results for a case of $We > 1$ are given in Figures 3.2 to 3.4 where the flow parameters correspond to the key parameters in two different breakup regimes observed by Huang. Comparisons with Huang's experiments will be made in Section 3.1e. The spatial amplification rate k_i is plotted against the wave number k_r at $\omega_i = 0$ and $We = 1.3$ and $Q = 0.0013$ in Figure 3.2. Both the upper and lower branches of the dispersion curves $\omega_i = 0$ in the region $k_r > 0$ are approached from the upper half plane where $\omega_i > 0$ and $k_i > 0$. Thus both branches are relevant to the downstream region because it

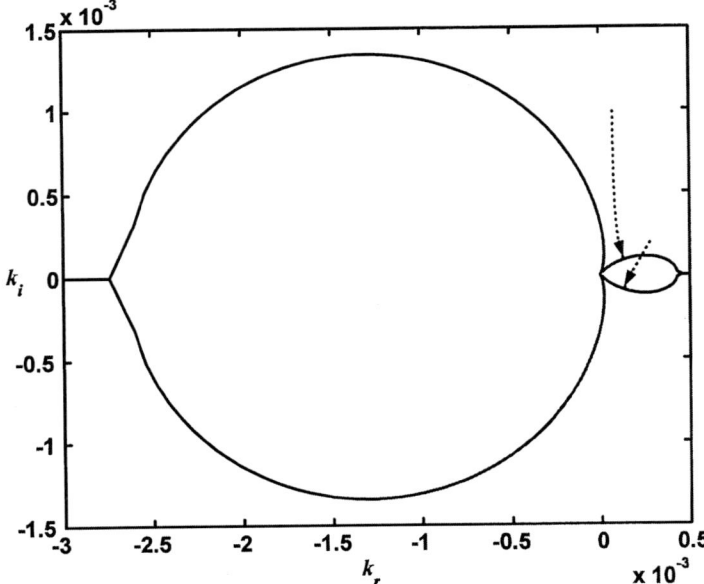

Figure 3.2. Spatial amplification curve. $Q = 0.0013$, $We = 1.3$, sinuous mode. (From Lin and Jiang, 2003)

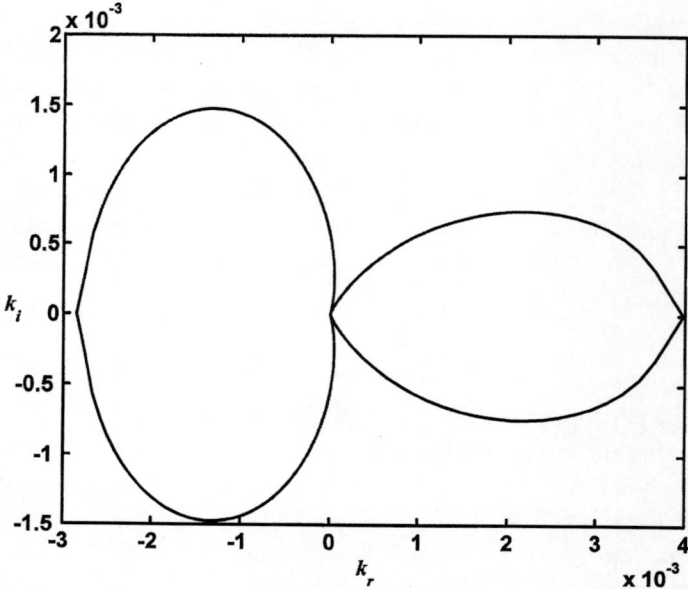

Figure 3.3. Spatial amplification curve. $Q = 0.0013$, $We = 3.8$, sinuous mode. (From Lin and Jiang, 2003)

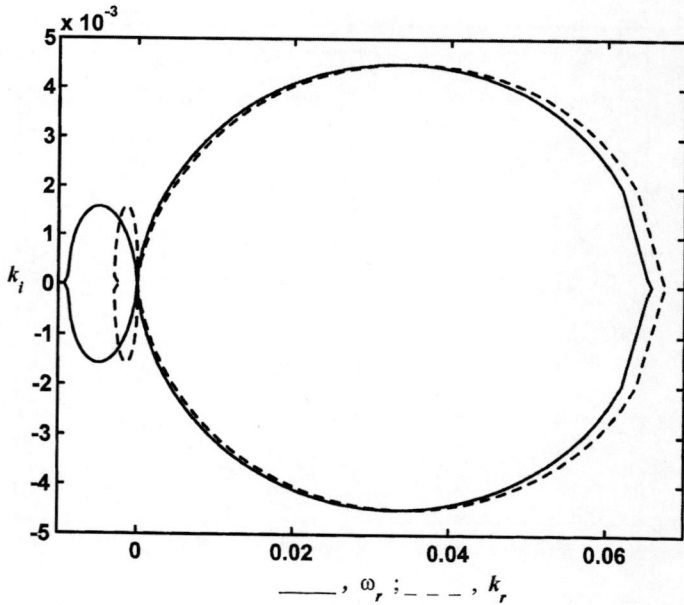

Figure 3.4. Amplification curve. $Q = 0.0013$, $We = 52.4$, sinuous mode. (From Lin and Jiang, 2003)

satisfies the causality condition that the disturbance is nonexistent when $t \rightarrow -\infty$ for all $0 \leq x \leq \infty$. Along both branches, it is found that $d\omega_r/dk_r > 0$. Therefore the wave packet consisting of all the Fourier components of the disturbance, regardless of whether introduced upstream or downstream of ξ_0, can only propagate in the downstream direction. The lower branch with $k_i < 0$ represents spatially growing waves, and the upper branch represents evanescent waves. For $k_r < 0$, the upper branch can be reached from $\omega_i > 0$ from the upper half of the k-plane where $k_i > 0$, and the lower branch can be reached from the lower half of the plane where $k_i < 0$. Thus the upper branch is relevant to the downstream region but the lower branch is relevant to the upstream region for the reasons just explained, $d\omega_r/dk_r > 0$ in both branches. Therefore the upper branch and the lower branch represent the downstream propagating wave packets which decay respectively in the downstream and upstream regions. The branches of amplification curves for $k_r < 0$ in Figures 3.2–3.4 possess the same properties and thus represent decaying waves. Hence the same explanations apply, and will not be repeated again for these figures. The statement of the fact that $d\omega_r/dk_r > 0$ is demonstrated in Figure 3.4. As explained in Chapter 2, the precise temporal and spatial growth of the disturbance must be determined from the double Fourier integral (2.28), except that (f', ψ_1, ψ_2) and x in (2.28) must be replaced respectively here by (h'_0, ψ'_e, ψ'_g) and r. It follows that the large time asymptotic behavior of the sinuous disturbance is given by (2.31) and (2.34) if $We > 1$. Thus for the case of $We > 1$, a sinuous disturbance decays in time at a given point in space but grows exponentially as one travels with the group velocity U_s corresponding to the frequency ω_s, which produces the maximum spatial amplification rate $-k_i(\omega_s)$. If $We \leq 1$ the sheet is absolutely unstable, and the sinuous disturbance grows algebraically according to (2.40) as it propagates in both downstream and upstream directions.

The limiting case of $Q = 0$ was considered in Exercise 2.6. As $Q \rightarrow 0$, the convectively unstable disturbance becomes stable and decays as fast as $\tau^{-1/2}$ locally, if $We > 1$. On the other hand if $We \leq 1$, the sheet instability changes from absolute to pseudo-absolute as $Q \rightarrow 0$ (Exercise 2.7).

3.1d. Varicose Modes

Stability analysis for the varicose mode can be similarly carried out based on (2.17a). Neither absolute nor pseudo-absolute instability is found. Convective instability remains for finite Q and can be enhanced by increasing Q as in the case of the sinuous mode. When $Q = 0$, the varicose disturbances are stable (Exercise 2.8). An example for finite Q is given in Figure 2.4. The upper branches represent downstream propagating evanescent waves and the lower branches represent downstream propagating convectively unstable disturbances. Both branches can be approached from the upper k-plane with $\omega_i > 0$. Notice that the amplification rates of the varicose mode and the sinuous mode are almost equal over the entire unstable wave spectrum for this case of $We = 1000$, $Q = 0.013$. This fact may have an important implication in the nonlinear evolution of the disturbance (Clark and Dombrowski,1972). However for the flow parameters in Figures 3.2 to 3.4, the growth rate of the varicose mode is two to three orders of magnitude smaller than that of the sinuous mode.

3.1e. Comparisons With Experiments

Huang carried out an extensive experiment to observe the breakup of a radially expanding liquid sheet formed by two equal vertical liquid jets impinging against each other. Two different regimes of sheet breakup were found. In regime I, the sheet breakup is along a nearly perfect circular edge. As the flow rate of the jet is increased, the sheet increases its radius but retains its smooth interface. In regime II, the breakup radius decreases with increasing jet flow rate, and the breakup is accompanied by a large-amplitude flaglike motion of the sheet. The broken edge becomes irregular (Villermaux and Clanet, 2002). Between these two regimes, there is a transition regime in which the Cardioid waves observed by Taylor (1959a) appear. They are gradually overshadowed by large-amplitude sinuous waves as the flow rate increases. Regimes I and II are reported to be, respectively, in the range $100 < We_d < 500$ and $2000 < We_d < 30,000$, where We_d is the Weber number based on the jet orifice diameter d,

$$We_d = \rho_l U^2 d / S,$$

U being the constant liquid velocity in the sheet. In regime I, the correlation between the breakup radius R_b and Weber number was found to be

$$R_b/(d/2) = 0.167 \, We_d. \tag{3.24}$$

The correlation in regime II depends on the gas to liquid density ratio. For a water sheet in air, it is given by

$$R_b/(d/2) = 1250 \, We_d^{-1/3}. \tag{3.25}$$

Thus R_b decreases as We_d increases. This was also observed earlier by Ostrach and Koestel (1965). The correlation in regime I was extended to $We_d = 800$ in the transition regimes, and that in regime II was extended back to $We_d = 1000$ by Huang.

Table 3.1. *Regimes of Sheet Breakup*

We_d	We_c	R_b (cm)	H_b ($\times 10^4$ cm)	$-k_i$	η/η_0	$d\omega_r/dk_r$ Experiments	$d\omega_r/dk_r$ Theory
100	1.0	1.33	31.8	4(E-6)	1.002	No wave	Cardioid
500	1.0	6.65	6.4	4(E-6)	1.04	No wave	Cardioid
800	1.0	10.64	4.0	4(E-6)	1.11	Cardi-Sin	Cardioid
1,000	1.3	9.92	4.1	1.1(E-4)	14.0	Sin-Cardi	0.12
2,000	3.3	7.88	5.2	6.4(E-4)	16,318	Sinuous	0.45
2,250	3.8	7.58	5.4	7.4(E-4)	32,448	0.55	0.49
4,450	9.5	6.04	6.8	1.6(E-3)	1.4(E6)	0.51	0.68
7,210	18.1	5.14	8.0	2.4(E-3)	5(E6)	0.48	0.76
10,300	29.1	4.57	9.0	3.3(E-3)	19(E6)	0.68	0.81
16,000	52.4	3.94	10.4	4.5(E-3)	25(E6)	0.73	0.86

To compare his experiments with our prediction, we relate We_d with We. The liquid discharge $\dot{Q}$ from the two orifices is given by

$$\dot{Q} = 2(C\pi d^2/4)U,$$

where C is the contraction coefficient given by Huang to be 0.67 in regime I and 0.65 in regime II. If $2H_b$ is the sheet thickness at $R = R_b$, then $\dot{Q} = 2\pi R_b(2H_b)U$. It follows that the local Weber number We_c at the sheet edge is

$$We_c = 0.167(d/2)We_d/R_b. \tag{3.26}$$

Similarly in regime II, we have

$$We_c = 0.1625(d/2)We_d/R_b. \tag{3.27}$$

The experimental values of We_d and R_b in the first three rows of Table 3.1 fall in the extended regime I. Substituting these values in (3.26) results in $We_c = 1$ for all We_d in regime I. Therefore regime I can be defined as the regime in which the onset of absolute instability occurs at the sheet edge. The condition that $We = 1$ at the broken edge was obtained by Taylor from the balance of inertial force and surface tension. Regime I was termed Savart–Bond regime by Huang because the linear dependence $R_d/(H_b/2)$ on We_d was found earlier by them (Savart, 1833; Bond, 1935). No waves were observed in regime I by Huang. He reported that Cardioid waves did not appear until the beginning of the transition regime at $We_d = 500$. On the other hand Taylor showed theoretically that Cardioid waves can be observed in the region $We > 1$ including regime I. The present theoretical prediction explains this discrepancy. The maximum spatial amplification rates k_i corresponding to the Weber number in the first two columns are given in the fifth column. Figures 3.2 to

3.4 illustrate how they are determined. The corresponding wave amplitudes normalized by the initial amplitudes are determined from $exp(-k_i R_b/H_b)$. They are given in the sixth column. The wave amplitudes including the initial amplitude η_0 were not recorded in Huang's experiments. We assume η_0 to be as small as 10^{-4} cm, which was measured in a radially expanding liquid sheet created by a fan spray nozzle in the experiments of Crapper et al. (1975). Then $\eta/\eta_0 = 1.04$ at $We_d = 500$ signifies that the wave amplitude η remains less than 4% greater than the initial amplitude $\eta_0 = 10^{-4}$ cm even near the edge of the sheet for $We_d < 500$. Note that $We > 1$ upstream of all the broken edges in Table 3.1. Thus the disturbances are all convectively unstable in the liquid sheet given in Table 3.1 except at the edges where $We = 1$. Therefore at any point in the sheet upstream of the broken edge the disturbance decays with time, but its amplitude grows as it is convected downstream and reaches its maximum observable magnitude near the edge. However, in regime I, it remains less than 1.04×10^{-4} cm, which is the threshold of optically observable magnitude. His experiments rely only on the reflection of regular light from the interface for wave detection. This may be why no waves were observed by Huang in regime I where $We_d < 500$. As We_d is increased to 800, the wave amplitude exceeds the threshold of optical detection by 11% and the Cardioid wave becomes observable. Wave motions in liquid sheets are described in Chapter 5. As explained by Taylor, the Cardioid wave is the loci of the components of the sinuous waves that are held stationary locally by the radial flow. In polar coordination (R, θ), the Cardioid is given by

$$\frac{r}{\bar{R}} = We^{-1} = [1 - cos(\theta - \theta_0)]/2, \tag{3.28}$$

where $\bar{R} = \rho U \dot{Q}/(4\pi S)$ and θ_0 is the angular position of an arbitrary point from which a Cardioid starts. In the range of $500 < We_d < 800$, a complete Cardioid can exist within the sheet with its outer edge tangent to the broken sheet edge according to (3.28), since $We \geq 1$ in this regime. This is consistent with Huang's observation that Cardioid waves predominate in this range. The dominated waves are the axisymmetric sinuous waves that cannot be held stationary and propagate radially downstream. This situation is indicated with Cardi-Sin in the seventh column. When We_d reaches 1000, according to Table 3.1, $We_c = 1.3$, and thus a complete Cardioid cannot exist within the sheet according to (3.28) since $\theta - \theta_0 < \pi$. Meanwhile the sinuous wave amplitude increases by one order of magnitude as indicated in the fourth row of Table 3.1. This is consistent with Huang's observation that "large amplitude antisymmetric (sinuous) waves appear and Cardioid waves diminish" across the line $We_d = 1000$. This is indicated by Sin-Cardi in the seventh column.

In this transition range of We_d, the amplitude of sinuous waves which break the edge with flaglike motion is quite large. The transition regime probably requires a nonlinear theory to describe. Nevertheless the numerical solution based on the present theory shows that the maximum spatial growth rate k_i increases as $We_d^{1/3}$ increases when Q is constant, and increases as Q increases when We_d is held constant. Thus if we assume that the breakup is to take place when the sinuous waves grow to be a certain amplitude, then the breakup radius decreases as $R_b \sim We_d^{1/3} Q^{-1}$. This may be compared with the criterion of Villermaux and Clanet $R_b \sim We_d^{1/3} Q^{-2/3}$. They obtain this criterion by assuming the breakup to be due to the Rayleigh-Taylor instability. Table 3.1 shows that as We_d exceeds 2000 the wave amplitudes become of order cm or greater. Comparisons of the results based on linear stability analysis with experiments at such large wave amplitudes are not legitimate. Moreover the waves exhibit nonaxisymmetric patterns at such a large Weber number (Villermaux and Clanet, 2002), and three-dimensional disturbances must be considered (Kim and Sirignano, 2000). Nevertheless the theoretical prediction appears to be consistent with Huang's observation that when We_d exceeds 2000, the large-amplitude antisymmetric waves predominate. Huang also measured wave speeds. The experimental values in the seventh column of Table 3.1 are read off from his Figure 3.5 near the broken edges. The theoretical wave speeds given in the last column are between 25% and 50% higher than the experimental values. The nonlinearity appears to have the effect of reducing the wave speed. We may conclude that while the breakup in regime I is due to the onset of absolute instability at sheet edge, the breakup in regime II is due to finite-amplitude convective instability associated with the sinuous mode. Finite-amplitude convective instability breakup has also been observed by Crapper et al. in a fan spray. They also offered a preliminary nonlinear theory, but comparisons with experiments were inconclusive.

3.2. Plane Liquid Sheet Thinning Under Gravity

3.2a. Basic Flow

A sketch of a plane liquid sheet of a constant width falling under the action of gravity is given in Figure 3.5. As the consequence of rapid fluid acceleration and conservation of mass, the local Weber number must increase in the flow direction. The effects of the Froude number in addition to the gas to liquid density ratio and the Weber number must be considered for a falling liquid sheet. The Froude number is already defined in Chapter 1. Here we take fluids 1 and 2 to be the liquid and gas respectively. The characteristic velocity U_0

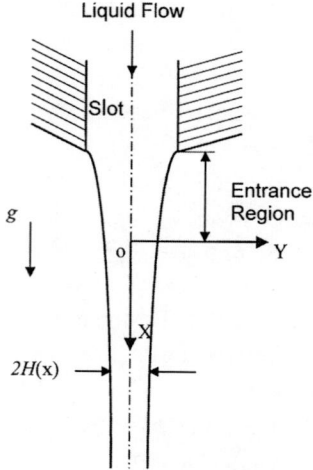

Figure 3.5. A falling liquid sheet.

and length h_0 are taken respectively to be the liquid velocity and the half sheet thickness at the nozzle exit. When the inertial force per unit mass of liquid is much greater than the gravitational force, then the sheet thins gradually as it falls. While the sheet thickness scales with h_0, the distance over which the thickness changes appreciably scales with U^2/g. Hence the Froude number is very large under this condition. Using $\varepsilon = 2Fr^{-1}$ as the small parameter, de Luca and Costa (1997a,b) obtained the basic flow velocity potential ϕ for an irrotational flow of an inviscid liquid falling vertically in a sheet,

$$\phi(y, \xi) = \frac{2}{3} \frac{[1 + (1 - Q_2)\xi]^{3/2}}{1 - Q_2} + O(\varepsilon^2), \qquad (3.29)$$

where $\xi = \varepsilon x$ and ϕ is related to the velocity components (v, w) in the Cartesian coordinates (y, x) by $v = \phi_{,y}$ and $w = \varepsilon\phi_{,\xi}$. The corresponding thickness variation is

$$h(\xi) = [1 + (1 - Q_2)\xi]^{-1/2}. \qquad (3.30)$$

It follows that the local Weber number $We = \rho U_0^2 h_0 \, w^2 h^2 / S$ increases with ξ. Note that the velocity does not change across the sheet thickness if the $O(\varepsilon^2)$ term is neglected.

3.2b. Local Instability

The length scale of ξ and x are so far apart that they can be treated as independent variables asymptotically. The effect of flow variation in space appears again only parametrically. The Weber number, disturbance wave number, frequency, and interfacial displacement are now functions of the slow variable ξ. Therefore the results of stability analysis for the uniform sheet in Chapter 2 can be applied locally to the present weakly nonparallel flow. Local stability analyses have been successfully used to describe the instability of other weakly nonparallel flows (Bouthier, 1972; Crighton and Gaster, 1976; Kulikovski, 1993; Lin, Phillips, and Valentine, 1993).

If $We > 1$ at the nozzle exit, then the entire falling sheet is supercritical because We increases in the downstream direction according to (3.29) and (3.30). Applying the results of Chapter 2 locally, we deduce that absolute instability will not occur in the present case. However the sheet is susceptible to convective instability with respect to sinuous and varicose mode disturbances if $Q_2 \neq 0$, according to (2.33). In Exercise 2.4 we show that when $Q_2 = 0$, the disturbance decays as fast as $\tau^{-1/2}$ locally as $\tau \to \infty$ in a sheet with $We > 1$ everywhere.

We will now further explain the transition from convective instability to stability as $Q_2 \to 0$. When $Q_2 = 0$, (2.17) can be rewritten as

$$\left[(k - \omega) \sqrt{k} \tanh(k) - k^2 \sqrt{We^{-1}} \right]$$
$$\times \left[(k - \omega) \sqrt{k} \tanh(k) + k^2 \sqrt{We^{-1}} \right] = 0 \qquad (3.31)$$

It can be shown readily that the two branches of the dispersion curves $\omega_i = 0$ coincide along the real k-axis and form a double root at $k = 0$, $\omega = 0$. However, this double root is not a pinch point singularity. Using Taylor's expansion shows that near this point

$$\omega_i = k_i(1 \mp We^{-1/2}) \mp We^{-1/2}(3k_r^2 - k_i^2)k_i/6 + \cdots, \qquad (3.32)$$
$$\omega_r = k_r(1 \mp We^{-1/2}) \mp We^{-1/2}(k_r^2 - 3k_i)k_r/6 + \cdots. \qquad (3.33)$$

It follows from these above equations that for the case of $We > 1$, which is under consideration, as the paths $\omega_i = C_\pm$ in the complex plane are lowered from $C_\pm > 0$ to $C_\pm = \varepsilon < 0$ the image of the path in the k-plane crosses over to $k_i = \varepsilon/(1 \mp W^{-1/2})$ in the lower half plane and coalesces as $\varepsilon \to -0$. Moreover $d\omega_r/dk_r > 0$ for both branches. Along these branches the wave packets propagate at different speeds but in the same direction. Therefore $(k, \omega) = (0, 0)$ is not a pinch point singularity. Branch points do not occur near $(0, 0)$ (c.f. Exercise 2.4). This leads to the sinuous disturbance decay rate $\tau^{-1/2}$ as $\tau \to \infty$.

However, when $We < 1$, (3.32) shows that as $C_\pm \to -\varepsilon$ the two images of $\omega_i = C_\pm$, one in the upper and the other in the lower k-plane, cross over into the opposite half planes. Moreover, according to (3.33) the group velocities corresponding to these two branches are in the opposite directions, and a pinch point singularity is formed at $(k, \omega) = (0, 0)$. The pinch point for the present case with $Q_2 = 0$ is of lower order than that discussed in Section 2.3 for the case of $Q_2 \neq 0$. For the present type of pinch point at which $\partial D/\partial k = \partial D/\partial \omega = \partial^2 D/\partial^2 k = \partial^2 D/\partial k \partial \omega = \partial^2 D/\partial^2 \omega = 0$, but not the higher derivatives, we find from the general formula (2.43) that the large time asymptotic behavior of the disturbance is

$$\sim exp\,(ik_0 x + \omega_0 \tau). \tag{3.34}$$

Thus the disturbance is nonvanishing but bounded for all time and all finite x. The corresponding instability is termed pseudo-absolute instability (Lin, Lian, and Creighton, 1990).

As Q_2 is raised from zero to a finite value, the instability becomes a full-fledged absolute instability when $We < 1$. The absolutely unstable sinuous disturbance grows algebraically as $\tau^{1/3}$ as $\tau \to \infty$ according to (2.40). In contrast to the case of the expanding sheet, absolute instability is now near the nozzle exit if it occurs. Because the disturbance in the absolute instability region can propagate in both the upstream and downstream directions, the disturbance originating near the nozzle exit will feedback to the upstream. The consequence of this feedback will be discussed later. Varicose mode disturbances can cause only convective instability if $Q_2 \neq 0$ (Exercise 2.3).

3.2c. *Initial Evolution of Long Waves*

Equations (2.17) and (2.17a) have been applied locally to dynamically different thinning inviscid liquid sheets. Values for k and ω must be chosen such that the governing equations with their boundary conditions are satisfied for given relevant flow parameters. For interfacial displacement of amplitude f and a wavelength that is long compared with the sheet thickness but shorter than ξ-scale, $tanh\,(k)$ in (2.17) can be approximated with k. In the absence of ambient gas, $Q_2 = 0$. The k and ω in the resulting equation arise respectively from the x- and τ-derivatives in the governing differential system. If we replace k and ω with the corresponding derivatives of f in the resulting approximate equation for long waves, we have

$$\left[\partial_\tau - (1 + We^{-1/2})\,\partial_x\right] \left[\partial_\tau - (1 - We^{-1/2})\partial_x\right] f = 0. \tag{3.35}$$

Hence along the characteristic lines $dx/dt = 1 \pm We^{-1/2}$, f remains constant. If a disturbance is introduced in the region where $We > 1$, two wave packets will travel downstream without attenuation at two different speeds $(1 \pm We^{-1/2}) > 0$. In Section 3.2a, we found for the case of $Q_2 = 0$ and $We > 1$ that indeed the disturbances all travel downstream, but they decay with time as fast as $\tau^{-1/2}$. The difference arises from the long-wave approximation in this section. With this approximation, the disturbance phase velocity relative to the local fluid particle velocity in the sheet is $\omega/k = \pm We^{-1/2}$ according to (2.17). Thus once a disturbance is introduced at a given point on the liquid sheet, it will split into two parts. One part will be convected downstream at velocity $1 + We^{-1/2} > 0$, but the other will be convected downstream at velocity $1 - We^{-1/2} > 0$ if $We > 1$. This is consistent with the prediction according to (3.35). However, the long-wave approximation makes the disturbance propagation velocity independent of wavelength or nondispersive. Consequently disturbances of different wavelengths travel together coherently, not allowing the disturbance energy to disperse. In Section 3.2b the disturbance including short wavelength is dispersive and allows its energy to disperse as fast as $\tau^{-1/2}$. Moreover the wave dispersion forces the present two packets to form one packet with the dominant component with $\omega = \omega_s$. This particular example illustrates a limitation of the long-wave approximation.

When a disturbance is introduced at a location where $We < 1$, a sinuous wave packet travels downstream at a velocity $1 + We^{-1/2} > 0$, but the other travels at a velocity $1 - We^{-1/2} < 0$ according to (3.35). Both packets travel without attenuation as predicted in Section 3.2a for the same case of $Q_2 = 0$. A numerical example is given in Figure 3.6. The long-wave approximation exhibits such a complete agreement with that predicted in Section 3.2b, where no such approximation was made, because the longest wave dominates the short-wave disturbance in causing absolute instability. In arriving at (3.35), we also assume that the sheet thickness and velocity vary slowly over a distance

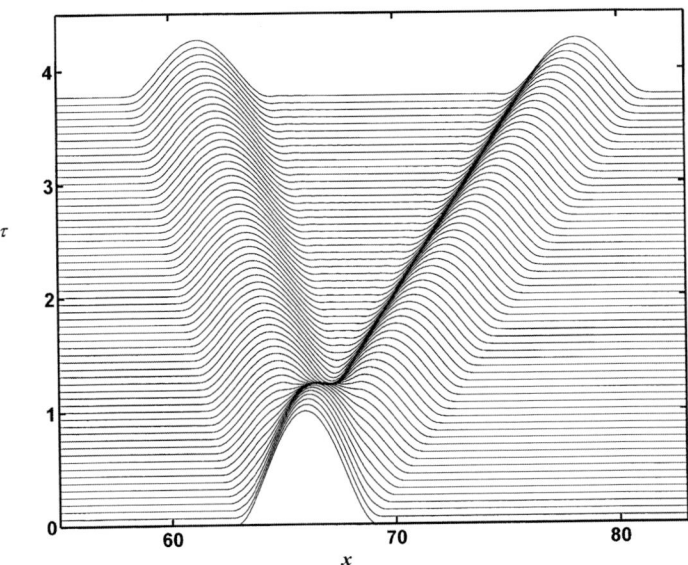

Figure 3.6. Sinuous wave, $We = 0.2$, $Q_2 = 0$.

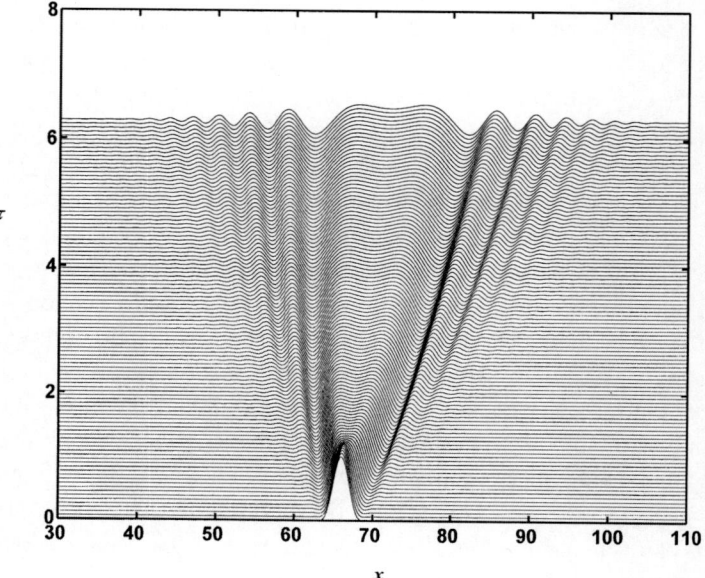

Figure 3.7. Convectively stable varicose waves. $We = 1.002$, $Q_2 = 0$.

of many wavelengths. If this distance is comparable with the wavelength, (3.35) is modified slightly, as discussed later in Exercise 3.1.

Similarly for varicose long waves we have (Exercise 3.2)

$$(\partial_\tau + \partial_x)^2 f + We^{-1} f_{,xxxx} = 0. \qquad (3.36)$$

The time evolution of f from a given initial displacement is given in Figure 3.7. The decay with time of a packet of dispersive varicose capillary waves as they are convected downstream on a liquid sheet is depicted in the figure to show the nature of convective stability. Note that the initial response to an impulsively introduced disturbance for both modes remains essentially the same as that predicted by the large time asymptotic theory given previously.

3.3. Curved Falling Sheet

Finnicum, Weinstein, and Ruschak (1993) showed that $We < 1$ can occur in a falling sheet without causing the sheet to disintegrate if the sheet is allowed to curve. Suppose that the occurrence of absolute instability in a falling liquid sheet deflects the sheet from its vertical position and the sheet eventually reaches a stationary position. Assume that in the first-order approximation, the flow in the sheet does not vary across the thickness. The time rate of change of momentum associated with the flow through the sheet of unit length must

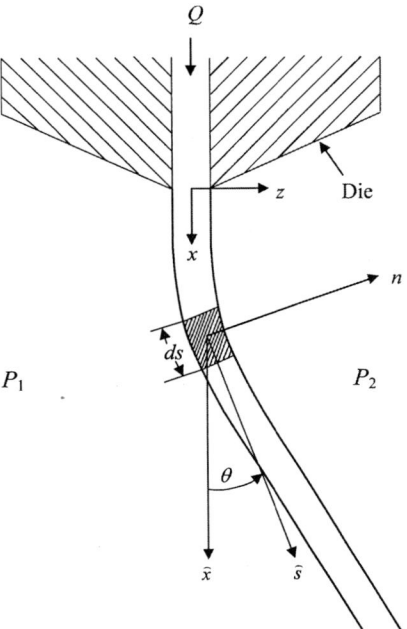

Figure 3.8. Curved falling liquid sheet.

be balanced by the total force acting on the segment. The balance equations
parallel and normal to the sheet are respectively given by (cf. Fig. 3.8)

$$\frac{d(\rho V^2 h)}{dX} = \rho g h, \tag{3.37}$$

and

$$\rho V^2 (2h) \frac{d\theta}{ds} = 2S \frac{d\theta}{ds} + (P_1 - P_2) - \rho g (2h) \left(\frac{dF}{ds}\right)\left(\frac{dX}{ds}\right), \tag{3.38}$$

where $2h$ is the local sheet thickness, X is the distance measured in the g-
direction, F is the distance measured from the X-axis to the sheet centerline,
s is the distance along the sheet, θ is the local tangent angle relative to the
X-axis, and V is the fluid velocity along the sheet under free fall,

$$V = \left(V_0^2 + 2gX\right)^{1/2}, \tag{3.39}$$

with V_0 the velocity at the die slot from which the sheet emanates. Using
geometric relations between θ, s, f, and X, nondimensionalizing the velocity
with the capillary velocity,

$$V_s = S/\rho V_0 h_0,$$

and all distances with $V_s^2/2g$, integrating nondimensionalized (3.38) once with respect to $\xi = X/(V_g^2/2g)$, and incorporating (3.37), we have

$$\frac{df/d\xi}{\left[1 + (df/d\xi)^2\right]^{1/2}} = \frac{\alpha\xi + C}{We - 1}, \tag{3.40}$$

where $f = F/(V_g^2/2g)$, C is an integration constant depending on the slope of f at $\xi = 0$, and

$$\alpha = \frac{(P_1 - P_2)\,S}{\rho^2 \dot{Q}^2 g}. \tag{3.41}$$

$\dot{Q} = 2V_0 h_0 = 2Vh$ is the constant volumetric flow per unit width of the sheet, and α is an additional parameter characterizing the imposed pressure difference across the sheet thickness. Upon substitution of (3.28) into the definition $We = \rho V^2 h/S$, we have

$$We = \left(We_0^2 + \xi\right)^{1/2}. \tag{3.42}$$

It follows from (3.40) and (3.42) that if $We_0 > 1$, (3.40) can be integrated with the initial condition $f = 0$ at $\xi = 0$ to determine the curtain shape. On the other hand when $We_0 < 1$, there will be a point downstream where $We = 1$. At this point (3.40) is singular unless C is properly chosen. $We = 1$ occurs at $\xi = \xi_s$. According to (3.42),

$$\xi_s = 1 - We_0^2. \tag{3.43}$$

Expanding ξ about ξ_s, we have

$$\frac{df/d\xi}{\left[1 + (df/d\xi)^2\right]^{1/2}} = \frac{\alpha\,(\Delta\xi + \xi_s) + C}{\frac{1}{2}\Delta\xi + O\left[(\Delta\xi)^2\right]},$$

where $\Delta\xi = \xi - \xi_s$.

Thus the right side of this equation is finite in the limit of $\Delta\xi \to 0$ if $C = -\alpha\xi_s$. This choice of C, however, dictates the deflection angle at $\xi = 0$ to be finite. It follows from (3.40) and (3.43) that θ at $\xi = 0$ is

$$\theta_0 = sin^{-1}\left[\alpha\,(1 + We_0)\right].$$

Finite θ_0 can be understood to be the consequence of the upstream propagating, absolutely unstable disturbance near the slot exit where $We < 1$. Note that when $P_1 = P_2$, $\alpha = 0$ and thus $\theta_0 = 0$.

3.4. Liquid Sheet Breakup Phenomena

It was pointed out at the end of Section 3.2 that if regions of absolute and convective instability coexist in a planar sheet, the former region must be located upstream of the latter region. The relative locations of these two regions are completely reversed in a radially expanding sheet. It was demonstrated that the upstream propagating, absolutely unstable sinuous waves collide with downstream propagating, convectively unstable sinuous waves at the circular edge where $We = 1$. The condition that $We = 1$ at the broken circular edge was obtained by Taylor from a balance between the liquid momentum change rate and the edge surface force. However, Huang observed that a broken edge can also exist at $We > 1$ because of convectively unstable sinuous waves in breakup regime II, the reason of which is explained in our theory.

Following the concept that global instability will result if the region of local absolute instability is sufficiently large in a flow (Chomaz, Huerre, and Redekopp, 1988; Monkewitz, Huerre, and Chomaz, 1993; Le Dizes, 1997), de Luca (1999) correlated the length of an accelerating liquid sheet, in which the sheet is locally absolutely unstable near the nozzle, with relevant flow parameters. It was argued that if a sufficiently large region of local absolute instability exists upstream of a convectively unstable region, a liquid sheet will rupture because of sinuous waves caused by global instability. However, theoretically if absolute instability occurs first at the nozzle exit, a sheet cannot be formed from the outset.

In contrast, in the axisymmetric expanding liquid sheet the Weber number decreases radially, and the region of absolute instability occurs downstream of the region of convective instability. The region of convective instability can at most extend only to a radius where $We = 1$. Local absolute instability is predicted in the rest of the flow field extending to $r \to \infty$ where the sheet thickness becomes infinitesimally thin. Hence the criterion that the region of local absolute instability must be sufficiently large in order for global instability to develop is amply satisfied. If global instability means that the flow is unstable in the entire flow region, then the present flow is globally unstable because only a small portion of an infinitely expanding sheet is convectively unstable. In fact, without the existence of the convectively unstable region, the absolutely unstable region could not have appeared in the flow. The present absolute instability can be viewed as the consequence of the convectively unstable disturbance entering from upstream. The existence of a convectively unstable region is necessary but not sufficient for absolute instability to occur. The spatial growth of the sinuous wave must be sufficiently small to allow the expanding sheet to reach the broken edge at $We = 1$.

The experimental results of de Luca seem to suggest that a falling plane liquid sheet cannot be formed if $We_0 < 1$. Paradoxically an intact falling plane liquid sheet with a large region of $We < 1$ was observed and predicted by Weinstein and coworkers (Finnicum et al., 1993; Weinstein et al., 1997;

Clarke et al., 1997; Weinstein, Hoff, and Ross, 1998). Their prediction of the existence of an intact subcritical region with $We < 1$ is based on a time-independent ordinary differential equation they derived for a gradually thinning inviscid liquid sheet falling under the action of gravity. A singularity arises at a location where $We = 1$ even if $We_0 > 1$. However, the singularity can be removed by allowing the sheet to adjust itself in response to the upstream propagating disturbances, in the presence or absence of the pressure drop across the liquid sheet. Experimentally, this is achieved by not using vertical guide wires as in the studies of de Luca and Costa and others. The liquid curtain is then guided by the side walls and assumes a curved shape. The shape of the curved liquid sheet depends on the location of $We = 1$ in a sheet of finite length in the experiment. It appears that the favorable capillary force associated with the imposed curvature allows the sheet to maintain an intact region in which $We < 1$. For the special case of zero pressure drop across the sheet, a vertical sheet with a large region of $We < 1$ near the nozzle was predicted and observed to exist by Finnicum et al. Therefore the onset of local absolute instability at $We = 1$ in an accelerating planar sheet does not appear to lead to global instability and sheet rupture as inferred by de Luca and Costa. Weinstein et al. (1997) and Clarke et al. (1997) also analyzed the wave motion by displacing the two interfaces symmetrically (varicose mode) and antisymmetrically (sinuous mode). They found that the sinuous wave can propagate only in the downstream direction if $We > 1$ everywhere in the sheet, but it can propagate in both the upstream and downstream directions in the local region where $We < 1$. The varicose wave can only propagate downstream. These waves were predicted by de Luca and Costa (1977a,b) to be the consequence of local convective and absolute instabilities. The same wave characteristics on a viscous liquid curtain were predicted earlier (Lin, 1981; Lin and Roberts, 1981; Lin and Lian, 1989; Lin et al., 1990; Teng, Lin, and Chen, 1997). While in the theory of de Luca and Costa, $We \leq 1$ signifies nonformability of the liquid sheet, it signifies in the work of Weinstein and his co-workers that the upstream propagating sinuous waves will eventually reach the nozzle exit and force a vertical liquid sheet to bulge at the nozzle exit or to leave at a finite angle with the nozzle orientation if a stationary sheet with finite region of $We < 1$ near the nozzle is to be maintained with the help of a pressure difference across the sheet. The appearance of an intact region with $We < 1$ could also be due to the stabilizing effect of the capillary meniscus formed via wetting along the side walls in the experiments of Weinstein et al. However the meniscus effect is not included in their analysis. If $We_0 > 1$ then $We > 1$ everywhere in the planar liquid sheet, and Weinstein et al. observed that no disturbance can reach the nozzle exit and the sheet

Figure 3.9. A liquid leaf formed by two equal jets making an angle $60°$.

can be issued vertically. On the other hand in the radially expanding axisymmetric liquid sheets of Savart (1833), Taylor (1959), and Huang (Regime I), even if $We_0 > 1$, the flow becomes critical at a downstream location where $We = 1$. Beyond this point the predicted basic flow is absolutely unstable, but the upstream propagating, absolutely unstable disturbance cannot penetrate this point to reach the nozzle exit.

A leaf of liquid sheet can be formed by impinging two equal coplanar but noncoaxial liquid jets on each other. An example is given in Figure 3.9. The sheet thickness decreases again inversely with the radical distance from the impinging jets, and the thickness variation along the polar angle is quite significant as demonstrated by Taylor (1960). Atomization occurs along the entire edge of the liquid leaf. It appears that the Weber number based on the local velocity perpendicular to the leaf edge reached the critical value of 1 along the entire leaf edge. Taylor (1960) used this fact to determine the shape of the leaf edge with the help of the measured thickness variation in the polar angle direction. Here we test the prediction based on the local stability analysis given in Section 3.1 by comparing the theoretical results with the observation given in Figure 3.9. By using of Equation (18), Figure 9 of Taylor (1960), and his estimate of the sheet velocity $U = 2gH$, where H is the water head he used to produce the sheet shown in Figure 3.9, we can calculate the Weber number near the observed waves to be $We = 5.4$. The experiment was carried out at 1 atmosphere, therefore $Q_2 = 0.0013$. The spatial amplification curve for this case is shown in Figure 3.10. The maximum amplification of the sinuous wave occurs at $k_i = 0.0035$. The corresponding wavelength is 1.8 cm, which compares quite well with the observed wavelength of 2 cm. Note that the region near the impinging jet is free of waves reflecting the convective instability nature of the liquid leaf.

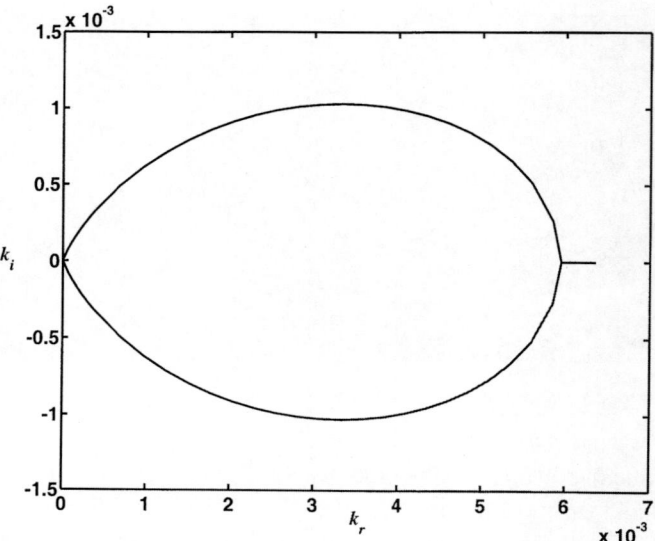

Figure 3.10. Spatial amplification curve. $We = 5.4$, $Q_2 = 0.0013$.

A round liquid jet impinging on a conical surface including the case of a 180° apex angle, with a sufficiently large momentum, may be reflected to form an axisymmetric sheet which closes at a downstream location to form a "water bell" (Savart, 1833a,b). Boussinesq (1869a,b) gave a catenary to the shape of Savart's water bell, from a balance between surface tension and centrifugal forces. The effect of liquid viscosity in the absence of gravity was recently considered by Clanet (2001). He also observed the instability of the water bell and gave physical explanation to it based on a model equation. Neglecting the ambient air and gravity, Taylor showed that the local Weber number cannot be less than one anywhere in the water bell, if it is to form. In the presence of gravity, Dumbleton (1969) showed that an upward pointing water bell against the gravitational pull will collapse to form a conical sheet which breaks up at an edge with $We = 1$. The breakup of water bells at $We = 1$ probably is related to the onset of absolute instability analogous to the situation in an axially expanding liquid sheet. The effect of feedback of disturbances to the nozzle exit considered by Yakubenko (1997) and Soderberg and Alfredsson (1998) is not observed in Taylor's experiment and is most likely unimportant in the radially expanding sheet with $We \gg 1$

Exercises

3.1. Weinstein et al. (1997) derived an equation for the free surface displacement whose length scale is comparable to that of the falling sheet thickness variation. Omitting the imposed pressure and using the normalization of Section 3.2, their equation can be written as

$$f_{,\tau\tau} + 2f_{,x\tau} + \frac{1}{u}[u(1 - We^{-1})f_{,x}]_{,x} = 0.$$

Show that if u varies appreciably only over a distance $\xi = \varepsilon x$ where $2Fr^{-1} = \varepsilon \ll 1$, the above equation is reduced to (3.35).

3.2. Obtain (3.36) from (2.17a) with $Q_2 = 0$ for long waves (Taylor, 1959a).

3.3. Show that in the absence of ambient gas the long sinuous waves at the onset of instability in a liquid sheet do not grow spatially (i.e., $k_i = 0$). The two group velocities $(1 - We^{-1/2})$ and $(1 + We^{-1/2})$ are equal to the phase velocities of its own constituent waves of all wavelength. Show that both wave groups satisfy the condition for the stationary phase.

3.4. For the case of $Q_2 = 0$, show that the varicose mode of disturbance does not cause absolute instability for all We in a liquid sheet.

3.5. A cylindrical jet of an inviscid liquid impinges at a mass flow rate $\dot{m}$ and velocity v_0 on a conical surface with a half apex angle ψ_0. The liquid is reflected by the conical surface and forms an axisymmetric bell-shaped liquid sheet that closes itself at the downstream side. Neglecting the ambient gas, show that the shape of the "water bell" in the cylindrical coordinates (r, θ, z), if it is formed, is given by (Lance and Perry, 1953)

$$R = \frac{e - \beta r}{\beta \sin\psi + r\cos\psi / e}, \tag{3.44}$$

where R is the local radius of curvature, $\beta = 4\pi S / \dot{m} v_0$, $\gamma = g / v_0^2$, g is the gravitational acceleration in the positive axial z-direction, $e = (1 + 2rz)^{1/2}$, and ψ is the angle between the local tangent plane and a horizontal plane. Neglecting gravity, show that the local Weber number We is related to ψ by

$$We^{-1} = 1 \mp (\sin\psi_0 / \sin\psi).$$

Show that for finite ψ_0, $(We)_{min}$, which is always greater than one, occurs at the maximum radius of the bell where the sheet thickness is the smallest. When $\psi_0 = 0$, the bell becomes a radially expanding liquid sheet in which $We \geq 1$.

3.6. Show by numerical integration of (3.44) that when a water bell is formed upward against gravity, the bell collapses to form a conical sheet with an edge located at $We = 1$ (Dumbleton, 1969).

References

Antoniades, M. G., Roberts, G., and Lin, S. P. 1980. *J. Coll. Interface Sci.* **77**, 583–587.

Baird, M. H. I., and Davidson, J. F. 1962. *Chem. Eng. Sci.* **17**, 467.

Bender, C. M., and Orszag, S. A. 1978. *Advanced Mathematical Methods for Scientists and Engineers.* McGraw-Hill.

Bers, A. 1975. Linear waves and instabilities. In *Physique des Plasmas* (Ed. C. De Witt and J. Peyraud). Gordon and Breach.

Bond, W. N. 1935. *Proc. Phys. Soc.* **47**, 549.

Boussinesq, J. 1869a. Théories des expérience de Savart sur la forme que prend une veine liquid e après choquée contre un plan circulaire. C. R. *Acad. Sci. Paris* **69**, 45.

Boussinesq, J. 1869b. Théories des expérience de Savart sur la forme que prend une veine liquid e après etre heurtée contre un plan circulaire (suite). C. R. *Acad. Sci. Paris* **69**, 128.

Bouthier, M. 1972. *J. Méc.* **4**, 599.

Brown, D. R. 1961. A study of the behaviour of a thin sheet of moving liquid. *J. Fluid Mech.* **10**, 297–305.

Chomaz, J. M., Huerre, P., and Redekopp, L. G. 1988. Bifurcations to local and global modes in spatially developing flows. *Phys. Rev. Lett.* **60**, 25–28.

Chubb, D. L., Calfo, F. D., Mcconley, M. W., Mcmaster, M. S., and Afjeh, A. A. 1994. Geometry of thin liquid sheet flows. *AIAA, J.* **32**, 1325–1328.

Clanet, C., and Villermaux, 2002. *J. Fluid Mech.* **462**, 307.

Clanet, C. 2001. Dynamics and stability of water bells. *J. Fluid Mech.* **430**, 111.

Clark, C. J., and Dombrowski, N. 1972. Aerodynamic instability and disintegration of inviscid liquid sheets. *Proc. R. Soc. Lond.* A **329**, 467–478.

Clarke, A., Weinstein, S. J., Moon, A. G., and Simister, E. A. 1997. Time dependent equation governing the shape of a two-dimensional liquid curtain. Part 2: Experiment. *Phys. Fluids.* **9**, 3637–3644.

Crapper, G. D., Dombrowski, N., and Jepson, W. P. 1975. Wave growth on thin sheets of non-Newtonian liquids. *Proc. R. Soc. Lond.* A **342**, 225–236.

Crapper, G. D., Dombrowski, N., and Pyott, G. A. D. 1975. Large amplitude Kelvin–Helmholtz waves on thin liquid sheets. *Proc. R. Soc. Lond.* A **342**, 209–224.

Crapper, G. D., Dombrowski, N., Jepson, W. P., and Pyott, G. A. D. 1973. A note on the growth of Kelvin–Helmholtz waves on their liquid sheets. *J. Fluid Mech.* **57**, 671–672.

Crighton, D. G., and Gaster, M. 1976. *J. Fluid Mech.* **77**, 397.

de Luca, L., and Costa, M. 1997a. Stationary waves on plane liquid sheets falling vertically. *Eur. J. Mech. B/Fluids* **16**, 75–88.

de Luca, L., and Costa, M. 1997b. Instability of a spatially developing liquid sheet. *J. Fluid Mech.* **331**, 127–144.

de Luca, L. 1999. Experimental investigation of the global instability of plane flows. *J. Fluid Mech.* **399**, 355–376.

Dombrowski, N., and Johns, W. R. 1962. The aerodynamic instability and disintegration of viscous liquid sheet. *Chem. Eng. Sci.* **18**, 203–214.

Dumbleton, J. H. 1969. Effect of gravity on shape of water bells. *J Appl. Phys.* **40**, 3950–3954

Finnicum, D. S., Weinstein, S. J., and Ruschak, K. J. 1993. The effect of applied pressure on the shape of a two-dimensional liquid curtain falling under the influence of gravity. *J. Fluid Mech.* **255**, 647–665.

Fraser, R. P., Eisenklam, P., Dombrowski, N., and Hasson, D. 1962. Drop formation from rapidly moving liquid sheets, *AIChE. J.* **8**, 672–680.

Hagerty, W. W., and Shea, J. F. 1955. A study of the stability of plane fluid sheets. *J. Appl. Mech.* **22**, 509–514.

Huang, J. C. P. 1970. The breakup of axisymmetric liquid sheets, *J. Fluid Mech.* **43**, 305–319.

Huerre, P., and Monkewitz, P. A. 1990. Local and global instabilities in spatial developing flows. *Ann. Rev. Fluid Mech.* **22**, 473–537.

Kim, I., and Sirignano, W. A. 2000. *J. Fluid Mech.* **410**, 147.

Koch, W. 1985. Local instability characteristics and frequency determination of self-excited wake flows. *J. Sound Vib.* **99**, 53–83.

Kulikovski, A. G. 1993. *J. Appl. Math. Mech.* **57**, 851.

Lance, G. N., and Perry, R. L. 1953. *Proc. Phys. Soc. Lond.* B **66**, 1067.

Le Dizes, S. 1997. Global modes in falling capillary jets. *Eur. J. Mech. B/Fluids* **16**, 761–778.

Li, X., and Tankin, R. S. 1991. On the temporal instability of a two-dimensional viscous liquid sheet. *J. Fluid Mech.* **226**, 425–443.

Li, X. 1993. Spatial instability of plane liquid sheets. *Chem. Eng. Sci.* **48**, 2973–2981.

Lienhard, J. H., and Newton, T. A. 1966. Effect of viscosity upon liquid velocity in axi-symmetrical sheets. *Z. Ang. Math. Phys.* **17**, 348–353.

Lin, S. P. 1981. Stability of a viscous liquid curtain. *J. Fluid Mech.* **104**, 111–118.

Lin, S. P., and Lian, Z. W. 1989. Absolute instability of a liquid jet in a gas. *Phys. Fluids* A **1**, 490–493.

Lin, S. P., and Roberts, G. 1981. Waves in a viscous liquid curtain. *J. Fluid Mech.* **112**, 443–458.

Lin, S. P., Lian, Z. W., and Creighton, B. J. 1990. Absolute and convective instability of a liquid sheet. *J. Fluid Mech.* **220**, 673–689.

Lin, S. P., Phillips, W. R. C., and Valentine, D. T. 1993. *Nonlinear Instability of Non-parallel Flows.* Springer-Verlag.

Lin, S. P., and Jiang, W. Y. 2003. Absolute and convective instability of a radially expanding liquid sheet. *Phys. Fluids* **15** (in print).

Matsumoto, S., and Takashima, Y. 1971. Studies of the standard deviation of sprayed drop size distribution. *J. Chem. Eng.* (Japan) **4**, 257–263.

Monkewitz, P. A. 1990. The role of absolute and convective instability in predicting the behavior of fluid systems. *Eur. J. Mech. B/Fluids* **9**, 395–413.

Monkewitz, P. A., Huerre, P., and Chomaz, J. M. 1993. Global linear stability analysis of weakly non-parallel shear flows. *J. Fluid Mech.* **251**, 1–20.

Nafeh, A. H., and Mook, D. T. 1979. *Nonlinear Oscillation*, John Wiley & Sons.

Ostrach, S., and Koestel, A. 1965. *AIChE J.* **11**, 294.

Savart, F. 1833a. Memoire sure le choc d'une veine liquid lancee contre un plan circulair. *Ann. Chim. Phys.* **LIX**, 55–87.

Savart, F. 1833b. Memoire sure le choc de deux veine liquides, animees de mouvements directement opposes. *Ann. Chim. Phys.* **LIX**, 257–310.

Soderberg, L. D., and Alfredson, P. H. 1998. Experimental and theoretical stability investigations of plane liquid jets. *Eur. J. Mech. B/Fluids* **17**, 189–737.

Squire, H. B. 1953. Investigation of the instability of a moving liquid film. *Brit. J. Appl. Phys.* **4**, 167–169.

Taylor, G. I. 1959a. The dynamics of thin sheets of fluid. II. Waves on fluid sheet. III. Disintegration of fluid sheets. *Proc. R. Soc. Lond.* A **253**, 296–321.

Taylor, G. I. 1959b. The dynamics of thin sheets of fluid. I. Water bells, *Proc. R. Soc. Lond.* A **253**, 289–295.

Taylor, G. I. 1960. Formation of thin flat sheet of water. *Proc. R. Soc. Lond.* A **259**, 1–17.

Teng, C. H., Lin, S. P., and Chen, J. N. 1997. Absolute and convective instability of a viscous liquid curtain in a viscous gas. *J. Fluid Mech.* **332**, 105–120.

Trefethen, L. N., Trefethen, A. E., Redy, S. C., and Driscoll, T. A. 1993. Hydrodynamic stability without eigenvalues. *Science* **261**, 578–584.

Villermaux, E., and Clanet, C. 2002. *J. Fluid Mech.* **462**, 341.

Weihs, D. 1978. Stability of thin, radially moving liquid sheets. *J. Fluid Mech.* **87**, 289.

Weinstein, S. J., Clarke, A., Moon, A. G., and Simister, E. A. 1997. Time-dependent equation governing the shape of a two-dimensional liquid curtain. Part 1. Theory. *Phys. Fluids.* **9**, 3625–3636.

Weinstein, S. J., Hoff, J. W., and Ross, D. S. 1998. Time-dependent equations governing the shape of a three-dimensional liquid curtain. *Phys. Fluids.* **10**, 1815–1818.

Yakubenko, P. A. 1997. Global capillary instability of an inclined jet. *J. Fluid Mech.* **346**, 181–200.

4

Viscous Liquid Sheet

In the previous chapter we mentioned that fluid viscosity might alter the critical Weber number that divides the parameter space into regimes of absolute and convective instability. The effects of gas and liquid viscosities are investigated separately in this chapter, not just to understand each individual effect but also to demonstrate the coupled effect, which is unexpected. In Chapter 3, stability analysis for an inviscid liquid sheet of uniform thickness was applied locally to investigate the stability of gradually thinning liquid sheets. The thinning was either due to axial expansion or gravitational acceleration. The local application of the inviscid theory for a uniform sheet to the two different cases of nonuniform sheets was made judiciously. Likewise the viscous theories given in this chapter can be applied judiciously to a gradually thinning viscous sheet whatever the cause of the thinning. The thinning may be caused by kinematic requirements, gravitational acceleration, or viscous extrusion. The breakup of a viscous liquid sheet in an inviscid gas is expounded in Section 4.1. The effect of gas viscosity is elucidated in Section 4.2. The effects of liquid and gas viscosities on the onset of sheet breakup are summarized in Section 4.3.

4.1. A Viscous Sheet in an Inviscid Gas

The basic flow attributed to G. I. Taylor is given in Section 4.1a, and its stability is analyzed in Section 4.1b. The physical mechanism of the sheet breakup is discussed in Section 4.1c, based on energy considerations.

4.1a. Basic Flow

Consider the steady flow in a Newtonian liquid sheet, which is extruded vertically downward by the viscous drag exerted by the horizontal moving plate, as shown in Figure 4.1. In film coating applications, the horizontal substrate moves as fast as 70 km/hr. Thus the fluid in the sheet can move faster

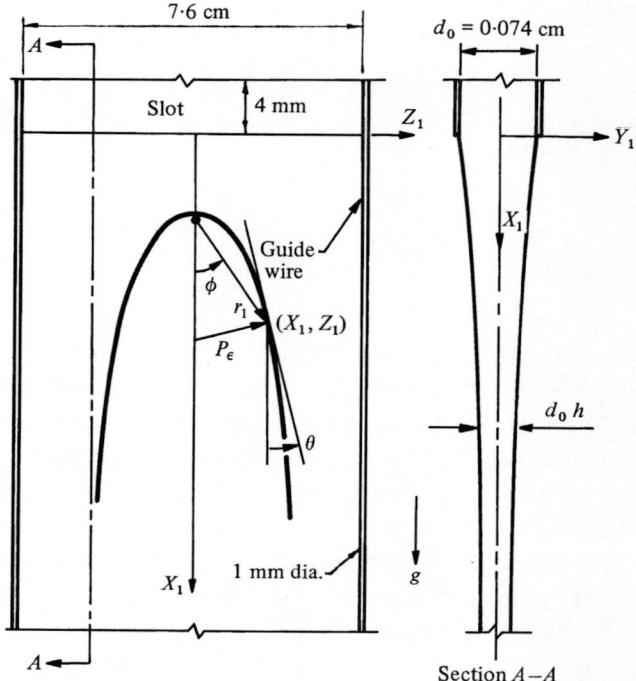

Figure 4.1. An extruded viscous sheet.

than that in a simple falling liquid sheet. Assuming that the flow is essentially two dimensional and that the effects of surface tension and the ambient gas as well as the normal stress variation across the sheets are negligible, Taylor derived the following dimensionless differential equations for the velocity variation in the liquid sheet according to Brown (1961):

$$(U,_X/U),_X - U,_X + U^{-1} = 0, \qquad U,_X + V,_Y = 0, \qquad (4.1)$$

where U and V are the dimensionless velocity components in the direction of the Cartesian coordinates X and Y respectively, and the subscripts denote differentiation. The derivation of (4.1) can be found in the work of Brown. The dimensionless quantities appearing in (4.1) are related to their dimensional counterparts (U_1, V_1) and (X_1, Y_1) by

$$(U_1, V_1) = (4\mu_1 g/\rho_1)^{1/3} (U, V),$$
$$(X_1, Y_1) = (4\mu_1/\rho_1)^{2/3} g^{-1/3} (X, Y),$$

where μ_1 and ρ_1 are respectively the dynamic viscosity and density of the liquid, and the distance X_1 is measured from the upper edge of the sheet in the direction of the gravitational acceleration g. Brown found good agreement between the numerical solution of (4.1) and his measured velocities,

$$U_1(X_1) = \left[(\dot{Q}/2H_0)^2 + 2gX_1 - 4(4vg)^{2/3} \right]^{1/2}, \tag{4.2}$$

except near the outlet of the liquid feeder. $\dot{Q}/2H_0 = U_0$, where $\dot{Q}$ in (4.2) is the volumetric flow rate per unit width of the sheet.

In the present stability analysis, the distance will be nondimensionalized with half of the maximum sheet thickness H_0, and the velocity by U_0. Thus the new dimensionless basic state velocity and the Cartesian coordinates are given respectively by $(\bar{u}, \bar{v}) = (U_1, V_1)/U_0$ and $(x, y) = (X_1, Y_1)/H_0$. It follows from the continuity equation that

$$\bar{u}_x = \delta U_X = -\bar{v}_y,$$

$$\delta = \left(\frac{Re_0}{Fr_0^2} \right)^{1/3} = \left(\frac{g^2}{4v} \right)^{1/3} \frac{H_0^2}{\dot{Q}},$$

$$Re_0 = \frac{\rho_1 U_0 H_0}{\mu_1} = \text{Reynolds number},$$

$$Fr_0 = \frac{U_0^2}{g H_0} = \text{Froude number}.$$

According to Figure 5 of Brown, $U_X = O(1)$. Thus for the case of thin sheets such that $\delta \ll 1$, the spatial variation in the basic-state velocity is much smaller than order one, according to (4.2). We consider here only the case of $\delta \ll 1$. This inequality is amply satisfied in many applications, including film coating. In the film coating application, the viscous liquid sheet is extruded by a rapidly moving substrate upon which the sheet falls perpendicularly. For any given feed rate $\dot{Q}$ the sheet thickness may become so thin, owing to the fast-moving substrate, that δ becomes much smaller than one. The condition $\delta \ll 1$ implies that the gravitational force is much smaller than the viscous force because δ represents the ratio of the former to the latter force. The inertial force and body force may be of the same order. Hence the basic flow velocity varies very slowly over a distance that is large compared with H_0, and thus the basic flow of the present problem is given by

$$\bar{u}_1 = U_1/U_0 = \left[1 + 2(4/Re_0 Fr_0)^{1/3}\xi - 4(4/Re_0 Fr_0)^{2/3} \right]^{1/2},$$

$$\bar{v}_1 = \bar{v}_2 = \bar{u}_2 = 0, \qquad \bar{p}_1 = \bar{p}_2,$$

where $\xi = \delta x$ is the stretched distance, and $\bar{p}_1$ and $\bar{p}_2$ are the pressures in the liquid and gas normalized with $\rho_1 U_0^2$. Note that $\bar{u}_x \to 0$ as $\delta \to 0$. In addition to $\delta \ll 1$, we impose the condition $(Re_0 Fr_0) = O(1)$ so that the stretched distance is physically consistent.

4.1b. Stability Analysis

Following Taylor's (1963) treatment of a related problem, we neglect the viscosity of the ambient gas based on the assumption that the instability is caused by capillary waves that may be assisted by the pressure fluctuation at the interface and on the assumption that gas viscosity is weakly stabilizing. This last assumption is based on the finding of Weber (1931) who showed that gas viscosity is stabilizing for a viscous liquid jet emerging into a quiescent gas. In Section 4.2 we will see that these assumptions are not entirely correct; gas viscosity can have a fundamental impact on absolute instability.

Substituting the perturbed velocity and pressure into the Navier–Stokes equations, subtracting out the basic state, then retaining only the terms of the first order in perturbation, we have the following governing equations of linear stability in a dimensionless form:

$$(\partial_\tau + \delta_{i1}\partial_x)\, v_i = -(\rho_i/\rho_1)\nabla p_i + (\delta_{i1}/Re)\nabla^2 v_i, \qquad (4.4)$$

$$\delta_{i1} = \begin{cases} 1, & i = 1, \\ 0, & i \neq 1, \end{cases}$$

$$\nabla \cdot v_i = 0 \qquad (i = 1,\, 2), \qquad (4.5)$$

where $i = 1$ and 2 stands for the liquid and gas, and the lower-case letters without over bars stand for perturbations of the corresponding basic flow quantities designated with over bars. In (4.4) and (4.5) the velocity and distance are normalized with the local velocity $U_0\bar{u}_1(\xi)$ and local half sheet thickness $h(\xi)$.

In linear theory, the boundary conditions need not be applied at the perturbed interfaces

$$\zeta = \pm 1.0 + \eta_\pm,$$

where η_+ and η_- are respectively the interfacial displacement from $y = 1.0$ and $y = -1.0$. We need only to expand all variables at the interface about $y = \pm 1.0$ by use of the Taylor series and retain only first-order terms in perturbations. In principle, the wave amplitude η must be even smaller than δ to allow the Taylor series expansion to be valid locally in ξ. However, in linear theory η is infinitesimal, and this requirement is satisfied. Hence we applied the expanded boundary conditions at $y = \pm 1.0$. From the balance of normal force per unit area of the interfaces at $\zeta = 1.0 + \eta_+$ and $\zeta = -1.0 + \eta_-$, we have

$$p_1 - p_2 \pm We^{-1}\,(\eta_\pm)_{,xx} - \frac{2v_{1,y}}{Re} = 0, \qquad (4.6a)$$

where $We = \rho_1 U_1^2 (\xi) h (\xi)/S = \rho_1 \dot{Q} U_0 \bar{u}_1 (\xi)/2S$ and

$$Re = U_1 (\xi) h (\xi)/v_1 = \dot{Q}/2v_1.$$

The balance of tangential stress at $y = \pm 1.0$ requires that

$$v_{1,x} + u_{1,y} = 0. \qquad (4.6b)$$

Because the interfaces are material surfaces, they must satisfy the kinematic boundary condition at $y = +1.0$,

$$v_i = \eta_{+,\tau} + \delta_{1i} \eta_{+,x} \qquad (i = 1, 2), \qquad (4.6c)$$

and at $y = -1.0$,

$$v_i = \eta_{-,\tau} + \delta_{1i} \eta_{-,x} \qquad (i = 1, 2). \qquad (4.6d)$$

There are altogether eight boundary conditions in (4.6a)-(4.6d).

We explained earlier that the normal mode solution will yield the same characteristic equation that can be obtained from representing the disturbance with Fourier integrals, and that only the properties of singularities associated with the characteristic equation are needed to describe the large time asymptotic behavior of the natural response of the flow to the disturbance.

The two-dimensional normal mode solution of the governing differential system will be sought in the form

$$(v_i, u_i, p_i, \eta_{\pm}) = [\hat{v}_i (y), \hat{u}_i (y), \hat{p}_i (y), \hat{\eta}_{\pm}] \, exp \, i \, (kx - \omega \tau),$$

where ω and k are respectively the complex frequency and wave number of disturbances. Weak dependence on ξ of all variables is understood and not explicitly indicated.

Taking the divergence of (4.4) and using the continuity Equation (4.5), we have

$$\nabla^2 p_i = 0.$$

The bounded normal mode solutions of this Laplace equation in the liquid sheet and the ambient gas yield the amplitudes of the pressure perturbations,

$$\hat{p}_1 = -[(k - \omega)/k] \, [A \, cosh \, (ky) + B \, sinh \, (ky)], \qquad (4.7a)$$

$$\hat{p}_2 = i Q_2 \omega F_+ \, exp \, (\mp ky), \qquad y > 0, \qquad (4.7b)$$

$$\hat{p}_2 = i Q_2 \omega F_- \, exp \, (\pm ky), \qquad y < 0, \qquad (4.7c)$$

where $Q_2 = (\rho_2/\rho_1)$, and A, B, F_+, and F_- are integration constants. The upper or lower sign in the exponents of $\hat{p}_2$ corresponds with $k_r > 0$ or $k_r < 0$. The amplitudes of the disturbance velocity components can be obtained from (4.4) with p_i given by (4.7a)-(4.7c). They are

$$\hat{v}_1 = -i\,[A\,sinh\,(ky) + B\,cosh\,(ky) + C\,sinh\,(my) + D\,cosh\,(my)],$$

$$\hat{u}_1 = \frac{1}{k}\,[kA\,cosh\,(ky) + mC\,cosh\,(my)$$

$$+ B\,sinh\,(ky) + mD\,sinh\,(my)], \tag{4.8}$$

$$m = [k^2 + iRe\,(k - \omega)]^{1/2}, \qquad \hat{v}_2 = \mp k F_\pm\,exp\,(\mp ky),$$

$$\hat{u}_2 = ik F_\pm\,exp\,(\mp ky) \tag{4.9}$$

Substitution of the solutions into the boundary conditions yields a system of eight linear homogeneous equations in eight unknown components A, B, C, D, F_+, F_-, η_+, and η_- of the eigenvector. These unknowns can be found directly from the eight equations. However, we shall exploit the fact that the odd and even eigen solutions are decoupled, thereby reducing the size of the system of equations.

Varicose Mode

For the odd solution of v_1, $B = D = 0$ in (4.8). The amplitudes of the interfacial displacements can be obtained from (4.6c), and (4.6d) with $i = 2$. They are

$$\hat{\eta}_\pm = \mp i\frac{k}{\omega}F_\pm\,exp\,(-k). \tag{4.10}$$

It follows from (4.6c), (4.6d), and (4.8) that we have for the odd mode

$$i\,(k - \omega)\,\hat{\eta}_\pm = \mp i\,[A\,sinh\,(k) + C\,sinh\,(m)]. \tag{4.11}$$

This equation states that the two interfaces of the liquid sheets are displaced in opposite directions for the odd mode, that is,

$$\hat{\eta}_+ = -\hat{\eta}_-. \tag{4.12}$$

Hence this mode is again termed the varicose mode. Substitution of (4.12) into (4.10) shows that

$$F_+ = F_- = F. \tag{4.13}$$

Substituting (4.7), (4.8), (4.10), (4.12), and (4.13) into the boundary conditions (4.6a), (4.6b), and (4.6d) with $i = 1$, we have respectively

$$-A\left[\frac{k-\omega}{k} - \frac{2ik}{Re}\right] \cosh(k) + iC\frac{2m}{Re} \cosh(m)$$

$$+ iF\left[-Q\omega + \frac{We^{-1}k^3}{\omega}\right] \exp(-k) = 0$$

$$A\lfloor 2k^2 \sinh(k)\rfloor + C(m^2 + k^2) \sinh(m) = 0,$$

and

$$A \sinh(k) + C \sinh(m) - i(\omega + k)\frac{k}{\omega} \exp(-k) = 0.$$

It is straightforward to show that the solution of this system is given by

$$A = \frac{k(m^2 + k^2)\exp(-k)F}{\omega Re \sinh(k)}, \qquad C = -\frac{2k^2 \exp(-k)Fr}{\omega Re \sinh(m)}, \qquad (4.14)$$

where the complex eigen frequency ω and complex wave number k are related by the characteristic equation for the varicose mode (Lin, 1981; Li, 1993)

$$D_1 = -Q_2\omega^2 + We^{-1}k^3 + (m^2 + k^2)^2 Re^{-2} \coth(k)$$

$$+ \left(\frac{2}{Re}\right)^2 k^3 m \coth(m) = 0. \qquad (4.15)$$

Sinuous Mode

For the even-mode solution, $A = C = 0$. It follows from (4.8), (4.6c), and (4.6d) that the fluid particle velocities as well as the interfacial displacements at the two interfaces are in unison, that is, $\hat{\eta}_+ = \hat{\eta}_-$, and $F_+ = -F_-$. The algebra required to yield the eigenvectors and the secular equations for this sinuous mode is identical to that for the varicose mode, except that hyperbolic sine and tangent functions must respectively be placed by the hyperbolic cosine and cotangent functions and vice versa. Thus the characteristic equation for the sinuous mode is (Lin, 1981; Li, 1993)

$$D_1 = -Q_2\omega^2 + We^{-1}k^3 + Re^{-2}(m^2 + k^2)^2 \tanh(k)$$

$$+ \left(\frac{2}{Re}\right)^2 k^3 m \tanh(m) = 0, \qquad (4.16)$$

and the eigenvector is given by

$$B = \frac{k(m^2 + k^2)\exp(-k)}{\omega Re \cosh(k)}, \quad D = -\frac{2k^3 \exp(-k)}{\omega Re \cosh(m)}, \qquad (4.17)$$

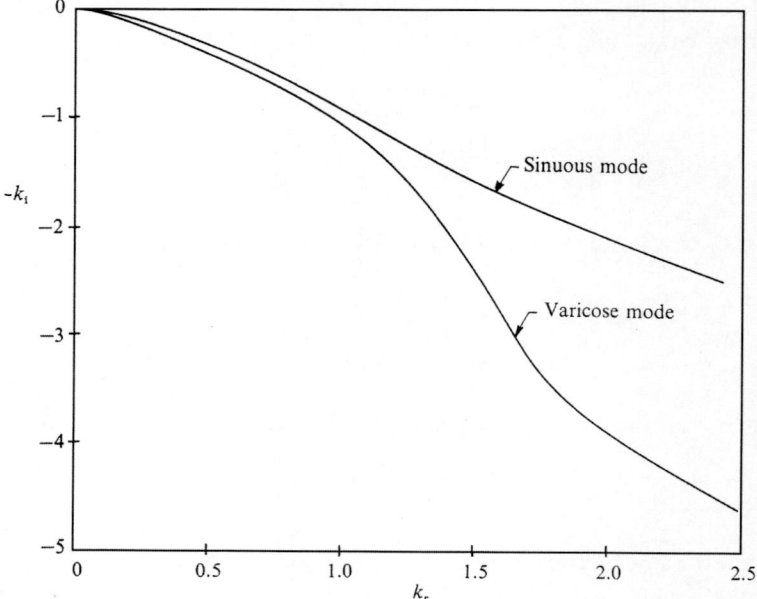

Figure 4.2. Stability of a liquid sheet. $Q_2 = 0$, $We = 5000$, $Re = 0.5$.

if F is put equal to 1 without loss of generality. When $k \to \infty$, $[coth \, (k)$, $coth \, (m)$, $tanh \, (k)$, $tanh \, (m)] \to 1$ and both (4.15) and (4.16) reduce to

$$-Q_2 \omega^2 + We^{-1} k^3 + \frac{[2k^2 + iRe\,(k - \omega)]^2}{Re^2}$$

$$+ k^3 \left(\frac{2}{Re}\right)^2 [k^2 + iRe\,(k - \omega)]^{1/2} = 0.$$

This is the characteristic equation obtained by Taylor (1963) in his investigation of the generation of ripples by wind over an infinitely deep viscous fluid. When $R \to \infty$, (4.16) and (4.15) reduce respectively to (2.17) and the corresponding varicose mode characteristic equation of Squire given in Exercise (2.1).

The spatial growth rate curves $\omega_i = 0$ for the sinuous and varicose disturbances in a liquid sheet at $Q_2 = 0$, $We = 5000$, and $Re = 0.5$ are given in Figure 4.2. Because ω_r decreases with k_r along these curves, the disturbance is convected downstream. Therefore, $k_i > 0$ on these curves signifies that the sheet is convectively stable below the slot at the head of the liquid sheet. An example of an unstable liquid sheet with $We > 1$ is given in Figure 4.3. Two branches of spatial amplification curves meet tangentially at the origin in the (k_r, k_i)-space of Figure 4.3(a). Figure 4.3(b) shows how

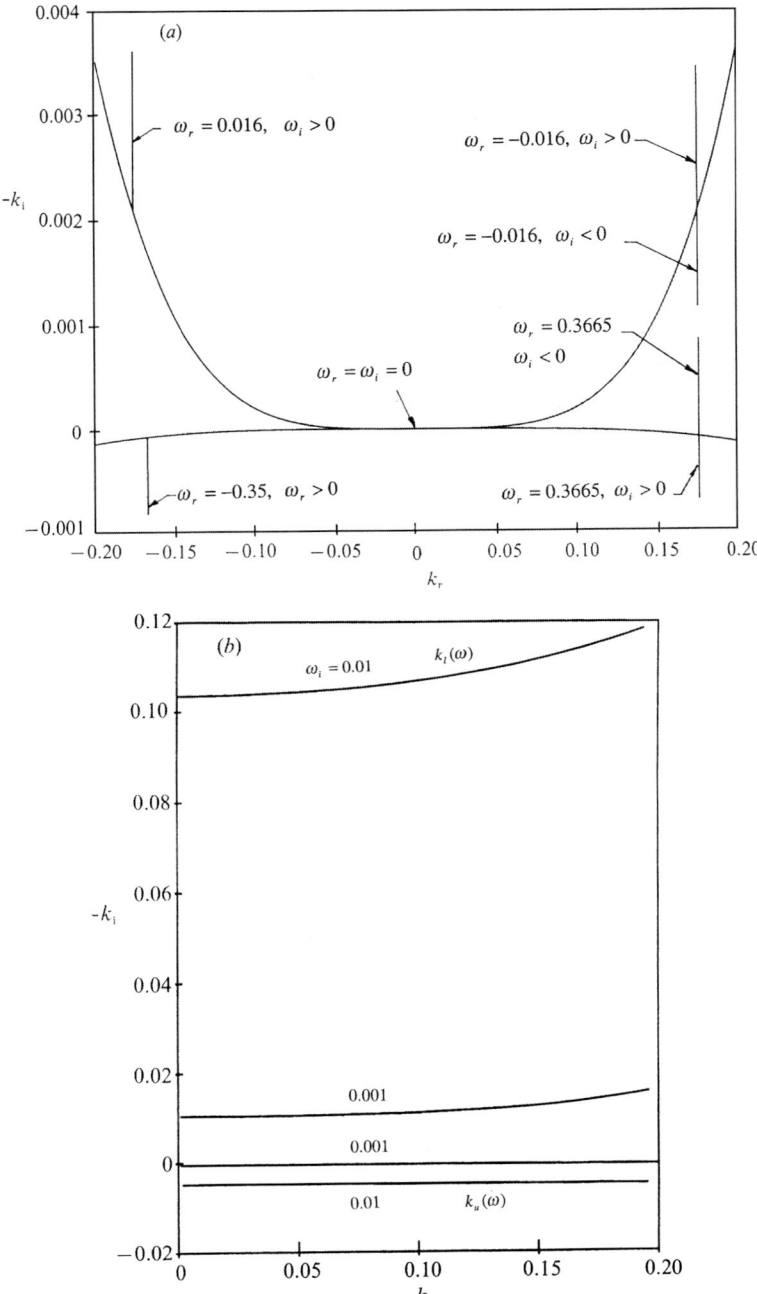

Figure 4.3. Pseudo-absolute instability caused by sinuous waves. $Q_2 = 0$, $We = 0.83$, $Re = 0.5$. (a) Amplification curve. (b) Formation of pinch point singularity.

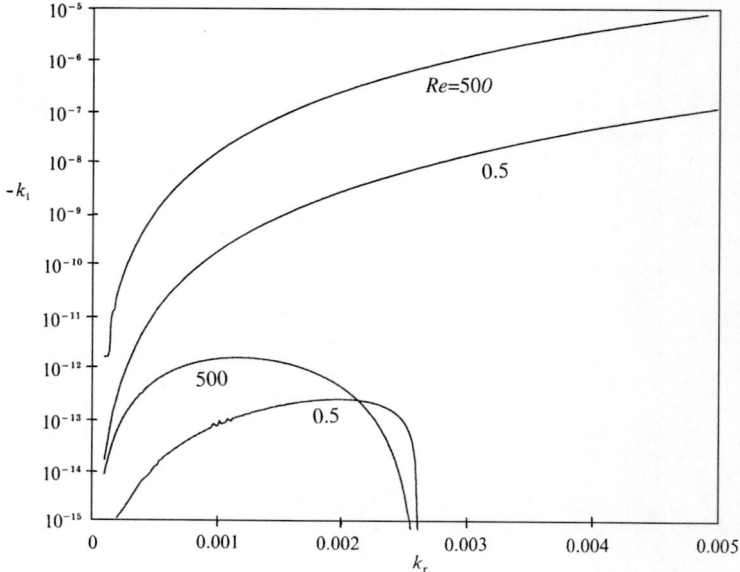

Figure 4.4. Sinuous mode. $Q_2 = 1.3 \times 10^{-3}$, $We = 0.998$.

the upper and lower branches of curves coalesce at the origin when ω_i is decreased toward zero from a positive finite value. At the origin in Figure 4.3, $D_1 = \partial D_1/\partial k = \partial D_1/\partial \omega = \partial^2 D_1/\partial k^2 = \partial^2 D_1/\partial k^2 = \partial^2 D_1/\partial k \partial \omega = 0$ but not the higher derivatives. Using the method that gives (2.40) or the formula for a general pinch point given in Exercise 2.4, it can be shown (cf. Exercise 2.5) that the large time asymptotic behavior of the disturbance amplitude is $G(\tau) \to \tau^0 exp(\omega_0\tau)$ for all x. Note that $d\omega_r/dk_r$ is positive along the upper branch and negative along the lower branch, and $\omega_0 = 0$. Thus the sinuous mode disturbance propagates both upstream and downstream. However, the amplitude remains nonvanishing, although it is not growing as it would in a full-fledged absolute instability. Since this type of instability satisfies the definitions of neither absolute nor convective instability, it is termed pseudo-absolute instability (Lin, 1981).

The presence of the ambient gas changes the nature of the singularity at $k = \omega = 0$ when $We < 1$, as demonstrated in Figure 4.4 for a wide range of Re. The branches of $\omega_i = 0$ in the region $k_r < 0$ are not shown in this figure. As in Figure 4.3, $\omega_r < 0(>0)$ for the upper (lower) branch of the amplification curve in the figure. Moreover, ω_r decreases (increases) along the upper (lower) branch with increasing k_r. Thus, the upper (lower) branch describes the upstream (downstream) propagating evanescent (spatially growing) waves.

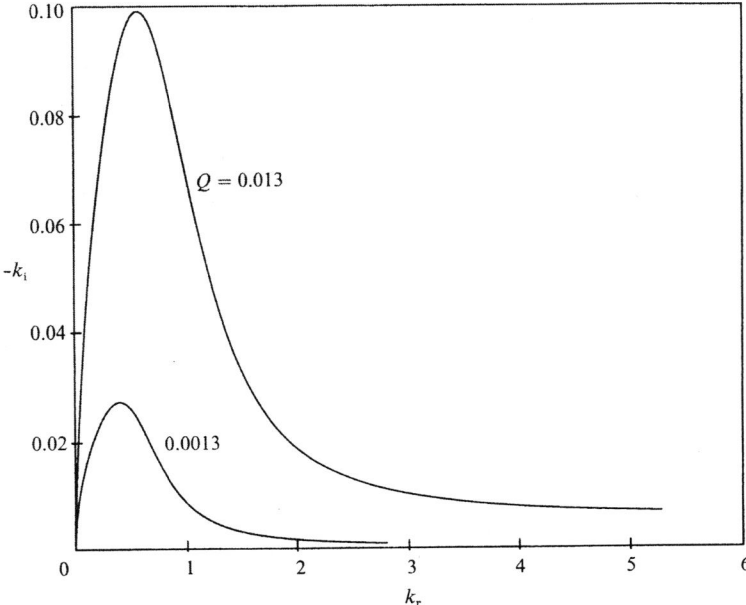

Figure 4.5. Effect of Q_2 on convective instability, sinuous mode. $We = 5000$, $Re = 0.5$.

The lower branch crosses over the real k-axis, and the two branches co-alesce at the origin where $\omega_r = 0$. For the case of $Q \neq 0$, $\partial D_1/\partial k = \partial D_1/\partial \omega = \partial^2 D_1/\partial k^2 = \partial^2 D_1/\partial k \partial \omega = 0$, but $\partial^2 D_1/\partial \omega^2$ and higher order derivatives do not vanish. Thus $(\omega - \omega_0)^2 \sim (k - k_0)^3$, and singularity at $(k_0, \omega_0) = (0, 0)$ behave as in (2.36). Therefore large time behavior of the disturbance is given by the algebraic growth indicated in (2.40).

$Q_2 = 0.0013$ and $Q_2 = 0.013$ in Figure 4.5 correspond respectively to the air-to-water density ratio at 1 and 10 times atmospheric pressure at room temperature. The spatial amplification curves $\omega_i = 0$ in this figure can be approached from $\omega_i > 0$ above the k_r-axis without violating the causality condition. Moreover, ω_r increases with increasing k_r along these curves. Thus these curves signify the spatial amplification of convectively unstable sinuous disturbances propagating in the downstream direction. Recall that it was shown in Figure 4.2 that the sheet is stable when $Q_2 = 0$, with the rest of the flow parameters being the same. Thus, the presence of ambient gas causes the convective instability. No pseudo-absolute instability has been found for the varicose mode in the presence or absence of ambient gas. In the absence of ambient gas, the varicose mode was found to be stable for a wide range of flow parameters. Figure 4.6 provides an example. However, in the presence of

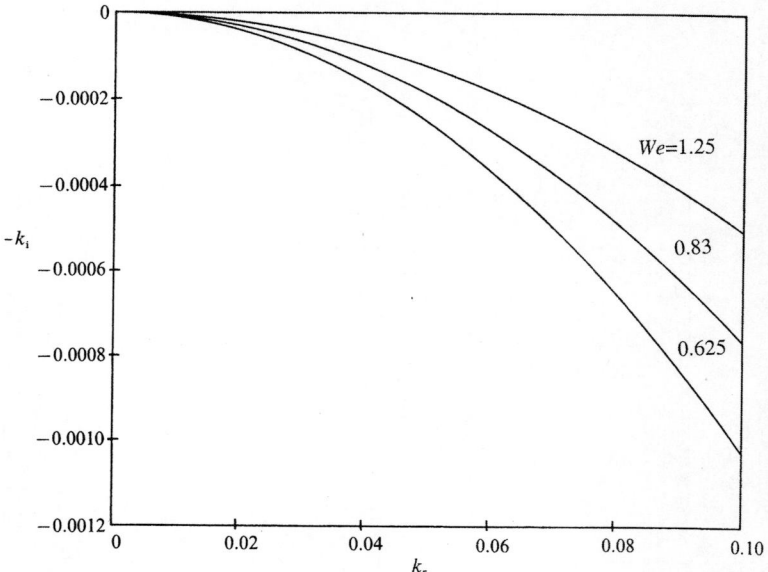

Figure 4.6. Asymptotic stability of varicose mode disturbances. $Q_2 = 0$, $Re = 0.5$.

the ambient gas, the varicose mode is found to be convectively unstable in the same wide range of parameters. Figure 4.7 gives such an example and shows that the surface tension is stabilizing because the amplification rates decrease as *We* decreases. The same dependence of growth rates on *We* is found for the sinuous mode when $We > 1$. Thus the convective instability is not caused by the surface tension. Figure 4.8 shows that the growth rates of convectively unstable varicose disturbances are slightly smaller than that of the sinuous mode near $k_r = 0$ but both modes have almost identical growth rates for short waves especially when *We* is much greater than one. Similar wave characteristics were found for three-dimensional temporal disturbance (Ibrahim and Akpan 1996), except when $We = O(10^3)$ the long varicose waves may dominate the sinuous waves. The critical Weber number above which instability is convective but below which it is pseudo-absolute is found to be 1.0 for *Re* ranging from 0.1 to 10000 and Q_2 from 0.00013 to 0.13.

4.1c. Energy Budget

To trace the energy sources of the absolute and convective instabilities, we balance the energy budget in a disturbed liquid sheet. Consider a control volume of liquid per unit sheet width over one wavelength, $\lambda = 2\pi/k_r$, of the disturbance. Forming the dot produce of (4.4) for liquid ($i = 1$) with v_1,

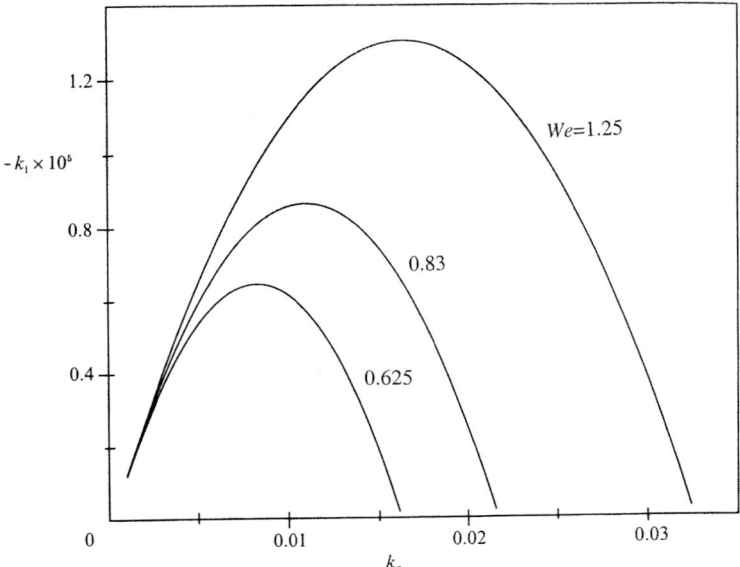

Figure 4.7. Stabilizing effect of surface tension on convectively unstable varicose mode disturbances. $Q_2 = 0.013, Re = 0.5$.

integrating over the control volume, and using (4.5) and the Gauss theorem to reduce some of the volume integrals to surface integrals, we arrive at the energy equation

$$\dot{K} = \dot{R} + \dot{I} + \dot{D}, \qquad (4.19)$$

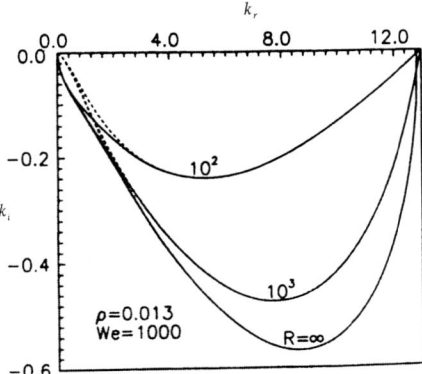

Figure 4.8. Spatial growth rate. $Q_2 = 0.013$, $We = 1000$. ———, sinuous mode; - - - -, varicose mode. (From Li, 1993).

where, omitting the subscript 1 of the liquid phase,

$\dot{K} \equiv$ time rate of changing of disturbance kinetic energy

$$= \int_0^\lambda (\partial_\tau - \partial_x) \, e \, dx, \quad e = \frac{1}{2} \mathbf{v} \cdot \mathbf{v},$$

$\dot{R} \equiv$ rate of reversible work

$$= -\int_0^\lambda [pv]_{y=-1}^{y=+1} \, dx - \int_{-1}^1 [pu]_{x=0}^{x=\lambda} \, dy,$$

$\dot{I} \equiv$ rate of irreversible work

$$= \frac{1}{R} \int_0^\lambda [u(u_y + v_x) + 2vv_y]_{y=-1}^{y=1} \, dx$$

$$+ \frac{1}{R} \int_{-1}^1 [v(u_y + v_x) + 2uu_x]_{x=0}^{x=\lambda} \, dy,$$

$\dot{D} \equiv$ rate of energy dissipation by viscosity

$$= \int_0^\lambda \int_{-1}^1 \left[(v_y)^2 + (u_x)^2 + \frac{1}{2}(u_y + v_x)^2 \right] dy dx,$$

where $[G(y)]_{y=b}^{y=a} = [G(a) - G(b)]$. In the above expressions, subscripts 1 for liquid velocity and pressure are omitted for clarity. The first term in the reversible work rate integral can be rewritten by applying the boundary condition (4.6a),

$$\int_0^\lambda [pv]_{y=-1}^{y=1} \, dx = \int_0^\lambda \left[v \left(p_2 \mp We \, (\eta_\pm)_{xx} + \frac{2v_y}{Re} \right) \right]_{y=-1}^{y=1} dx.$$

Note that in (4.19) the last term in the above integrand cancels the second term of the first integral in the expression of $\dot{I}$. Moreover, $(u_y + v_x)$ in the first integral of $\dot{I}$ vanishes by virtue of the boundary condition (4.6b). Henceforth it is understood that $\dot{I}$ consists of only the second integral in the above expression of $\dot{I}$, and $\dot{R}$ is now redefined as

$$\dot{R} = -\int_0^\lambda [p_2 v_1]_{y=-1}^{y=1} dx - \int_{-1}^1 [pu]_{x=-\lambda}^{x=0} dy + We \int_0^\lambda [\pm v \, (\eta_\pm)_{xx}]_{y=-1}^{y=1}.$$

The first integral represents the power $\dot{P}$ exerted by the gas pressure on the liquid sheet and the second integral represents the flow work rate at the top and bottom of the control volume. The third integral in $\dot{R}$ gives the surface tension work rate $\dot{S}$. Equation (4.19) can now be rewritten as

$$\dot{K} = \dot{P} + \dot{P}_f + \dot{S} + \dot{I} + \dot{D}, \tag{4.20}$$

Table 4.1. *Pseudo-Absolute Instability, Re = 100, We = 0.38, k_r = 0.2*

$Q \times 10^4$	$k_i \times 10^8$	ω_1	$\dot{p}$	$\dot{p}_f$	$\dot{s}$	$\dot{t}$	$\dot{d}$
130	45891	−0.2478	1.24	0	184.55	−0.10	−85.68
1.3	49169	−0.2566	0.01	0	179.20	−0.10	−79.20

where

$$\dot{P} = -\int_0^\lambda [p_2 v_1]_{y=-1}^{y=1} \, dx, \quad \dot{P}_f = -\int_{-1}^1 [pu]_{x=0}^{x=\lambda} \, dy,$$

$$\dot{S} = We \int_0^\lambda [\pm v \, (\eta_\pm)_{xx}]_{y=-1}^{y=1} \, dx.$$

The upper and lower signs in $\dot{S}$ are associated respectively with the surface at $y = 1.0$ and $y = -1.0$. It is understood that the pressure and velocity components appearing in all integrals in this section are the real parts of their counterparts in Section 4.1b. To serve as a check on the final accuracy, all integrals on the right-hand side of (4.20) are added and the results are compared with the independently evaluated $\dot{K}$. The error is mostly less than 1%. For a better appreciation of the relative importance of each item the energy budget will be presented in the variables defined below:

$$(\dot{p}, \dot{p}_f, \dot{s}, \dot{t}, \dot{d}) = [(\dot{P}, \dot{P}_f, \dot{S}, \dot{I}, \dot{D})/\dot{K}] \times 100.$$

Table 4.1 gives a typical energy budget for the case of pseudo-absolute instability. Tables 4.2 and 4.3 give typical energy budgets for the convective instability of the sinuous mode and the varicose mode respectively. The values of k_r, k_i, and ω_i are all taken from the calculated spatial amplification curves $\omega_r = 0$. In particular, these values in Tables 4.2 and 4.3 correspond to the maximum spatial growth rate on $\omega_r = 0$. While the positive dominant item in the pseudo-absolute instability case is the surface tension work term, that in the convective instability case are the terms due to the gas pressure work at the interface. Moreover, the surface tension work terms in the convective instability are negative. Thus while the surface tension work is responsible for the pseudo-absolute instability, the gas pressure work is responsible for the convective instability of both the varicose and sinuous modes. It has been

Table 4.2. *Convective Instability, Sinuous Mode; Re = 5000, Q = 0.0013*

$We \times 10^4$	k_r	k_i	ω_1	$\dot{p}$	$\dot{p}_f$	$\dot{s}$	$\dot{t}$	$\dot{d}$
769.2	0.508	0.0255	0.505	204.92	0.03	−104.35	0.01	−0.61
7692	6.249	0.1261	0.242	312.43	0.55	−195.24	0	−17.74

Table 4.3. *Convective Instability, Varicose Mode;* $Re = 5000$, $Q = 0.0013$

$We \times 10^4$	k_r	k_i	ω_1	$\dot{p}$	$\dot{p}_f$	$\dot{s}$	$\dot{\tau}$	$\dot{d}$
13	0.74	0.0079	0.7397	403.84	0.32	−298.81	0	−5.35
1.3	6.30	0.1256	6.2927	317.67	0.59	−200.14	0	−18.11

pointed out that the spatial growth rate of the varicose mode is smaller than that of the sinuous mode for the same flow parameters. Comparing the viscous dissipation and the surface tension terms in Tables 4.2 and 4.3, we find that the reason for this is that the varicose mode entails larger viscous dissipation and surface tension work by disturbances. It can be seen that in Tables 4.2 and 4.3 k_r scales with Q_2/We. The same situation has been found in other parameter ranges.

4.2. A Viscous Sheet in Viscous Gas

4.2a. Basic Flow

Consider a viscous liquid sheet of constant thickness $2h_0$. The liquid sheet is surrounded by a viscous gas that is bounded by two vertical walls located at equidistance L from the centerline of the liquid sheet. For the liquid sheet to maintain a constant thickness, the dynamic pressure gradient in the liquid and gas flows must remain constant. In the presence of gravity, such parallel flows of incompressible liquid and gas that satisfy the Navier–Stokes equations are given by

$$U_i = \frac{1}{2} K_i \, y^2 \pm A_i \, y + B_i \qquad (i = 1, 2,), \qquad (4.21)$$

where U is the velocity normalized with the centerline velocity of the liquid sheet U_0 in the direction of gravitational acceleration, the subscript i stands for the liquid or the gas phase depending on if it is 1 or 2, y is the horizontal distance measured from the centerline to the right divided by h_0, and $+$ or $-$ signs in front of A_i are for the domain $1 \leq y \leq 1+l$ or $-1-l \leq y \leq -1$ respectively, l being $(L - h_0)/h_0$. The integration constants A_i, B_i, and K_i, which are related to the dynamic pressure gradient, are given by

$K_1 = Re(K - Fr^{-1})$, $\qquad A_1 = 0$, $\qquad B_1 = 1$,

$K_2 = (Re/\mu_r)(K - \rho_r \, Fr^{-1})$, $\qquad A_2 = (K_1/\mu_r - K_2)$,

$B_2 = -K_2(1 + l)^2/2 - (K_1/\mu_r - K_2)(1 + l)$,

$K = Fr^{-1}(\mu_r + 2l + \rho_r l^2)/(\mu_r + 2l + l^2) - 2\mu_r/[Re(\mu_r + 2l + l^2)]$,

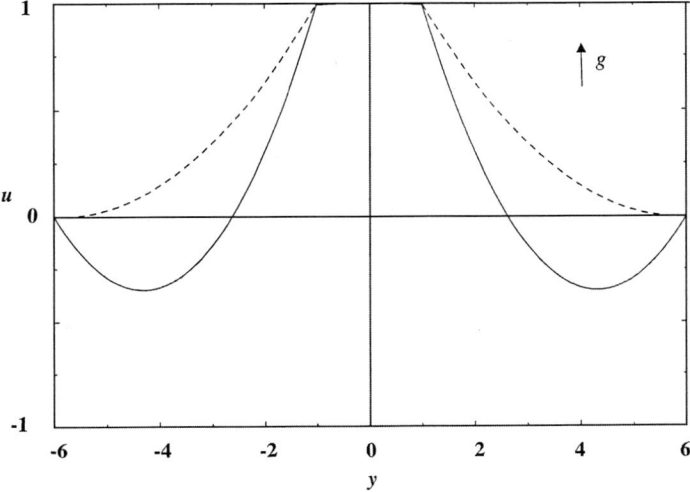

Figure 4.9. Basic flow velocity profile. $Re = 52.72$, $\rho_r = 0.0013$, $\mu_r = 0.017$, $l = 5$.
———, $Fr^{-1} = 0.00340625$; - - -, 0.00016.

where $\mu_r = \mu_2/\mu_1$, $\rho_r = \rho_2/\rho_1$, and $l = (L - h_0)/h_0$. Note that in the notation of Chapter 2, $\mu_r = M_2$ and $\rho_r = Q_2$. The flow is in the direction of
gravitational acceleration g, that is, in the x-direction. Some representative
velocity distributions are given in Figure 4.9. Note that the gas flow may or
may not have counter flows depending on the flow parameters. The values of
Q_2 and μ_r given in this figure correspond to a water curtain in air at 1 atm.
pressure and 20°C.

4.2b. Stability Analysis

Let us now consider the stability of the basic flow described by (4.21) with
respect to two dimensional disturbances. The governing equations of the onset
of instability are again given by (4.4) and (4.5), except that δ_{1i} on the left side
of (4.4) is now replaced by $U_i(y)$ and that on the right side is replaced by
$N_i' = M_i/Q_i$. Equation (4.5) enables us to represent the perturbation velocity
components (u_i, v_i) in the Cartesian coordinates (x, y) by the Stokes stream
function ψ_i as

$$u_i = \psi_{i,y}, \qquad v_i = -\psi_{i,x}. \qquad (4.22)$$

The normal mode representation of ψ_i is given by

$$\psi_i = \phi(y) e^{i(kx - \omega\tau)}, \qquad (4.23)$$

where $\phi(y)$ is the amplitude, which varies across the flow. Substituting (4.22)
with (4.23) into the modified (4.4) and eliminating the pressure terms by cross

differentiation, we have

$$\left[-i\omega - (N_i'/Re)D^2\right]D^2\phi_i\,(y) + ikU_i\,D^2\phi_i - ik\left(d^2\,U_i\right)\phi_i = 0, \quad (4.24)$$

$$D^2 = d^2 - k^2, \qquad d^2 = d^2/dy^2.$$

This is the well-known Orr–Sommerfeld equation, which governs the onset of instability in parallel flows (Drazin and Reid, 1985).

Since the interface is a material surface, the interfacial displacement $\eta_\pm$ relative to the unperturbed interface at $y = \pm 1$ must satisfy

$$\eta_{\pm,\tau} + U_1\,\eta_{\pm,x} = -\psi_{1,\,x}.$$

Other interfacial kinematic conditions are the continuity of the y- and x-components of the velocity across the interfaces given respectively by

$$[\psi_{i,x}]_2^1 = [\psi_{1,x} - \psi_{2,x}]_{y=\pm 1} = 0, \qquad [\psi_{i,y} + U_{i,y}\,\eta_\pm]_2^1 = 0,$$

where η_+ or η_- is to be used for the boundary conditions at $y = 1$ or -1. The balancing of forces per unit area of the interface in the tangential and normal directions leads, respectively, to the dynamic conditions at $y = \pm 1$,

$$[M_i(\eta_\pm U_{i,yy} + \psi_{i,yy} - \psi_{i,xx})]_2^1 = 0,$$

$$[p_i + (2/Re)M_i\psi_{i,xy}]_2^1 \pm We^{-1}\eta_{\pm,xx} = 0,$$

where p_i is the perturbation pressure and $M_i = \mu_i/\mu_1$. Thus, We signifies the ratio of the inertial force to the surface tension force per unit area of the interface. The boundary conditions at the walls $y = \pm(1 + l)$ are

$$\psi_{2,x} = 0, \qquad \psi_{2,y} = 0.$$

The normal mode pressure and interfacial displacement are written as

$$[p_i\eta_\pm] = [\zeta_i\,(y),\,\xi_\pm]\,e^{i(kx-\omega\tau)}. \tag{4.25}$$

We substitute (4.21), (4.23), and (4.25) into the above boundary conditions, and rewrite them in the same order of appearance

$$ik\,\phi_1 + [-i\omega + ikU_1]\xi_\pm = 0, \tag{4.26a}$$

$$[\phi_i]_2^1 = 0, \tag{4.26b}$$

$$[\phi_{i,y} + \xi_\pm U_{i,y}]_2^1 = 0, \tag{4.26c}$$

$$\left\{M_i\left[(d^2 + k^2)\phi_i + U_{i,yy}\,\xi_\pm\right]\right\}_2^1 = 0, \tag{4.26d}$$

$$[\zeta_i + (2ik/Re)M_i\phi_{i,y}]_2^1 \mp k^2We^{-1}\xi_\pm = 0, \tag{4.26e}$$

$$\phi_2\left[\pm(1 + l)\right] = 0, \tag{4.26f}$$

$$\phi_{2,y}\left[\phi\,(1 + l)\right] = 0. \tag{4.26g}$$

The pressure amplitude discontinuity in (4.26e) obtained from the linearized Navier–Stokes equations is found to be

$$(-ik)[\zeta_i]_2^1 = \left[Q_i\{(-i\omega + ikU_i)\phi_{i,y} - ikU_{i,y}\phi_i\} - M_i D^2 \phi_{i,y}/Re\right]_2^1,$$

where $Q_i = \rho_i/\rho_1$.

Nontrivial solutions of (4.24) with its boundary conditions (4.26a)-(4.26g) for given flow parameters Re, Fr, We, Q_2, M_2, l, and k exist only for certain eigenvalues ω.

4.2c. Solution by Chebyshev Expansion

The solution of the formulated problem will be expanded in Chebyshev polynomials. The gas domains $y \in [1, 1 + l]$ and $y \in [-1 - l, -1]$ are mapped into $y_+ \in [-1, 1]$ and $y_- \in [-1, 1]$, respectively, by use of the linear transformations

$$y = 1 - \frac{1}{2}l(y_+ - 1), \qquad y = -1 - \frac{1}{2}l(y_- + 1).$$

In the new variables $y_\pm$, the expressions of the Orr–Sommerfeld equation and its boundary conditions remain the same except the n-th derivatives appearing in these equations must be multiplied by $(-2/l)^n$, and the basic flow in the gas region is written as

$$U_{2\pm} = Ey_\pm^2 \pm Fy_\pm + H, \tag{4.27}$$

where

$$E = \left(K_2 l^2/8\right), \qquad F = -[l(1 + l/2)K_2/2 + lA_2/2],$$

$$H = (1 + l + l^2/4)K_2/2 + (1 + l/2)A_2 + B_2.$$

The corresponding perturbation amplitude can be expanded in the series of Chebyshev polynomials of n-th degree

$$\phi_1(y) = \sum_{n=0}^{N_1} a_n T_n(y) \qquad (-1 \le y \le 1),$$

$$\phi_{2+}(y_+) = \sum_{n=0}^{N_2} b_n T_n(y_+) \qquad (-1 \le y_+ \le 1), \tag{4.28}$$

$$\phi_{2-}(y_-) = \sum_{n=0}^{N_2} c_n T_n(y_-) \qquad (-1 \le y_- \le 1).$$

By substituting (4.28) into the Orr–Sommerfeld equation and performing the Galerkin projection with T_m, where $0 \le m \le N_1$ in liquid domain and

Viscous Liquid Sheet

$0 \leq m \leq N_2$ in each gas domain, we obtain $N + 1$ equations in liquid domain, and $N_2 + 1$ equations in each gas domain. The Galerkin projections of ϕ_1, $\phi_{2\pm}$ as well as their derivatives and their products with powers of y and $y_{\pm}$ can also be expanded in series of Chebyshev polynomials in each domain (Gottlieb and Orszag, 1977). The Orr–Sommerfeld equation in terms of a_n, b_n, and c_n are given in Appendix A.

Since the governing system of equations is linear and homogeneous, the solution can be separated into even and odd functions of y. For the even solution, $\phi_1(1) = \phi_1(-1)$. It follows from (4.26a) that $\xi_+ = \xi_- = \xi$, since $U_1(1) = U_1(-1)$. The displacements at the two liquid-gas interfaces are in phase for the even solution, which is again termed the sinuous mode. For the odd solution, $\phi_1(1) = -\phi_1(-1)$ and $U_1(1) = U_1(-1)$, and therefore it follows from (4.26a) that $\xi_+ = -\xi_-$. Hence the odd solution is termed, as before, the varicose mode. The even and odd solutions are again decoupled.

For the sinuous mode, we retain only the even Chebyshev polynomials in ϕ_1. Using the relation $T_n(\pm 1) = (\pm 1)^n$, the corresponding boundary conditions at $y = y_+ = 1$ are written as

$$[-i\omega + ikU_1(1)]\,\xi + ik\sum_{n=0}^{N_i} a_{2n} = 0, \tag{4.30a}$$

$$\sum_{n=0}^{N_1} a_{2n} - \sum_{n=0}^{N_2} b_n = 0. \tag{4.30b}$$

$$[U_{1,y} - U_{2,y}]\xi + \sum_{n=0}^{N_1}(2n)^2\,a_{2n} + (2/l)\sum_{n=0}^{N_2} n^2\,b_n = 0, \tag{4.30c}$$

$$\sum_{n=0}^{N_1}[4n^2(4n^2 - 1)/3 + k^2]\,a_{2n} - \mu_r\sum_{n=0}^{N_2}[(2/l)^2 n^2(n^2 - 1)/3 + k^2]\,b_n$$

$$+ (U_{1,yy} - \mu_r U_{2,yy})\,\xi = 0, \tag{4.30d}$$

$$\sum_{n=0}^{N_1}\left[-\left(\frac{3\,k^2}{Re} + ikU_1\right)(4n^2) + ikU_{1,y} + (2n)^2\,(4n^2 - 1)\right.$$

$$\left. \times (4n^2 - 4)/(15\,Re) + i\omega\,(2n)^2\right]a_{2n} - \sum_{n=0}^{N_2}\left[ik(3ik\mu_r/Re\right.$$

$$- U_2\rho_r)\left(-\frac{2}{l}\right)n^2 + ikU_{2,y}\,\rho_r + \mu_r\left(-\frac{2}{l}\right)^3 n^2(n^2 - 1)$$

$$\left. \times (n^2 - 4)/(15\,Re) + i\omega\rho_r\left(-\frac{2}{l}\right)n^2\right]b_n - ik^3\,We^{-1}\,\xi = 0, \tag{4.30e}$$

Table 4.4. *Chebyshev Approximation for* $Re = 15711.77$, $Fr^{-1} = 7.67 \times 10^{-6}$, $We = 8880.9947$, $\rho_r = 0.0013$, $\mu_r = 0.017$, $l = 5.0$, $k_r = 5.0$, $k_i = 0$

N_1	N_2	ω_i	$-\omega_r$
15	30	0.01668370	4.71282368
20	40	0.01487745	4.71166567
30	62	0.01477907	4.71171634
31	64	0.01477890	4.71171627
32	66	0.01477908	4.71171503
33	68	0.01478001	4.71171482
34	70	0.01478024	4.71171521
35	72	0.01478075	4.71171620

$$\sum_{n=0}^{N_2} (-1)^n b_n = 0, \tag{4.30f}$$

$$\sum_{n=0}^{N_2} (-1)^{n+1} b_n = 0. \tag{4.30g}$$

For the varicose mode, the system of equations remains the same except that the a_{2n+1} terms are retained in the expansion of ϕ_1.

The unknowns of the system are ξ, b_n $(n = 0$ to $N_2)$ and a_{2n} or a_{2n+1} $(n = 0$ to $N_1)$ depending on whether we are dealing with the sinuous or varicose mode. To render the number of equations the same as the number of unknowns $(N_1 + N_2 + 3)$, we used the method of Lanczos (1956) to truncate the Galerkin projection of the Orr–Sommerfeld equation. We choose $0 \le m \le N_1 - 2$ for the liquid domain and $0 \le m \le N_2 - 4$ for the gas domain. Hence the seven equations from the boundary conditions and the $N_1 + N_2 + 4$ equation from the Orr–Sommerfeld equations allow us to determine the eigenvector up to an arbitrary multiplication factor for each eigenvalue of the system if the flow parameters are given.

The numbers of terms retained in ϕ_1 and $\phi_{2\pm}$ are systematically increased until five significant digits are obtained for the eigenvalue. A typical example of the convergence test is given in Table 4.4.

4.2d. Numerical Results

The spatial amplification curves for convectively unstable varicose disturbances are given in Figure 4.10 for various values of *We* and the set of other flow parameters specified in the caption. Along these curves, ω_r increases

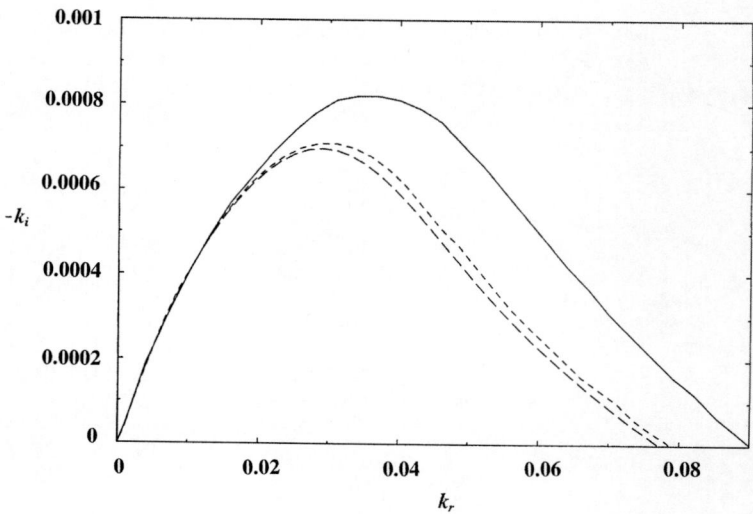

Figure 4.10. Varicose mode amplification rate. $Re = 52.72$, $Fr^{-1} = 3.40625 \times 10^{-4}$, $\rho_r = 0.0013$, $\mu_r = 0.017$, $l = 5$. ____, $We = 2.0$; _ _ _, 1.1111; _ _ _, 1.0204.

with k_r, and thus the packet of disturbances is convected in the downstream direction. It is seen that the amplification rate decreases as *We* is decreased. The converse is true for the sinuous mode with the same set of flow parameters, as shown in Figure 4.11. Note that $-k_i$ of the sinuous mode is two orders

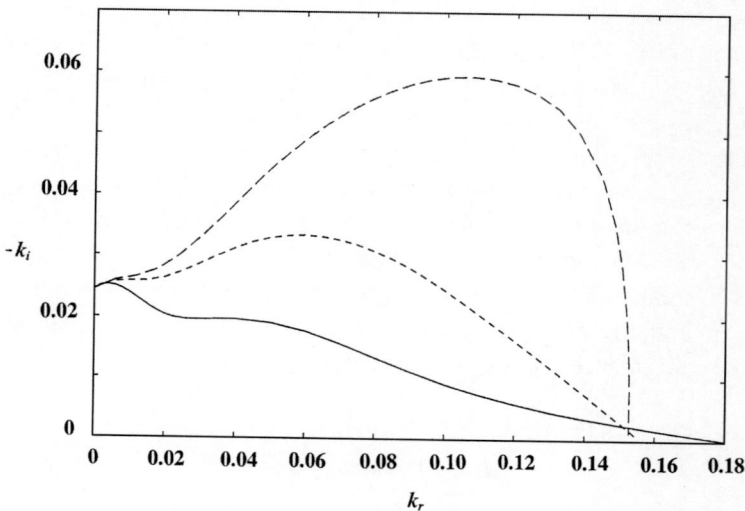

Figure 4.11. Sinuous mode amplification rate. $Re = 52.72$, $Fr^{-1} = 3.40625 \times 10^{-4}$, $\rho_r = 0.0013$, $\mu_r = 0.017$, $l = 5$. ____, $We = 2.0$; _ _ _; 1.1111; _ _ _, 1.0204.

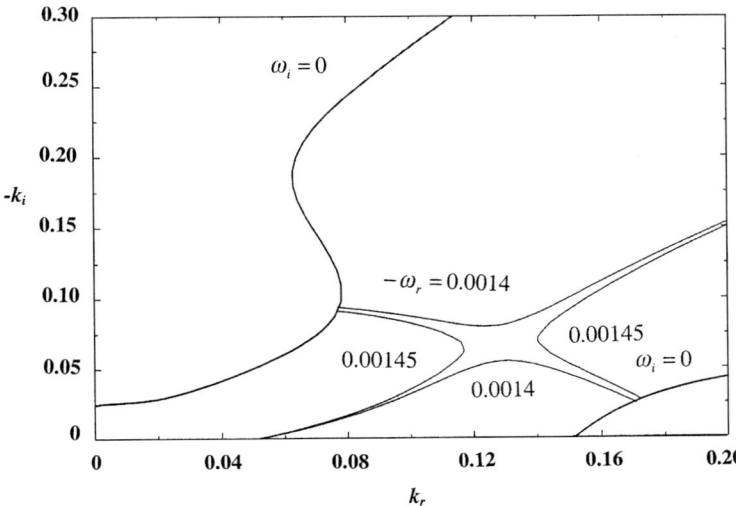

Figure 4.12. Saddle point in k-plane. $Re = 52.72$, $Fr^{-1} = 3.40625 \times 10^{-4}$, $\rho_r = 0.0013$, $\mu_r = 0.017$, $l = 5$, $We = 1.0$.

of magnitude larger than that of the varicose mode. The convectively unstable sinuous mode becomes absolutely unstable when We is decreased slightly below the smallest Weber number given in this figure. The emergence of a saddle point when We is reduced to 1, while the rest of the flow parameters remain the same as in the caption of Figures 4.10 and 4.11, is shown in Figure 4.12. The wave frequency of the saddle point lies between -0.014 and -0.0145. Two sets of curves of the same ω_r are shown in this figure. A member of each set originates from $k_i < 0$, and the other member from $k_i > 0$. Both members originate with $\omega_i > 0$. As ω_i is decreased toward zero, each member of the same set approaches a different branch of $\omega_i = 0$. Along the branch with wave numbers, $d\omega_r/dk_r > 0$; and along the other branch, $d\omega_r/dk_r < 0$. Hence, the disturbances propagate both upstream and downstream. Moreover, the emergence of saddle point at k_0 signifies $(\partial D/\partial k)_{k_o=0} = 0$, where D is the determinant of the coefficient matrix of the system (4.30a)-(4.30g). Furthermore the image of the saddle point in the ω-plane is at $\omega_0 = \omega_{0r} + i\omega_{0i}$ with $\omega_{0i} > 0$ and $\partial D/\partial \omega \neq 0$. Hence according to the formula given in Exercise 2.4 the large time asymptotic behavior of the disturbance is that of an absolute instability $\mathbf{G}(x, \tau) \to \tau^{-1/2} \exp(\omega_o \tau)$. The transition from convective instability to absolute instability occurs at $We = 1.01$ for the rest of the flow parameters specified in the caption of Figure 4.12. An example of the transition Weber number as a function of Re is given in Figure 4.13. The transition curve (We, Re) shown in Figure 4.13 corresponds to the basic flows

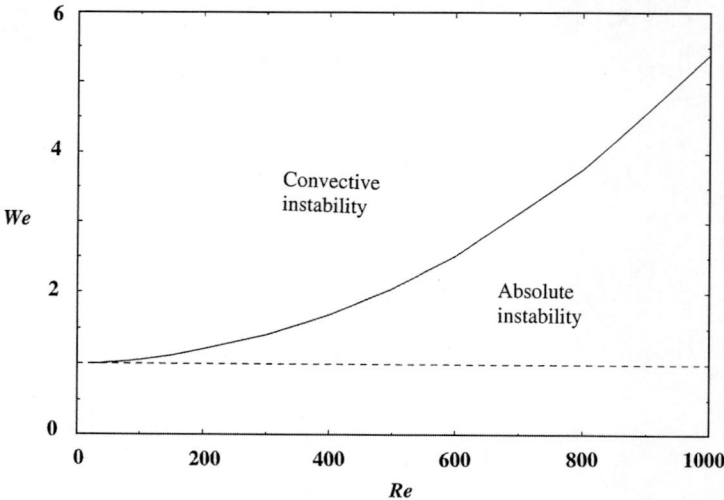

Figure 4.13. Critical Weber number for absolute instability. $Fr^{-1} = 0.000340625$, $\rho_r = 0.0013$, $\mu_r = 0.017$, $l = 5$.

with counterflows similar to that shown by the solid line in Figure 4.9. Huerre and Monkewitz (1985) demonstrated that, in the limit of $Re \rightarrow \infty$, if the velocity of the counterflow is greater than 13.6% of the mainstream velocity in free shear layers, absolute instability will occur. However, the absolute instability reported here is not due to counterflows because the absolute instability can be changed to convective instability by increasing the Weber number above We_c without reducing the maximum velocity of the counterflow. In fact the basic flow without counterflow as shown in the dashed line in Figure 4.9 can become absolutely unstable when We becomes smaller than 0.998. Emergence of a saddle point for this flow as We is reduced from 0.999 to 0.998 is shown in Figure 4.14. The dashed lines in this figure correspond to the $\omega_i = 0$ curves in Figure 4.12, and the solid lines with smaller and larger values of $-k_i$ correspond, respectively, to the downstream amplification and the upstream decay of the convected disturbances. It is found that the transition Weber number increases with Re for a given set of flow parameters, and approaches 1 as $Re \rightarrow 0$. This seems to correspond to the experimental observation of Lin and Roberts (1981) and Antoniades and Lin (1980) for the case of small Re. They found that the liquid curtain tends to rupture at a Weber number slightly greater than one. For $We > 1$, they observed both sinuous and varicose waves. They also found that both waves are downstream propagating, but sinuous waves have much higher amplification rates. Detailed discussions on wave motion in a liquid sheet will be given in the next chapter.

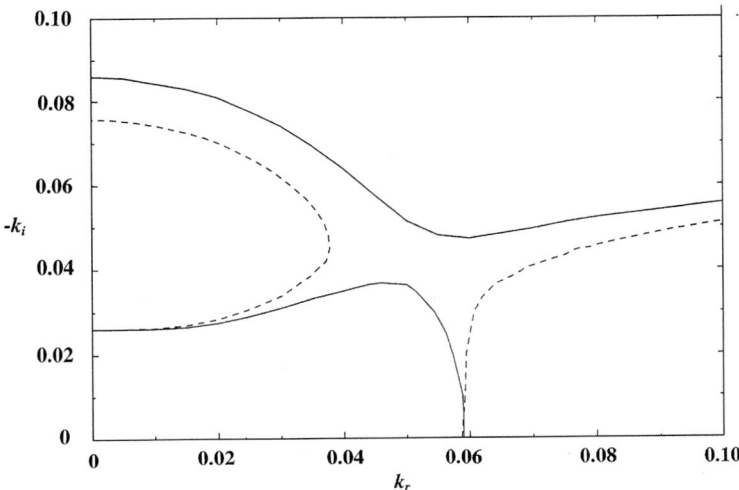

Figure 4.14. Emergence of saddle point. $Re = 52.72, Fr^{-1} = 1.6 \times 10^{-4}, \rho_r = 0.0013,$ $\mu_r = 0.017, l = 5.$ ——, $We = 0.999$; ---, 0.998.

Although the surface tension is shown to be responsible for the absolute instability, it actually reduces the amplification rate of disturbances in the regime of convective instability when $We \gg 1$, as shown in Figure 4.15. We see that the amplification rate of the sinuous mode is higher than that of the varicose mode for relatively long waves, but it is practically the same for shorter waves. For both modes, the amplification rate as well as the cutoff wave number above which the disturbance is damped are reduced when the surface tension is increased. Thus the surface tension is not a cause of convective instability. As expected, the effect of surface tension becomes relatively insignificant as $k_r \to 0$ because the surface tension cannot exhibit its effect in long waves with insignificant curvature.

The effect of gravity is shown in Figure 4.16. As the gravitational force is decreased relative to the inertial force, the cutoff wave number as well as the amplification rates of both sinuous and varicose modes of convectively unstable disturbance are reduced. The reduction of the amplification rate is quite significant for long sinuous waves. This suggests that the origin of long sinuous waves is the gravitational force. Figure 4.17 shows that this is indeed the case. As the gravity vanishes, the long sinuous waves of $k_r = 0.03$ become stable. However, the varicose mode remains unstable. Nevertheless, the sinuous mode remains dominant for the rest of the unstable range of wave numbers. Figure 4.18 shows that the growth rates of both modes increase with Re. The growth rate of the sinuous mode is much higher than that of

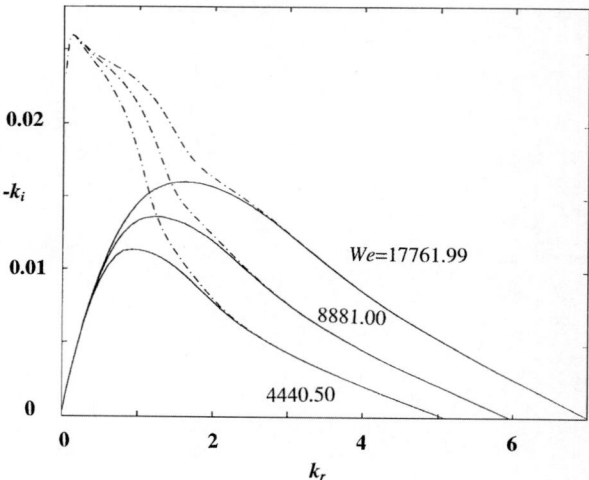

Figure 4.15. Effect of *We* on convective instability. $Re = 15711.77, Fr^{-1} = 3.835 \times 10^{-6}, \rho_r = 0.0013, \mu_r = 0.017, l = 5.$ _ _ _, sinuous mode; _____, varicose mode.

the varicose mode in the range $k_r < 1$. The growth rate of both modes are practically identical as predicted earlier when the gas viscosity is neglected (Fig. 4.8). However in Figure 4.8, the maximum growth rates of both modes occur at the same wavelength, while in Figure 4.18 the maximum growth rate of the sinuous mode occurs at a much longer wavelength. Moreover the cutoff wave number decreases with a decrease in *Re* when the gas viscosity is taken

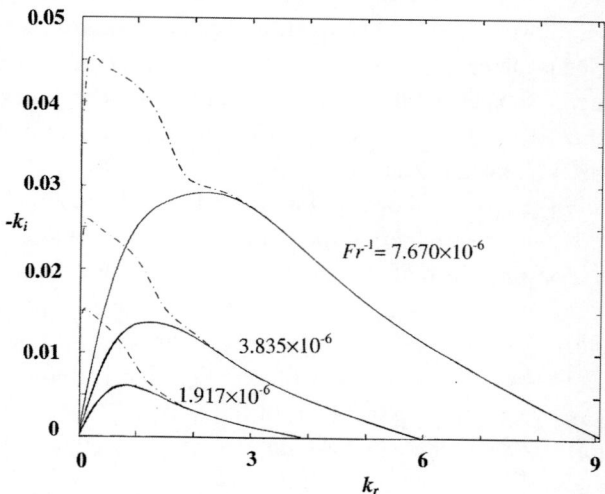

Figure 4.16. Effect of *Fr* on convective instability. $Re = 15711.77, We = 8881.0, \rho_r = 0.0013, \mu_r = 0.017, l = 5.$ _ _ _, sinuous mode; _____, varicose mode.

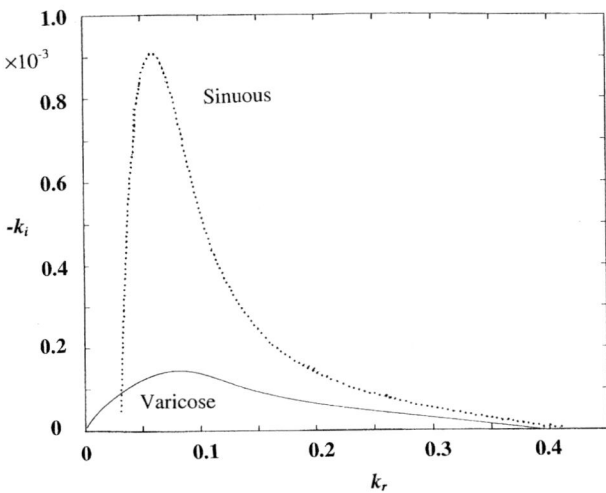

Figure 4.17. Stabilization of long sinuous waves. $Fr^{-1} = 0, Re = 7855.89, \rho_r = 0.0013,$ $\mu_r = 0.017, We = 2220.25, l = 5.$

into account, while it remains the same when the gas viscosity is neglected as in Figure 4.8. Thus in atomization applications of viscous sheets in viscous gases without selective external forcing, the sheet will breakup due to the amplification of the sinuous wave. This finding is contrary to the conclusion reached when the gas viscosity is ignored. Figures similar to Figure 4.18,

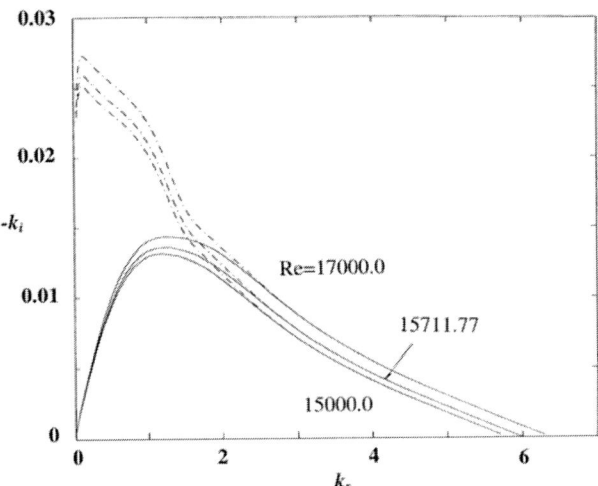

Figure 4.18. Effect of Re on convective instability. $Fr^{-1} = 3.835 \times 10^{-6}, We = 8881.0,$ $\rho_r = 0.0013, \mu_r = 0.017, l = 5.$ _ _ _ _, sinuous mode; _____, varicose mode.

depicting the effects of other flow parameters on the onset of sheet instability, can be obtained (Teng, Lin, and Chen, 1997). These effects will be described without displaying the relevant numerical results. Figure 4.18 can be viewed as a demonstration of the effects of liquid viscosity. The gas viscosity has a similar effect except that its effect is more pronounced for long waves. The enhancement of the convective instability can be achieved by increasing the gas inertial force through an increase in the gas density. The amplification rates of convectively unstable disturbances can be reduced by reducing the gap filled by gas.

4.3. Axially Expanding Viscous Sheets

When the experimental results for the axially expanding liquid sheet described in Chapter 3 were compared with theories, it was assumed that the theory for rectilinear flow could be applied locally to a radially expanding sheet. Without this assumption, Weihs (1978) derived the following dimensional equation to describe the dynamics of sinuous motion of the centerline of a radially expanding viscous Newtonian liquid sheet in an otherwise quiescent invisicid gas

$$2\rho_2 k U^2 yr + uKy_{,rrt} + 2S(y_{,r} + ry_{,rr}) - \rho_1 Ky_{,tt} = 0, \quad (4.31)$$

where y is the distance between the centerlines of the perturbed and unperturbed sheet, k is the wave number associated with the radially traveling disturbance in the gas, U is the liquid velocity, which is constant as explained in Chapter 3 where K is shown to be related to the local sheet thickness h by $K = 2rh$. As $r \to \infty$, (4.31) is reduced to the equation of Dombrowski and Johns (1963) for the case of a rectilinear sheet when their nonlinear term $h_t y_t$ is consistently neglected. The solution of (4.31) has the form

$$h = R(r)T(t), \quad (4.32)$$

$$T(t) = C_3 \exp\left[\frac{\lambda t}{(\rho_1 K)^{1/2}}\right] + C_4\left[\frac{-\lambda t}{(\rho_2 K)^{1/2}}\right],$$

$$R(r) = C_5 \exp(i\alpha^{1/2}r)M\left[\frac{1}{2} + i\frac{\Omega\alpha + \lambda_1^2}{2\alpha^{1/2}}, 1, -2i\alpha^{1/2}(r + \Omega)\right],$$

where M is Kummer's function of the first kind (Erdelyi et al., 1953) and

$$\alpha = \frac{\rho_2 U^2 k}{S}, \qquad \Omega = \frac{\mu}{2S}\left(\frac{k}{\rho_1}\right)^{1/2}, \qquad \lambda_1 = \frac{\lambda^2}{2S}.$$

The temporal amplification factor λ and the wave number k appear in the first parameter of the confluent hypergeometric function M. Hence the wave form and the amplification both change during the spatial development of instability according to (4.32). This makes the quantitative description of the dependence of wave behavior on the flow parameters quite difficult. A few qualitative descriptions of the behavioral change of the wave motion in the radial direction were made by Weihs. In particular, he showed that a temporally stable disturbance may actually grow in amplitude spatially in the radial direction, which is reminiscent of convective instability. Although, no single wavelength for the maximum growth rate was found for the entire sheet, local maximum growth rates were found in the far field where the sheet thickness variation along the radial direction could be neglected. The local maximum growth rate was found to depend on the viscosity. This is in contrast to the results of Crapper et al. (1975) who considered the linear stability of a parallel flow in a rectilinear viscous sheet under viscous gas. They found that liquid viscosity had no effect on the initial wave growth of sinuous waves, and that no maximum growth rate could be found for finite viscosity.

4.4. Discussion

No absolute instability is found in Section 4.1 when the viscosity of gas is neglected. When the gas viscosity is taken into account in Section 4.2, we find absolute instability occurring at a transition Weber number that depends on the Reynolds number with the rest of the flow parameters fixed. The transition Weber number approaches one from above, as $Re \to 0$. Lin and Roberts (1981) carried out an experiment using 12% gelatin solution for the test fluid. The fluid is Newtonian for an experimental shear rate that is much smaller than the limit of 2000/sec. Experimental measurements are given in Table 4.5 for comparison with calculated quantities. δ_1 in this table is given by $\delta_1 = (Re/4Fr^2)^{1/3}$. The test fluid has viscosity and surface tension

Table 4.5. *Velocity Variation*

	U_1(cm/s)					
X_1(cm)	Measured	Eq.(4.2)	Free fall	$2H(\mu m)$	We^{-1}	δ_1
1.47	48.7	51.7	56.1	448	1.22	0.082
2.27	62.7	65.1	68.6	348	0.95	0.049
4.73	93.1	95.2	97.6	235	0.64	0.023
5.56	102.3	103.4	105.6	213	0.58	0.019
10.16	139.6	140.4	142.0	156	0.42	0.010
10.99	147.5	146.0	147.7	148	0.40	0.009

respectively given by 35 CP and 45 dynes/cm. $Re = 3.1$ and $We = 0.82$ at $X_1 = 1.47$ cm. This We is below the transition Weber number in Figure 4.13 and thus the sheet should be absolutely unstable between $X = 0$ and at least $X_1 = 1.47$ cm. It appears that absolutely unstable in a vertical liquid sheet does not necessarily imply rupture. Other conditions are required for rupture to occur. It should be pointed out, however, that the sheet flow configurations in Lin and Roberts and Teng et al. are not exactly the same. In particular the thinning of the liquid sheet between the nozzle exit and $X_1 = 1.47$ cm is not gradual in the experiment of Lin and Roberts. In fact at $X_1 = 0$, $U_1 = 16.2$ cm/sec and $h_0 = 0.0678$ cm, and thus $We = 0.27$ and $\delta = 0.7476$, which is not much smaller than 1. In Chapter 2, we find that the transition from convective to absolute instability occurs at $We = 1$ in an inviscid liquid sheet for which $Re = \infty$. Therefore if one extends the numerical computation beyond the range given in Figure 4.13, the transition curve should turn around and approach one again as $Re \to \infty$ takes place.

Finally we point out some shortcomings of neglecting the gas viscosity. This omission results in missing the absolute instability caused by a sinuous wave. It causes failure to predict that at slightly larger Weber numbers than the transition Weber number, the surface tension suppresses the varicose waves but promotes the sinuous waves, which are the vehicle of absolute instability.

Nevertheless, the theories that neglect either the liquid viscosity or the gas viscosity or both have captured the essentials of the breakup phenomena. However, for quantitative understanding of the onset of instability, it is essential to include viscosities. We summarize below some main findings that may have considerable practical significance when gas viscosity is present. It was found that the spatial growth rate of convectively unstable disturbances decreases as μ_r is increased or ρ_r, Fr, Re, We, and l are decreased separately with the rest of the flow parameters kept constant. This implies that the inertia force relative to the body and viscous forces is destabilizing. Therefore in order to retard on earth the convectively unstable wave packet in the curtain coating of a given liquid, one should coat at high speed in low-density high-viscosity gas surrounding the liquid in a narrow gap. The absolute and convective instability of a liquid curtain is shown to be due to the surface tension effect. It is not caused by the interfacial shear associated with the counterflow effect discussed by Huerre and Monkewitz.

Exercises

4.1. Using $2Fr^{-1}$ as the small parameter, obtain the velocity variation in a slowly thinning viscous falling sheet in an inviscid still air by means of the regular perturbation method.

4.2. Obtain the thickness variation in a axisymmetrically expanding viscous liquid sheet, using an appropriate small parameter. What is the effect of liquid viscosity on the sheet breakup?

4.3. Using δ defined in Section 4.1 as the small parameter, find the velocity variation in an extruded viscous liquid curtain.

4.4. Obtain the eigenvectors from the eigenvalues obtained in Section 4.2, and construct energy budgets to elucidate the effects of interfacial shear exerted by the ambient gas on the liquid sheet.

4.5. When the two fluids are of the same density and viscosity and the interfacial tension vanishes, recover the neutral curves for the Poiseuille flow, obtained by Orszag (1971) by applying the results given in Appendix A and (4.30a)-(4.30g).

4.6. Using the results in Appendix A and (4.30a)-(4.30g), show that the results of Renardy (1987) for the case of three-layered vertical Poiseuille flow can be recovered.

References

Antoniades, M. G., and Lin, S. P. 1980. *J. Colloid Interface Sci.* **77**, 583.

Brown, D. R. 1961. *J. Fluid Mech.* **10**, 297.

Crapper, G. D., Dombrowski, N., and Jepson, W. P. 1975. *Proc. R. Soc. Lond.* A **342**, 225.

Dombrowski, N., and Johns, W. R. 1963. *Chem. Eng. Sci* **18**, 203.

Drazin, P. G., and Reid, W. H. 1985. *Hydrodynamic Stability.* Cambridge University Press.

Erdelyi, A., Magnus, W., Oberhettinger, F., and Tricomi, F. E. 1953. Higher Transcendental Functions. McGraw-Hill.

Gottlieb, D., and Orszag, S. G. 1977. *Numerical Analysis of Spectral Methods; Theory and Applications.* CBMS-NSF Regional Conference Series in Applied Mathematics, Vol. 26, SIAM.

Hurre, P., and MonKewitz, P. 1985. J. Fluid Mech. **159**, 151.

Ibrahim, E. A., and Akpan, E. T. 1996. *Atomiz. Sprays* **6**, 649.

Kelly, R. E., Goussis, D. A., Lin. S. P., and Hsu, F. K. 1989. *Phys. Fluids A* **1**, 819.

Lanczos, C. 1956. *Applied Analysis.* Prentice–Hall.

Li, X. 1993. *Chemi. Eng. Sci.* **48**, 2973.

Lin, S. P. 1981. *J. Fluid Mech.* **104**, 111.

Lin, S. P., and Lian, Z. W. 1989. *Phys. Fluids A* **1**, 490.

Lin, S. P., and Roberts, G. 1981. *J. Fluid Mech.* **112**, 443.

Lin, S. P., Lian, Z. W., and Creighton, B. J. 1990. *J. Fluid Mech.* **220**, 673.

Muller, D. E. 1956. *Mathematical Tables and Aid to Computation*, Vol. 10, pp. 208–230. National Research Council. Washington D.C.

Orszag, S. A. 1971. *J. Fluid Mech.* **50**, 689.

Renardy, Y. Y. 1987. *Phys. Fluids* **30**, 1638.

Rombers, S. 1984. *Problem Solving Software for Mathematical and Statistical FOR-TRAN Programming.* I & II. IMSL.

Taylor, G. I. 1959a. *Proc. R. Soc. Lond. A* **253**, 289.

Taylor, G. I. 1959b. *Proc. R. Soc. Lond. A* **253**, 296.

Taylor, G. I. 1959c. *Proc. R. Soc. Lond. A* **253**, 313.

Taylor, G. I. 1963 *The Scientific Papers of G. I. Taylor.* Vol. 3, No. 25. Cambridge University Press.

Teng, C. H., Lin, S. P., and Chen, J. N. 1997. *J. Fluid Mech.* **332**, 105.

Weber, O. 1931. Z. Angent. Math. Mech. **11**, 136.

Weihs, D. 1978. *J. Fluid Mech.* **87**, 289.

5

Waves on Liquid Sheets

5.1. Generation of Plane Waves

In the stability analyses given in the previous chapters, we examined the dynamic response of liquid sheets to a disturbance introduced impulsively along a local line perpendicular to the flow direction. In some certain flow parameter space, the disturbance grows in time as it is convected downstream. However, a long time after the wave caused by the disturbance passes by a given location in space, the wave appears to decay at that location. To maintain a sustained wave motion in such a liquid sheet, an external source of forcing must be introduced.

Consider a periodic line source given by $\delta(x - x_0)e^{-i\omega_e\tau}$, where δ is the Dirac delta vector function of a constant amplitude across the sheet width, and ω_e is the constant real frequency of the external forcing. The double Fourier transformation of the governing differential system (4.4) to (4.6) with this forcing is

$$\mathbf{H} = \int\!\!\int \frac{\mathbf{D}_{adj}}{D(k,\omega)} \frac{dx}{2\pi} \frac{d\tau}{2\pi} \left[\delta(x - x_0)e^{-i\omega_e\tau}e^{-ikx+i\omega\tau}\right],$$

$$\sim \mathbf{I}\frac{e^{-ikx_0}}{(\omega - \omega_e)D(k,\omega)}. \tag{5.1}$$

Here $\mathbf{D}_{adj} \cdot \delta$ is the constant amplitude of the disturbance vector. Because it is determined only up to an arbitrary multiplication vector in the linear theory, we put its arbitrary multiple to be the unit vector $\mathbf{I}$. The inverse transformation of (5.1) gives the dynamic response

$$\mathbf{G}(x,\tau) = \mathbf{I}\int_F\int_L \frac{e^{i[k(x-x_0)-\omega\tau]}}{(\omega - \omega_e)D(k,\omega)}\frac{d\omega}{2\pi}\frac{dk}{2\pi}$$

$$= \mathbf{I}\int_F i\frac{dk}{2\pi}\frac{e^{i[k(x-x_0)-\omega_e\tau]}}{D(k,\omega_e)}$$

$$= \mathbf{I}\sum_{k_d}\frac{-H(x-x_0)}{(\partial D/\partial k)_{k_d}}e^{i[k_d(\omega_e)(x-x_0)-\omega_e\tau]}. \tag{5.2}$$

In (5.2) we assume that only downstream propagating, convectively unstable disturbances existed before the introduction of the external forcing and that after the external forcing only the downstream propagating waves are amplified. The forced disturbance no longer decays with time at a fixed point in space unlike the case for an impulsively introduced disturbance. The forcing selects a natural frequency to resonate with its own frequency and force the wave to pass the same point with fixed frequency without decaying in time. Moreover, since $k_i < 0$ in a convectively unstable sheet, the wave amplitude increases downstream of the source. Actually a wave motion can also be sustained in a limited region of a convectively stable liquid sheet.

It was shown in Chapter 2 that in the absence of the ambient gas, the sinuous disturbances in Exercises 2.5 and 2.6 are convectively stable. Even this decaying disturbance can be forced to amplify according to (5.2) because $\omega_{ei} = +\varepsilon \to 0$ and thus $k_{di} \to +\varepsilon$ (cf. Exercise 2.4). Therefore the wave train amplitude decays in space in the downstream direction. However, at a fixed point in space the disturbance is forced to maintain its amplitude while oscillating at the forced frequency according to (5.2).

We have demonstrated how a one-dimensional wave train can be sustained in a moving liquid sheet. We will now apply the results to describe the stationary wave patterns created by a point source in a liquid sheet and thereby confirm the theories developed in the previous chapters. In Sections 5.2 to 5.4 we will treat the case of inviscid liquid sheets, and in Section 5.5 we will consider the case of very viscous sheets.

5.2. Waves on Uniform Inviscid Sheets

Consider a simple case of wave motion on an inviscid plane liquid sheet of uniform thickness in the absence of gravity and ambient gas. A line source spanning across the width of the sheet at $x = 0$ generates plane sinuous waves of lengths much longer than the sheet thickness. The sustained frequency is given approximately by (3.31) and (3.32). According to (3.22) two wave trains can be resonated, one with phase velocity $1 - We^{-1/2}$, the other with phase velocity $1 + We^{-1/2}$. In these two phase velocities, 1 arises from the fluid particle velocity and $\pm We^{-1/2}$ represents the wave speed relative to the moving fluid particle. The plus and minus signs refer to the relative wave speed in the downstream and upstream directions, respectively. Hence only when $We = 1$ can the plane wave be held stationary in space.

If instead of a line source, a point forcing is introduced at $x = 0$, we may be able to create a two-dimensional stationary wave pattern. The fluid particle that is perturbed at $x = 0$ is convected down stream along the x-axis, carrying

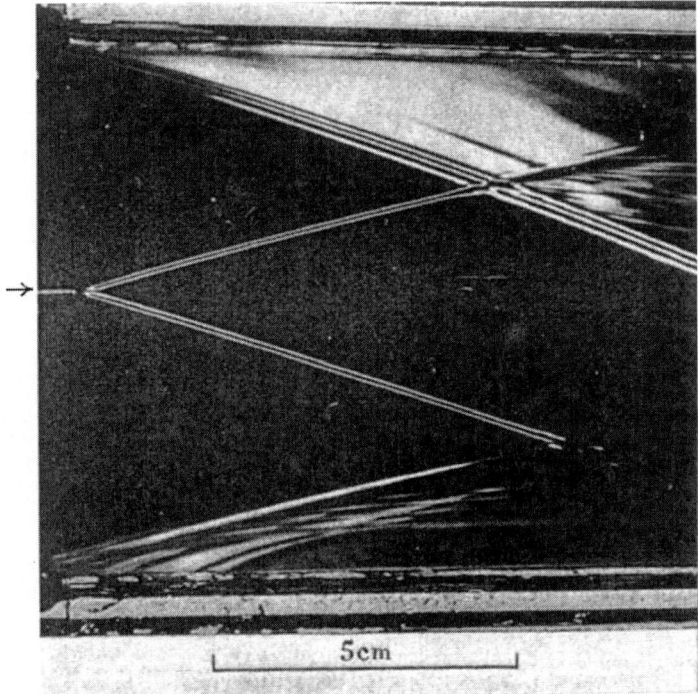

Figure 5.1. Sinuous waves in a sheet of uniform thickness. The arrow marks the position of the glass capillary that directs the air jet. (From Taylor, 1959).

with it the band of frequency imparted by the original source. The string of excited fluid particles along the x-axis in turn serves as the sources emitting two-dimensional centered waves. The wave front perpendicular to the x-axis cannot be held stationary if $We \neq 1$. However, in a certain radial direction the wave speed will be just balanced at some point by the fluid velocity in the same direction. In this particular example, the balance is given by (cf. Fig. 5.1),

$$U_0 \sin(\pm\theta) = U_0 \, We^{-1/2}, \qquad (5.3)$$

where θ is the angle between the tangent at the point of balance and the x-axis. The loci of point of balance form a set of lines of the constant wave phase. According to (5.3) the loci consist of two straight lines making angles $\pm \sin^{-1}(We^{-1/2})$ with the x-axis (Taylor, 1959). Taylor provided experimental evidence to support his prediction.

Equation (5.3) can be rewritten as

$$\mathbf{K}_r \cdot \mathbf{U} + \Omega_0 = 0, \qquad (5.4)$$

where $\mathbf{K}_r$ is the local wave number vector pointing in the direction of the wave propagation with frequency $\Omega_0 = U|\mathbf{K}_r|We^{-1/2}$ in a quiescent fluid. Equation (5.4) is a generally valid statement of the kinematics of a stationary wave. It states that the frequency of a wave in a moving fluid is the sum of its frequency Ω_0 in a quiescent fluid and its Doppler shift $\mathbf{K}_r \cdot \mathbf{U}$ due to the motion of the fluid, and the frequency is zero for a stationary wave.

5.3. Waves on a Falling Liquid Sheet

In the falling liquid sheet discussed in Section 3.2, the sheet thickness decreases and the fluid velocity increases in the flow direction. These changes are significant only over a distance much larger than the sheet thickness when $F_r \gg 1$. For this case (5.4) can be applied locally, and written in a dimensionless form

$$\mathbf{k}_r(\xi) \cdot \mathbf{u}(\xi) + \omega_0 = 0, \tag{5.5}$$

where nondimensionalization is achieved with the same characteristic quantities used in previous chapters. In Section 5.2 we consider only waves of wavelength much larger than the sheet. Here we include short waves whose lengths change appreciably only over a large distance $\xi = O(1)$. Then the condition (5.5) can be used with the aid of some general kinematic relations of wave motion to describe the stationary wave pattern (Ursell, 1960; Whitham, 1974).

The phase angle θ_1 of any Fourier component is given by

$$\theta_1 = \mathbf{k}_r(\omega, \xi, T) \cdot \mathbf{r} - \omega(k, \xi, T)\,\tau, \tag{5.6}$$

where the wave number vector and frequency are allowed to vary over a large distance ξ and a large time T, and $\mathbf{r}$ is the position vector. It follows from (5.6) that

$$\omega = -\frac{\partial \theta_1}{\partial \tau}, \tag{5.7}$$

$$\mathbf{k}_r = \nabla \times \theta_1. \tag{5.8}$$

Elimination of θ_1 between (5.7) and (5.8) yields

$$\frac{\partial \mathbf{k}_r}{\partial \tau} + \nabla\omega = 0, \tag{5.9}$$

$$\nabla \times \mathbf{k}_r = 0. \tag{5.10}$$

It follows from (5.10) that the wave number vector is irrotational.

In the present application of (5.10) to two-dimensional waves,

$$\frac{\partial \beta}{\partial x} - \frac{\partial \alpha}{\partial y} = 0, \qquad (5.11)$$

where $\mathbf{k}_r = (\alpha, \beta)$ in Cartesian (x, y) coordinates. For the stationary waves, α and β are related, according to (5.5), by

$$G(\alpha, \beta) = \alpha u + \omega_0(\alpha, \beta) = 0. \qquad (5.12)$$

The solution of (5.12) can be written, in general, as

$$\alpha = F(\beta). \qquad (5.13)$$

Substituting (5.13) into (5.11), we have

$$\frac{\partial \beta}{\partial x} - \frac{\partial F}{\partial \beta} \frac{\partial \beta}{\partial y} = 0.$$

It follows that along the characteristic line

$$\frac{dy}{dx} = -\frac{dF}{d\beta}, \qquad (5.14)$$

(α, β) is constant. Thus along the characteristics passing through a source at (x_0, y_0) we have

$$\frac{y - y_0}{x - x_0} = -F_{,\beta},$$

$$\frac{y - y_0}{x - x_0} = \frac{G_{,\beta}}{G_{,\alpha}}. \qquad (5.15)$$

Thus (5.12) and (5.15) can be solved for α and β as functions of $\mathbf{x} = (x, y)$, and the phase $\theta(\mathbf{x})$ along each characteristic can be determined from

$$\theta_1 = \int_{\mathbf{x}_0}^{\mathbf{x}} \mathbf{k} \cdot dx \qquad (5.16)$$

along any path between $\mathbf{x}_0$ and $\mathbf{x}$, since $\mathbf{k}$ is irrotational. However, it is convenient to choose characteristics as the integration paths along which $\mathbf{k}$ are constant. A set of constant phase lines can then be traced by assigning constant values, $\theta_1 = 2n\pi$ for example.

For the special example of an inviscid liquid sheet falling in a vacuum, $Q_2 = 0$ and ω_0 according to (2.17), and (5.2) is given by

$$\omega_0 = \left[We^{-1} k_r^3 \tanh(k_r) \right]^{1/2}, \qquad (5.17)$$

for sinuous waves. For varicose waves, it is given by (Exercise 2.1)

$$\omega_0 = \left[We^{-1} k_r^3 \coth(k_r) \right]^{1/2}.$$

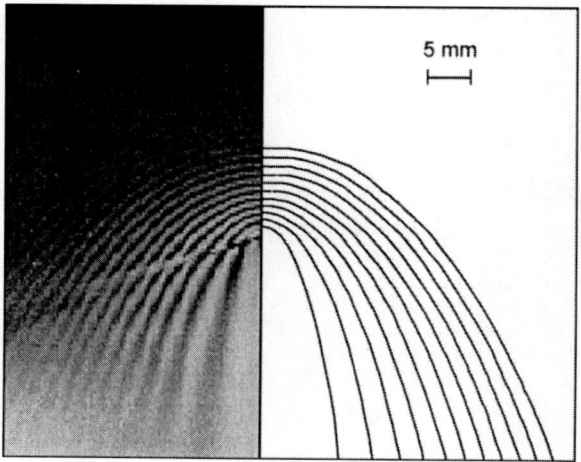

Figure 5.2. Experimental and computed varicose wave patterns on pure water, nozzle-to-obstacle distance: 2.5 cm. (From de Luca, 1997).

Note that We and k_r are slowly varying functions of ξ. With ω_0 given, the wave pattern can be obtained by using method described in the previous paragraph. An example of a varicose wave pattern in the case of $We_0 = 0.75$ is given in Figure 5.2. The calculated lines of constant phase on the right half of the figure compare favorably with the observed wave pattern created by placing an obstacle 2.5 cm below the sheet nozzle (de Luca and Costa, 1997). However, it should be pointed out that near the upstream stagnation point on the obstacle, $We < 1$ and the sheet is absolutely unstable. The disturbance created by the obstacle can propagate upstream. It appears that these upstream propagating disturbances serve as the source of the wave pattern upstream of the obstacle. Because the absolutely unstable upstream region seems to allow the existence of stationary waves, the observation appears to offer experimental evidence that local absolute instability need not lead to sheet rupture. Stationary varicose and sinuous waves were also observed earlier by Taylor. Instead of using an obstacle, he directed a weak thin air jet at an angle to the sheet so that the flow in the sheet was little retarded. The stationary waves seemed to all stand downstream of the forcing air jet in his experiments, testifying that $We > 1$ throughout the liquid sheet. (Figure 5.1).

Exercise

5.1. Obtain the stationary sinuous and varicose wave patterns on a falling liquid sheet in air with $Q_2 = 0.0013$, $We_0 = 10$, $Fr = 1000$. Include the effect of spatial amplitude growth.

5.4. **Waves on an Expanding Sheet**

The axisymmetric expanding liquid sheet described in Section 3.1 is convectively unstable in air, and stable in the absence of air. In either case, stationary waves can be sustained by external forcing according to the mechanism described in Section 5.1. Assuming long sinuous waves are the least attenuated, we may locally apply (5.3) to find

$$sin\,\theta = We^{-1/2}, \tag{5.18}$$

where θ is now the angle between the tangent to the wave crest at a point where a radius and the wave crest intersect. *We* is now defined in terms of the local velocity and thickness, that is, by (3.13). Because $(ruh)^2$ in (3.13) is the square of the constant dimensionless volumetric discharge rate, and We_0 is a given initial Weber number, (5.18) can be written as

$$sin\,\theta = \frac{rh^{1/2}}{R_0}, \tag{5.19}$$

where $R_0 = We_0^{1/2}\,(ruh)$ is a constant. To the first-order approximation, $(rh)^{1/2} = 1$ according to (3.12). Thus (5.19) can be rewritten as

$$sin\,\theta = r^{1/2}/R_0. \tag{5.20}$$

In polar coordinates (r, ϕ), the standing wave defined by (5.20) can be described by

$$\frac{dr}{rd\phi} = cot\,\theta = \frac{[1 - (r\,R_0)]^{1/2}}{(r/R_0)^{1/2}}. \tag{5.21}$$

The solution of this differential equation with the initial condition that $r = 0$ at $\phi = \phi_0$ yields

$$\frac{r}{R_0} = \frac{1}{2}\,[1 - cos\,(\phi - \phi_0)], \tag{5.22}$$

which is the equation of a Cardioid (Taylor, 1959). Taylor created eight cardioids on an expanding liquid sheet by impinging a jet on an impact disc with eight grooves. The experimental observation agrees well with the theoretical prediction. The calculated wave pattern is given in Figure 5.3.

Exercise

5.2. Obtain the stationary sinuous and varicose wave patterns on a radially expanding liquid sheet in a gas with $We = 10$ and $Q_2 = 0.013$.

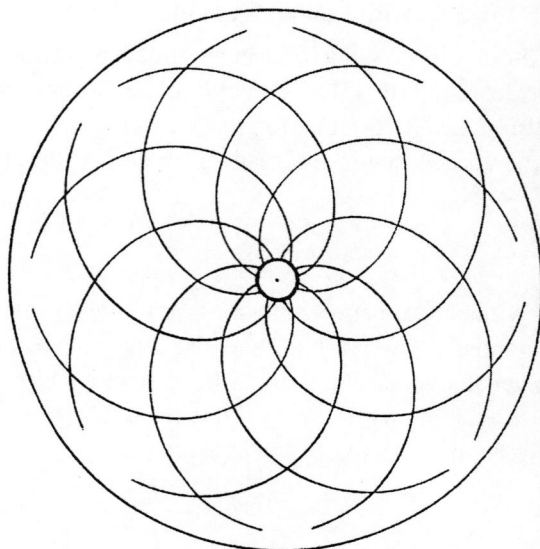

Figure 5.3. Eight cardioids. (From Taylor, 1959).

5.5. Waves on Viscous Sheets

5.5a. Sinuous Waves

Consider the wave motion in the very viscous liquid curtain discussed in Section 4.1. The long wave expansion of the sinuous mode characteristic (4.16) allows the solution

$$(\omega_r/k_r) = 1 \pm We^{-1/2}\left[1 - \frac{1}{8}k_r^2 + O\left(k_r^4\right)\right], \tag{5.23}$$

$$\omega_i = -\frac{-2}{3\,Re}k_r^4\left[1 + \frac{1}{3}k_r^2 + O\left(k_r^4\right)\right], \tag{5.24}$$

$$k_i = k_r^4 \Big/ \left\{6\,Re\left[1 \pm We^{-1/2}\left(1 - \frac{3}{8}k_r^2\right)\right]\right\}, \tag{5.25}$$

where k_r, ω, We, and Re are functions of slow variable ξ. The same notations were used in the previous chapters. Therefore a long sinuous wave at a given point in the viscous curtain with $We > 1$ decays with time at rate ω_i and decays spatially at rate k_i while it is convected downstream at a speed ω_r/k_r according to the above three equations. Note that both decay rates increase with the viscosity. When $We < 1$, the disturbance can propagate both upstream and downstream, while it grows in time as $\omega_i \to -0$. These asymptotic results for $k \to 0$ are consistent with the numerical results given in Section 4.1 for finite k.

To the first-order approximation, (5.23) provides the frequency relative to the stationary fluid. Thus

$$\omega_0 = We^{-1/2}(\xi),$$

and the angle between the tangent to the constant wave phase line with the x-axis is given by

$$\sin(\pm\theta) = We^{-1/2}(\xi). \tag{5.26}$$

This is the same as (5.3) except We is now a function of ξ. In the Cartesian coordinates (x, z) measured in the unit of h_0, (5.26) can be rewritten as

$$\frac{dz}{dx} = \pm\left(\frac{2S}{\rho\dot{Q}U_1 - 2S}\right)^{1/2},$$

where U_1 is given in Section 4.1a and $\dot{Q} = 2U_1 h = 2U_0 h_0$ is the constant volumetric flow rate per unit width of the curtain. The solution of this differential equation with the boundary condition $z = 0$ at $x = m$ is

$$z = We_0^{-1/2}\left(\frac{\dot{Q}^2}{2gh_0^3}\right)\left[\frac{2}{3}\{F^{3/2}(x) - F^{3/2}(m)\}\right.$$

$$\left. + 2We_0^{-1}\{F^{1/2}(x) - F^{1/2}(m)\}\right], \tag{5.27}$$

where

$$F(x) = \left[1 - 16\left(\frac{h_0}{\dot{Q}}\right)^2(4vg)^{2/3} + 8g\left(\frac{h_0}{\dot{Q}}\right)^2 xh_0\right]^{1/2} - We^{-1},$$

and m is the shortest distance between the z-axis and the constant phase line. This distance is taken to be the distance from the z-axis to the obstacle placed in the curtain to introduce the disturbance. A plot of (5.26) is given in Figure 5.4b for comparison with the corresponding experimental observation given in Figure 5.4a.

5.5b. Varicose Wave

Let us assume again that viscous damping is more effective on shorter waves and that longer waves are more easily observable in a viscous sheet of relatively low Reynolds numbers. The long wave solution of (4.15) is given

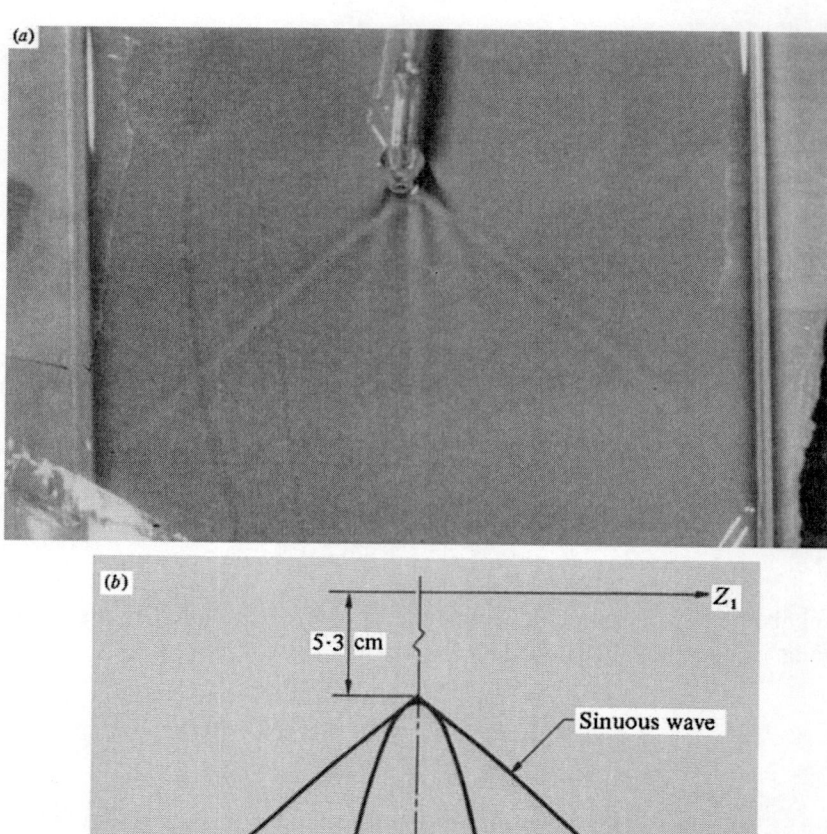

Figure 5.4. Stationary waves in a gelatin solution at 40.6°C, flow rate 21.3 cm³/s; the distance between the lower edge of the coating lip and the obstacle is 5.3 cm. (a) Photograph. (b) Theoretical curve drawn on the same scale as the photograph.

by

$$\omega_0 = \pm k_r^2 \left[We^{-1} - \left(\frac{2}{Re}\right)^2 \right]^{1/2} \qquad \text{and} \qquad \omega_i = -\frac{2}{Re} k_r^2,$$

$$\text{if } We^{-1} > \left(\frac{2}{Re}\right)^2,$$

or

$$\omega_0 = k_r \qquad \text{and} \qquad \omega_i = -\frac{2}{Re} k_r^2 \pm k_r^2 \left[\left(\frac{2}{Re}\right)^2 - We^{-1} \right]^{1/2},$$

$$\text{if } We^{-1} < \left(\frac{2}{Re}\right)^2.$$

The second case of the above two cases cannot satisfy the stationary wave condition (5.5). Therefore we consider only the case of $We^{-1} > (\frac{2}{Re})^2$. For this case the damping rate of long varicose waves is proportional to the first power of viscosity and inversely proportional to the second power of the wavelength. Thus we expect the longest varicose wave to be the most observable in experiments.

Because varicose waves are more dispersive than the sinuous waves, we use the method of Rayleigh (1893) to obtain curves of constant phase.

The varicose waves are held stationary for the same reason the sinuous waves are. Thus, along a given phase of the stationary varicose waves, we have

$$U_1 \sin(\pm\theta) = \pm U_1 \left(\frac{2\pi h_0}{\lambda}\right) \left[We^{-1} - \left(\frac{2}{Re}\right)^2 \right]^{1/2} \qquad (0 \le \theta \le \pi).$$

$$(5.28)$$

The stationary wave is the envelope of varicose waves of different wavelengths λ but of the same phase $\varepsilon - \varepsilon_0$, where ε_0 is a constant reference phase. λ and $\varepsilon - \varepsilon_0$ are related by (cf. Fig. 4.1),

$$\lambda = 2\pi P_\varepsilon / (\varepsilon - \varepsilon_0), \qquad (5.29)$$

where P_ε is the distance measured normally from any point on the curve of constant phase to the line of sources that produce waves. The line of wave source in the present problem is the centerline of the wake behind the obstacle shown in Figure 4.1. The phase at this centerline will be referred to as ε_0. In the polar coordinates (r, ϕ), as shown in Figure 4.1, P_ε is given by

$$P_\varepsilon = d_0 r \sin\phi / \cos\theta, \qquad (5.30)$$

where r is the radial distance r_1 measured from the center of the obstacle divided by d_0. Substituting (5.29) and (5.30) into (5.28), we have

$$r \sin\phi \tan\theta = (\varepsilon - \varepsilon_0) \left[We^{-1} - \left(\frac{2}{Re} \right)^2 \right]^{1/2}.$$

This equation can be written in the Cartesian coordinate system as

$$z \frac{dz}{dx} = (\varepsilon - \varepsilon_0) \left[We^{-1} - \left(\frac{2}{Re} \right)^2 \right]^{1/2}. \tag{5.31}$$

Consider a set of n curves of constant phase each corresponding to a stationary wave crest. Then $\varepsilon - \varepsilon_0 = 2n\pi$ and (5.31) can be integrated to give

$$z^2 = 4n\pi \int_{x_n}^{x} \left[We^{-1} - \left(\frac{2}{Re} \right)^2 \right]^{1/2} dx, \tag{5.32}$$

where x_n is the distance measured from the origin along the x-axis to the apex of each curve of constant phase. Note that (5.32) may not be an accurate description in a region upstream of the obstacle where $\theta \to \frac{1}{2}\pi$ and $\phi \to \pi$. It follows from the definition of k_r, (5.29), and (5.30) that in this region

$$\alpha = \frac{h_0}{P_\varepsilon} (\varepsilon - \varepsilon_0) \sim \frac{\varepsilon - \varepsilon_0}{r}.$$

Thus if the stationary wave crests cut the x-axis at small r such that $2n\pi/r$ is not small, then the condition $k_r \ll 1$, which was used to arrive at (5.32), is violated. In this region the complicated full-dispersion relation (4.15) must be used. However, short waves are so rapidly damped that it would be difficult to observe in practice. Note that if the curtain thickness had been uniform, the integrand in (5.32) would have been a constant and the resulting curves of constant phase would have formed a set of parabola of focal lengths $n\pi [We^{-1} - (2/Re)^2]^{1/2}$. A theoretical curve obtained from (5.32) is given in Figure 5.4 for comparison with experiments.

Comparing the expressions of ω_i for the varicose and sinuous modes, we see that although the viscous damping rates of both modes are proportional to the first power of viscosity, the long varicose mode damps more quickly than the sinuous wave. This is seen clearly in Figure 5.4.

5.5c. Local Surface Tension Variation

Surface tension is very sensitive to temperature and concentration of surface contaminants (Probstein, 1989). Even a pure solution may have concentration

Table 5.1. *Dynamic Surface Tension*

X_1(cm)	U_1(cm/s) Measured	U_1(cm/s) Eq. (4.2)	$d(\mu m)$	θ(degree), Eq. (5.26) Using Static T	θ(degree), Measured	T(dynes/cm) Eq. (5.26)
1.7	53.5	55.7	410	59.8	No wave	—
5.3	100.0	100.8	220	39.2	50	66
9.4	134.5	134.9	163	33.1	42	68

variations along the free surface of a liquid sheet, especially when it is rapidly extended. As the slope of the constant phase line depends on the local surface tension value, the local slope may be measured and the local surface tension calculated by use of (5.26), for example. If the surface tension as a function of the concentration is given, then the concentration can also be inferred.

An example of local surface tension measurements is given in Table 5.1. The test fluid was a 12% gelatin solution of water at 40.6°C, which is a Newtonian fluid for shear rates much smaller than 2000/sec. The static surface tension measured with a Wilhelmy surface tensionmeter was 45 dynes/cm. The fifth column gives the value of θ calculated from (5.26) using the static surface tension value. The sixth column gives the measured values of θ. The last column gives the dynamic surface tension calculated from (5.26) using the measured values of θ. It is seen that the dynamic surface tension differs considerably from the static surface tension (Lin and Roberts, 1981; Antoniades and Lin, 1980; Thomas and Potter, 1975).

5.6. Broken Sheets

When a nonwetable obstacle is placed in a steady liquid sheet, its two free surfaces meet and form a stationary wedge as shown in Figure 5.5. The shape of the free edge can be determined by applying the momentum principle. Consider a small volume of fluid enclosed by a segment of the free edge, a plane parallel to it, and the two side planes perpendicular to these planes (cf. Fig. 5.5). The surface force perpendicular to the free edge must balance the time rate of change of momentum in the same direction in order for the flow associated with this volume to be stationary. Thus simply

$$sin\,\theta = \frac{S}{\rho U_1^2\,(\xi)\,h\,(\xi)\,H_0}.$$

This is the same as (5.26).

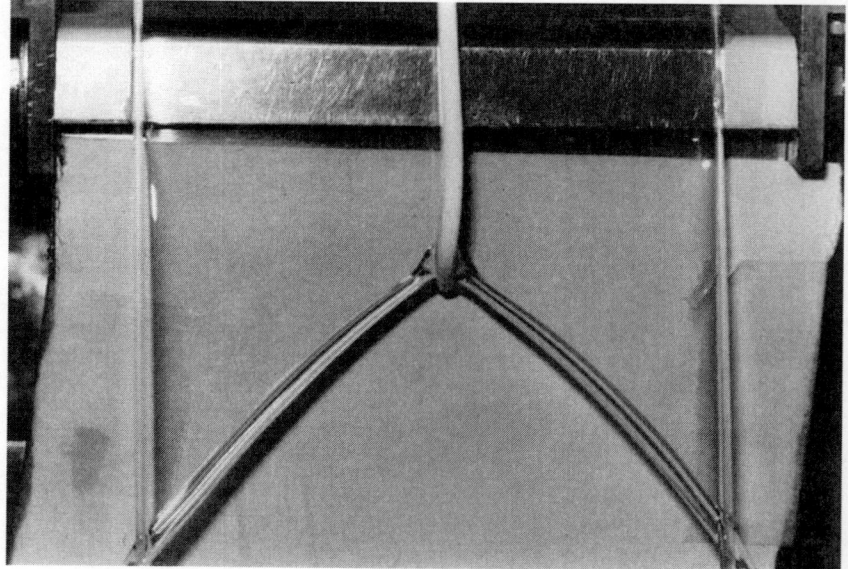

Figure 5.5. Free edge created by an obstacle. $Re = 3.1$. (From Lin and Roberts, 1981).

Since $U_1 H_0 h$ is constant and U_1 increases downward in a falling liquid sheet, θ should decrease in the downstream direction as can be seen in Figure 5.5. In this figure the flow along the viscous sheet edge is laminar, Re being equal to 3.1. The same angle has been observed for turbulent edge in a water sheet by Taylor (1959). For the case of an radially expanding sheet, $2\pi R_1 U_1 (2H_0) = $ constant and the sheet thickness decreases radially, and therefore θ must increase radially along the free edge (Taylor, 1959). It may be worth pointing out that the Weber number based on the velocity perpendicular to the edge is the transition Weber number from convective to absolute instability in the limit of $Re \to 0$ (see Fig. 4.13).

5.7. Summary

This chapter showed how stationary sinuous and varicose waves can be generated by introducing external forcing into the liquid sheets that are predicted in Chapter 4 to be either convectively stable or unstable. The wave properties predicted on the basis of linear stability analyses are consistent with experimental observation. The neglected nonlinear effect appears to stabilize the unstable linear disturbances. We have shown how we can apply our knowledge of wave properties to predict the dynamic surface tension (see also Antoniades and Lin, 1980). Other applications are conceivable. Note that

Figure 5.6. Swirl atomizer spray. (From Taylor, 1959).

although theories of plane waves have been successfully applied locally to construct the observed wave patterns in this chapter, there are wave patterns in liquid sheets that require a three-dimensional wave theory. An example is given in Figure 5.6. Nonaxisymmetric three-dimensional wave patterns are clearly seen in a conical liquid sheet produced by a swirl atomizer.

References

Antoniades, M. G., and Lin, S. P. 1980. *J. Colloid Interface Sci.* **77**, 583.

Brown, D. R. 1961. *J. Fluid Mech.* **10**, 297.

de Luca, L. 1997. *European J. Mech, B/Fluids*, **16**, 75

de Luca, L., and Costa M. 1996. *J. Fluid Mech.* **331**, 127.

de Luca, L., and Meola, C. 1995. *J. Fluid Mech.* **300**, 71–85.

Lin, S. P., 1981. *J. Fluid Mech.* **104**, 111.

Lin, S. P., and Roberts, G. 1981. *J. Fluid Mech.* **112**, 443.

Probstein, R. F. 1989. *Physicochemical Hydrodynamics*. Butterworth.

Rayleigh, L. 1893. *Proc. Lond. Math. Soc.* **15**, 69.

Savart, F. 1833. *Ann. Chim. Phys.* **59**, 55 and 113.

Taylor, G. I. 1959. *Proc. R. Soc. Lond.* A **253**, 289.

Thomas, W. D. E., and Potter, L. 1975. *J. Colloid Interface Sci.* **50**, 397.

Ursell, F. 1960. Steady wave patterns on a non-uniform steady fluid flow. *J. Fluid Mech.* **9**, 333–346.

Whitham, G. B. 1974. *Linear and Nonlinear Waves*. John Wiley & Sons.

6

Phenomena of Jet Breakup

A liquid jet emanating from a nozzle or orifice exhibits richly varied phenomena that depend on the orifice geometry, the inlet condition before the jet is emanated, and the environmental situation into which the jet is issued. A liquid jet cannot escape the ultimate fate of breakup because of hydrodynamic instability. The breakup possesses two major regimes: large drop formation and fine spray formation. These two regimes are controlled by distinctively different physical forces, and between them there exist intermediate regimes. All the regimes arise from a subtle dynamic response of the jet to the disturbances.

6.1. Geometry of Liquid Jets

Citing the experiment of Bidone, Rayleigh (1945, p. 355) stated, " Thus in the case of an elliptical aperture, with major axis horizontal, the sections of the jet taken at increasing distances gradually lose their ellipticity until at a certain distance the section is circular. Further out the section again assumes ellipticity, but now with major axis vertical." This statement is illustrated in Figure 6.1, which was taken from Taylor (1960), who also carried out the experiment. The phenomenon was understood as the vibration of a jet enclosed in an envelope of constant tension about its equilibrium configuration with a circular cross section. However, Taylor (1960) demonstrated that the phenomenon can still be predicted without the surface tension in the absence of gravity. With gravity, if the jet is issued vertically downward, it will accelerate. For the case of a small gravitational force compared with the inertial force, Tuck (1976) showed that the governing differential equation of an elliptic inviscid liquid jet issued into a vacuum possesses two fixed points. One corresponds to a circular jet and the other to an infinitely thin sheet. The geometries of the liquid jet issued from an orifice with an opening other than elliptic are quite unexpected, as described in Rayleigh's book. For example if the orifice is in the form of an equilateral triangle, the jet resolves itself into three sheets. The

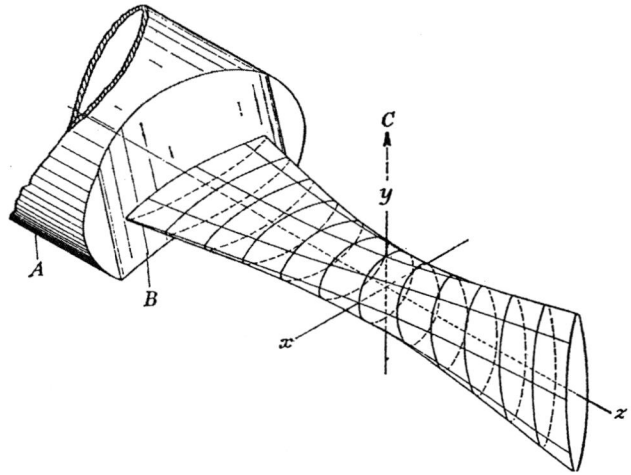

Figure 6.1. Projection of jet with elliptic sections. (From Taylor, 1960).

planes of the sheet, being perpendicular to the sides of the orifice, intersect along the jet axis. Similarly if the aperture is a regular polygon of any number of sides, a corresponding number of sheets develop perpendicular to the sides of the polygon.

The most commonly encountered orifice has a circular aperture. The circular cylindrical jet issued from it can be made to breakup in a controlled manner for various applications including jet cutting, ink jet printing, and spray formation. To be able to control the breakup process reasonably well, one needs to know precisely the physical mechanisms involved in fairly complex breakup processes.

6.2. Regimes of Jet Breakups

The following experimental observation has been frequently cited (Chieger and Reitz, 1996; Lin, 1996; Lin and Reitz, 1998) to bring out in stages the important factors involved in the breakup of a liquid jet. A liquid jet is issued from a nozzle with a circular cross section. If the liquid momentum is not sufficiently large, a continuous jet cannot be formed. One observes liquid dripping from the nozzle tip as from a water faucet that has not been tightly turned off. When the liquid discharge is increased beyond a certain rate, the intermittent release of drops is replaced by a steady streaming of a liquid jet. Near the nozzle tip the jet appears smooth and unperturbed, but after some distance downstream wavy disturbances become discernable at the liquid surface. The surface wave is seen to increase in amplitude as it is propagated

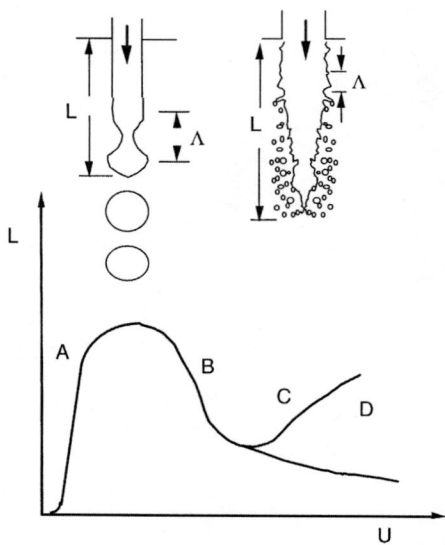

Figure 6.2. Schematic diagram of the jet breakup length L vs. jet velocity U. (From Lin and Reitz, 1998).

downstream. When the wave amplitude becomes equal to the jet radius, the first drop of liquid is pinched off from the jet. We term the distance between the nozzle tip to the point just before the first drop is formed the intact length or breakup length. This unbroken jet length increases almost linearly at first as the jet velocity increases. The rate of increase then becomes less than linear before the intact length reaches its maximum, corresponding to point between A and B in Figure 6.2. Subsequent increases in the jet velocity result in a decrease in the intact length. Point A in Figure 6.2 marks the end of dripping and the beginning of jet formation. The drops formed below the intact length between points A and B have radii almost twice that of the liquid jet. The drop radius below the intact length between points B and C are approximately the same as that of the jet. Beyond point C the droplets appear to be stripped off from the surface rather than pinched off by segments. The jet may be said to be locally atomized. The depth of surface stripping grows deeper and the average droplet radius become smaller. As the jet velocity increases beyond point D in Figure 6.2, the entire jet is completely atomized except near the nozzle tip, where a small core of liquid in the shape of a sharp tongue remains. A completely atomized jet is commonly called a spray. The average radius of the droplets in the spray decrease with the inlet jet velocity.

The appearance of a jet in different breakup regimes corresponding to A, B, C, and D in Figure 6.2 is shown in Figure 6.3. The development of

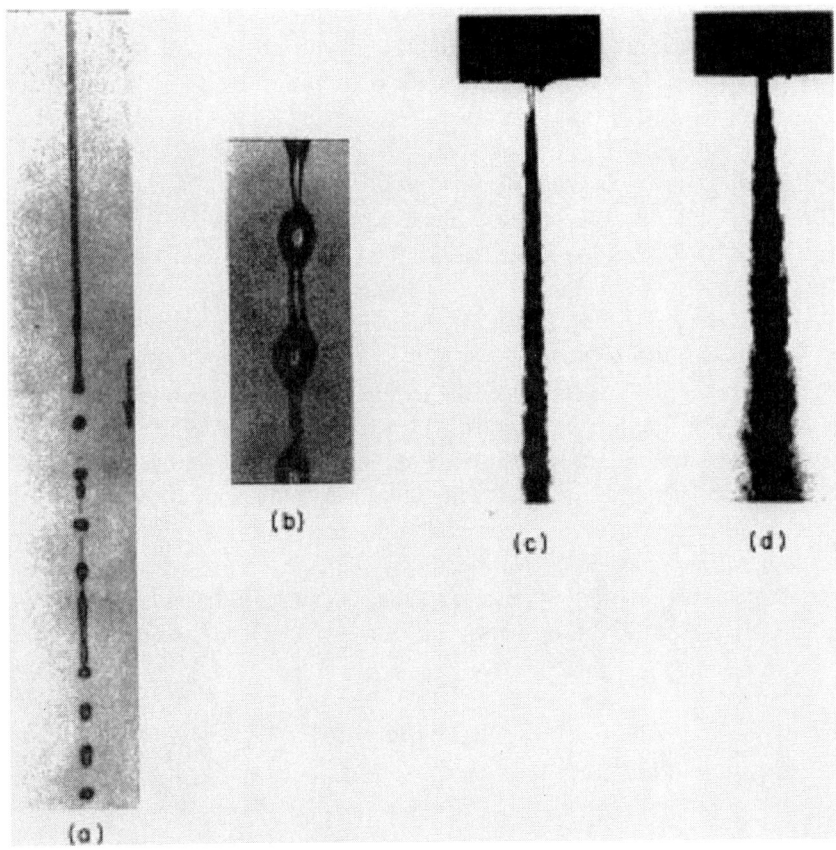

Figure 6.3. Four jet breakup regimes. (From Chieger and Reitz, 1996).

atomization beyond point D is quite complex and can be greatly influenced by the inlet condition. For example, Hiroyasu, Arai, and Shimizu (1991) showed that the intact length can be made to grow again after point D with a specially designed orifice edge geometry. Figure 6.2 reveals that for a given inlet condition, different physical mechanisms are active in different breakup regimes. They do not need to evolve from one to another. In each regime different physical forces play a dominant role in the breakup. The relevant physical forces include the surface tension, the inertial forces in the liquid and gas, the viscous forces, and the body force. The relative importance of these forces are characterized by the Weber number, Reynolds number, Froude number, Mach number, density ratios, and velocity of the fluids involved. The relative importance of these forces varies in different regions of the parameter space and can shift as the nonlinear evolution of the initial instability progresses.

Figure 6.2 is plotted in terms of two-dimensional quantities. A universal relation between the intact length and the rest of the physical parameters is not known. The intact length varies not only with the flow parameters but also with the nozzle inlet condition. The factors influencing the inlet condition include possible cavitation and flow fluctuation inside the nozzle, the nozzle length, the exit edge shape, etc. A review of the influence of nozzle condition on jet breakup can be found in several review articles (Chieger and Reitz, 1996; Lin and Reitz, 1998). Experimental observations of different breakup regimes and theoretical attempts including numerical simulation (Tchavdarov et al., 1988) to confirm the observations can be found in several review articles listed in Chapter 3. In the subsequent chapters, we will concentrate on the fluid dynamics of the jet breakup arising from the interfacial instability, considering the nozzle effects as given. Instability that leads to the formation of drops comparable in size to the jet diameter shall be termed the Rayleigh (1897) mode and instability that leads to the formation of droplets much smaller than the jet diameter shall be termed the Taylor mode. This is because the author, following Taylor and Rayleigh, views the formation of drops and droplets as the consequence of interfacial instability, although Savart (1833), Plateau (1873), and others studied the problem from different points of view.

References

Chieger, N., and Reitz, R. D. 1996. In *Recent Advances in Spray Combustion: Spray Atomization and Drop Burning Phenomena* (Ed. K. K. Kuo), Vol. 1, pp. 109–135. AIAA.

de Luca, L., and Costa, M. 1997. *J. Mech. B/Fluids* **16**, 75.

Grant, R. P., and Middleman, S. 1969. *AIChE J.* **12**, 669.

Hiroyasu, H., Arai, M., and Shimizu, M. 1991. Breakup length of a liquid jet and internal flow in a nozzle. *Proc. ICLASS* **91**, Paper 26.

Lin, S. P. 1996. In *Recent Advances in Spray Combustion: Spray Atomization and Drop Burning Phenomena* (Ed. K. K. Kuo), Vol. 1, pp. 137–160. AIAA.

Lin, S. P., and Reitz, R. D. 1998. *Ann. Rev. Fluid Mech.* **30**, 85.

McCarthy, M. J., and Molloy, N. A. 1974. *Chem. Eng. J.* **7**, 1.

Plateau, J. 1873. Statique Experimental et Theorque des Liquids Soumle aux Seule Force Moleculaire. Vols. 1, 2. Gauthier Villare.

Rayleigh, L. 1945. *The Theory of Sound.* Vol. 2, 2nd ed. republication. Dover Publications.

Rayleigh, L. 1897. *Lond. Math. Soc.* **10**, 4.

Reitz, R. D., and Bracco, F. V. 1986. In *Encyclopedia of Fluid Mechanics.* (Ed. N. Cheremisnoff). Vol. 3, 233–249. Gulf Publishing.

Sterling, A. M., and Sleicher, C. A. 1975. *J. Fluid Mech.* **68**, 477.

Savart, F. 1833. *Ann. Chim.* **53**, 337.

Taylor, G. I. 1960. *Proc. R. Soc.* A **259**, 1.

Taylor, G. I. 1963. *The Scientific Papers of Geoffery Ingram Taylor*. Vol. 3, No. 25. Cambridge University Press.

Tchavdarov, B., Radev, S., Gospodinov, P., Iijima, T., and Asanuma, T. 1988. Numerical prediction of jet breakup length considering the velocity profiles relaxation. *Proceedings of the Faculty of Engineering, Tokai University* **14**, 23–38.

Tuck, E. O. 1976. The shape of free jets of water under gravity. *J. Fluid Mech.* **76**, 625.

7

Inviscid Jets

Stability analysis of inviscid liquid jets issued into a vacuum or into another inviscid gas clearly brought out the physical origin of jet breakup. However, the results predicted by inviscid theories require quantitative and sometimes even qualitative modifications when the effects of viscosity are included. The known basic flows of inviscid circular cylindrical jets are inherently unstable although they satisfy exactly the governing differential system. The origin of instability can be analyzed by examining its onset.

The effects of surface tension, inertial and body forces, and fluid compressibility on the onset of instability in infinite and semi-infinite inviscid liquid jets are examined in this chapter. The governing differential system that includes all these factors except body force and jet swirl are derived in the next section. Each individual effect is isolated in the subsequent sections.

7.1. Stability Analysis of Inviscid Jets

Consider the onset of instability of a jet of an inviscid compressible fluid emanating from a circular nozzle of inner radius R_0 into an unbounded inviscid compressible fluid. The conservation equations of momentum, mass, and energy and the equation of state are, respectively,

$$\rho_i \left(\partial_t + \mathbf{V}_i \cdot \nabla \right) \mathbf{V}_i = -\nabla P_i, \tag{7.1}$$

$$\partial_t \rho_i + \nabla \cdot \left(\rho_i \mathbf{V}_i \right) = 0, \tag{7.2}$$

$$\left(\partial_t + \mathbf{V}_i \cdot \nabla \right) e_i - P_i \left(\partial_t + \mathbf{V}_i \cdot \nabla \right) \left(\rho_i^{-1} \right) = 0,$$

and

$$P_i = P_i \left(\rho_i, T_i \right) \qquad (i = 1, 2),$$

where the subscripts $i = 1, 2$ denote, respectively, the jet fluid and the ambient fluid, ρ is density, t is time, $\mathbf{V}$ is the velocity vector, P is dynamic pressure,

108

e is internal energy per unit mass, and T is temperature. Note that the flow is assumed to be adiabatic and the body force is negligible. The basic state that exactly satisfies (7.1) and (7.2) is given by

$$\mathbf{V}_1 = (\bar{U}_1, \bar{V}_1, \bar{W}_1) = (0, 0, W_1), \qquad \bar{\rho}_1 = \text{constant},$$
$$\bar{P}_1 = \text{constant} \qquad (0 < R < R_0),$$
$$\mathbf{V}_2 = (\bar{U}_2, \bar{V}_2, \bar{W}_2) = (0, 0, W_2), \qquad \bar{\rho}_2 = \text{constant},$$
$$\bar{P}_2 = \bar{P}_1 - S/R_0 \qquad (R_0 < R < \infty),$$

where the overbar denotes the basic state quantities, U, V, and W are, respectively, the radial, azimuthal, and axial components of velocity in the cylindrical coordinate system (R, θ, Z), W_1 and W_2 are the constant velocities in the liquid and gas respectively, and S is the surface tension. Z is positive in the flow direction.

We investigate the stability of this basic state with respect to the disturbance defined by

$$\mathbf{V}_i = \bar{\mathbf{V}}_i + \mathbf{V}'_i, \qquad \rho_i = \bar{\rho}_i + \rho'_i, \qquad P_i = \bar{P}_i + P'_i, \qquad R = R_0 + d, \tag{7.3}$$

where primes designate the perturbations of the variables. Substituting (7.3) into (7.1) and (7.2) and retaining only linear terms produces

$$(\partial_t + W_i \partial_z) \mathbf{V}'_i = -(1/\bar{\rho}_i) \nabla P'_i, \tag{7.4}$$
$$(\partial_t + W_i \partial_z) \rho'_i = -\bar{\rho}_i \nabla \cdot \mathbf{V}'_i. \tag{7.5}$$

By taking the divergence of (7.4) and then using (7.5), we have

$$(\partial_t + W_i \partial_z)^2 \rho'_i = \nabla^2 P'_i. \tag{7.6}$$

Assuming the change of pressure with respect to density is an isentropic process, and expanding $P_i [\rho_i, s (\rho_i, T_i)]$ about $\bar{\rho}_i$, we have $P'_i = (\partial P_i / \partial \rho_i)_s \rho'_i = c_i^2 \rho'_i$, where s is the entropy. Thus (7.6) can be written as

$$(\partial_t + W_i \partial_z)^2 \rho'_i = c_i^2 \nabla^2 P'_i. \tag{7.7}$$

With respect to moving fluids, (7.7) is the wave equation with the speed of propagation of pressure fluctuation c_i. The speed of sound c_i depends on the specific form of the equation of state and the fluid properties. It follows from the energy equation that

$$\frac{de_i}{dt} = P_i \rho_i^{-2} \frac{d\rho_i}{dt},$$

$$\frac{dh_i}{dt} = \frac{d\,(e_i + P_i/\rho_i)}{dt} = \rho_i^{-1} \frac{dP_i}{dt},$$

where (d/dt) stands for the substantial derivative, and $h_i\,[T_i,\,s_i\,(\rho_i,\,T_i)]$ is the enthalpy. Hence, for the isentropic process of sound propagation we have,

$$\left(\frac{\partial P_i}{\partial \rho_i} \right)_s = \frac{P_i}{\rho_i} \left(\frac{\partial h_i/\partial T_i}{\partial e_i/\partial T_i} \right)_s .$$

For the ideal gas surrounding the jet, $P_2 = \bar{R}\rho_2 T_2$, $\partial h_2/\partial T_2 = C_p$, $\partial e_2/\partial T_2 = C_v$, and thus $c_2 = (\gamma \bar{R} T_2)^{1/2}$, where $\bar{R}$ is the ideal gas constant, and $\gamma = C_p/C_v$ is the ratio of the constant pressure specific heat C_p to the constant volume specific heat C_v. For the jet liquid, $P_1\,[(\partial h_1/\partial T_1)/(\partial e_1/\partial T_1)] = K$ is the adiabatic bulk modulus of elasticity, and $c_1 = (K/\rho_1)^{1/2}$. The corresponding linearized boundary conditions to (7.7) at the interface $R = r_0 + d$ are the dynamic force balance

$$P_1' - P_2' = S\left(R_0^{-2} - R_0^{-2}\partial_{\theta\theta} - \partial_{zz} \right) d \qquad (7.8)$$

and the kinematic conditions

$$(\partial_i + W_i)^2\, d = -\bar{\rho}_i^{-1}\partial_R P_i'. \qquad (7.9)$$

Note that P_i' in (7.8) and (7.9) are to be evaluated at $R = R_0$.

The normal mode solution of the differential system (7.7)-(7.9) will be sought in the form

$$\begin{bmatrix} P_1' \\ P_2' \\ d \end{bmatrix} = \begin{bmatrix} \bar{\rho}_1 W_1^2 p_1\,(r) \\ \bar{\rho}_1 W_1^2 p_2\,(r) \\ R_0 f \end{bmatrix} exp\,[i\,(kz + n\theta - \omega\tau)], \qquad (7.10)$$

where $(r,\,z,\,f)$ and τ are dimensionless variables defined by

$$(r,\,z,\,f) = (R,\,Z,\,d)R_0^{-1} \qquad \text{and} \qquad \tau = t(W_1/R_0);$$

and $(k,\,n)$ and ω are the dimensionless wave number vector and wave frequency, respectively.

Substitution of (7.10) into (7.7)-(7.9) yields the governing equation for the dimensionless perturbation pressure amplitudes

$$\left\{ D^2 + r^{-1}D - \left[n^2 r^{-2} + k^2 - M_1^2\,(k - \omega)^2 \right] \right\} p_1 = 0, \qquad (7.11)$$

$$\left\{ D^2 + r^{-1}D - \left[n^2 r^{-2} + k^2 - (W_1/W_2)^2\, M_2^2\,(kW_2/W_1 - \omega)^2 \right] \right\} p_2 = 0. \qquad (7.12)$$

and the corresponding boundary conditions at $r = 1$,

$$p_2 - p_1 + We^{-1}(k^2 + n^2 - 1)f = 0, \tag{7.13}$$

$$(k - \omega)^2 f - Dp_1 = 0, \tag{7.14}$$

$$Q (kW_2/W_1 - \omega)^2 f - Dp_2 = 0, \tag{7.15}$$

where $D = d/dt$, $We = \dfrac{\bar{\rho}_1 W_0^2 R_0}{S}$, $Q = \dfrac{\bar{\rho}_2}{\bar{\rho}_1}$, $M_1 = \dfrac{W_1}{c_1}$, and $M_2 = \dfrac{W_2}{c_2}$.

Note that We is the Weber number, M_1 is the Mach number of the jet, and M_2 is the Mach number of the ambient fluid with respect to a frame of reference moving with the jet at a speed W_0. The bounded solutions of (7.11) and (7.12) are

$$p_1 = AI_n (\lambda_1 r), \qquad \lambda_1 = \left[k^2 - M_1^2 (k - \omega)^2\right]^{1/2},$$

$$p_2 = BK_n (\lambda_2 r), \qquad \lambda_2 = \left[k^2 - (W_1/W_2)^2 M_2^2 (kW_2/W_1 - \omega)^2\right]^{1/2},$$

where A and B are the integration constants and I_n and K_n are, respectively, the modified Bessel function of the first and second kinds. Substitution of these solutions into (7.14) and (7.15) yields

$$A = (k - \omega)^2 f/I_n' (\lambda_1) \lambda_1,$$

$$B = Q (kW_2/W_1 - \omega)^2 f/[K_n' (\lambda_2) \lambda_2],$$

where primes now denote the differentiation $d/d (\lambda_i r)$. It follows from (7.13) that

$$
\begin{aligned}
D_0 &\equiv (k - \omega)^2 I_n (\lambda_1)/\lambda_1 I_n' (\lambda_1) \\
&\quad - Q (kW_2/W_1 - \omega)^2 K_n (\lambda_2)/[K_n' (\lambda_2) \lambda_2] \\
&\quad - We^{-1}(k^2 + n^2 - 1) = 0.
\end{aligned}
\tag{7.16}
$$

For a given set of four flow parameters We, Q, M_1, and M_2, this is the secular equation the solution of which gives the characteristic wave frequency ω and the wave number k.

7.2. An Infinite Jet

Rayleigh (1878) investigated the instability of an incompressible liquid jet extending to infinity in the upstream and downstream directions along the jet axis. The effect of the surrounding gas is neglected. In the case $Q = 0$,

$c_1 \to \infty$, $c_2 \to 0$, $\lambda_1 \to k$, $\lambda_2 \to \infty$, and (7.16) is reduced to

$$I_n(k)(k - \omega)^2/I_n'(k) + k(1 - k^2 - n^2)We^{-1} = 0. \tag{7.17}$$

Because the jet is homogeneous along the infinitely long jet axis, the Galeri transformation can be applied. With respect to a coordinate system (R, θ, X) moving with W_0, the liquid appears stationary. X is related to Z by $Z = X + W_0 t$. It follows from (7.10) that (7.17) is still valid if ω is replaced by $k + \omega_1$, where ω_1 is the dimensionless frequency with respect to the moving reference frame. With this Doppler's shift, (7.17) is reduced to

$$\omega_1^2 We = -\frac{kI_n'(k)}{I_n(k)}(1 - k^2 - n^2). \tag{7.18}$$

The left side of this equation can be written as

$$\omega_1^2 We = \left(\frac{\Omega}{W_0/R_0}\right)^2 \left(\frac{\rho_1 W_0^2 R_0}{S}\right), \tag{7.19}$$

where Ω is the dimensional frequency. In a stationary column of liquid, W_0 is not the characteristic velocity, but the capillary velocity $(S/\rho R_0)^{1/2}$ is. Replacing W_0 with the capillary velocity, yields

$$\omega_1^2 We = \left[\frac{\Omega}{(S/\rho_1 R_0^3)^{1/2}}\right]^2 = \omega_R^2. \tag{7.20}$$

Rayleigh treated ω_R as the complex frequency the imaginary part of which gives the temporal growth rate for the given real wave number appearing on the right side of (7.18). Note that $kI_n'(k)/I_n(k)$ is positive for all real k. Hence if the disturbance is nonaxisymmetric $n \neq 0$, then the exponential growth rate $\omega_{Ri} = 0$ according to (7.18). For the axisymmetric disturbance $n = 0$ and $\omega_R^2 < 0$ in $-1 < k < 1$, but $\omega_R^2 > 0$ in $-1 \geq k \geq 1$. Therefore the axisymmetric disturbance grows in time with an exponential rate ω_{Ri} in the range $-1 < k < 1$ and oscillates neutrally in $-1 \geq k \geq 1$. A plot of the growth rate is given in Figure 7.1. The disturbance of wavelength longer than that corresponding to the cut-off wave number $k_r = 1$ is amplified. The maximum amplification rate occurs at $k = 0.697$. Rayleigh equated the volume of the jet cylinder over one wavelength of the most amplified wave with that of a sphere of radius R to give an estimate of the size of the drop resulting from the jet breakup,

$$R = 1.89R_0. \tag{7.21}$$

The mechanism of instability can be understood from the expression of (7.18). The first term in the parenthesis, 1, arises from the curvature associated

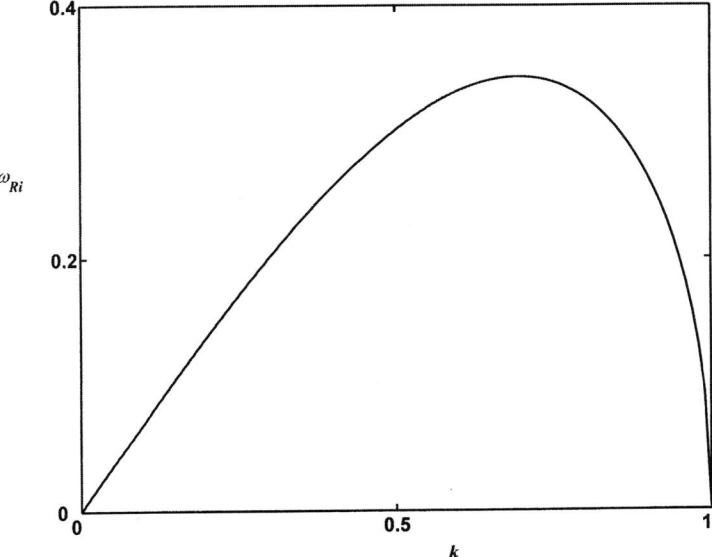

Figure 7.1. Temporal growth rate.

with the jet radius. The second term $-k^2$ arises from the curvature associated with the axial direction. They are of opposite sign. The former represents the capillary pinching in the radial direction; the latter arises from the curvature associated with the other principal direction and represents the capillary force that opposes the capillary pinching. For long waves with $k < 1$, the capillary pinching dominates over the restoring force. For shorter waves with $k \geq 1$, the favorable surface curvature becomes sufficiently large to overcome the capillary pinching. When the restoring force exceeds the capillary pinching, the excess surface energy associated with $k > 1$ is quickly dispersed by the disturbance of different wavelengths.

Temporal stability analysis identified the main mechanism of capillary instability of a liquid jet. However, the Galeri transformation did not allow us to compare the relative importance of the surface and inertial forces in causing the instability, as W_0 disappears from the growth rate ω_{Ri}.

7.3. Semi-Infinite Inviscid Jets

Consider a semi-infinite inviscid incompressible liquid jet emanating from a nozzle into a vacuum. A disturbance is impulsively introduced at a certain location in the jet. We wish to observe the natural response of the jet to this disturbance. The disturbance may excite waves that propagate and amplify

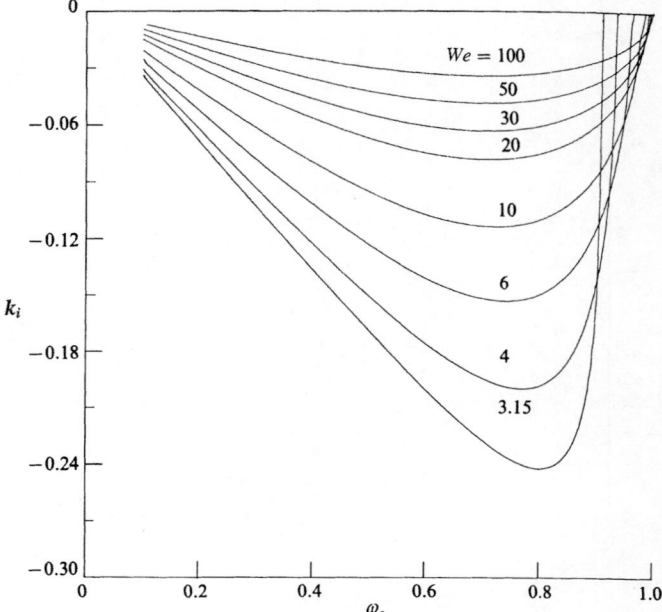

Figure 7.2. Growth rate for convectively unstable waves. (From Leib and Goldstein, 1986a).

spatially as well as temporally. Hence we allow both k and ω to be complex. The relevant dispersion equation is (7.17). The right side of (7.17) appears in the denominator of the integrand in the Fourier superposition of all frequencies and wavelength of the dynamic response, as we learned in previous chapters.

A representative spatial growth rate, $-k_i$, at the onset of instability $\omega_i = 0$ is obtained from (7.17) and plotted in Figure 7.2 to show the effect of the Weber number. Note that as the inertial force relative to the surface force becomes larger, the wave number corresponding to the maximum spatial growth rate steadily decreases toward the value of 0.697, which corresponds to the maximum temporal growth rate reported by Rayleigh. This trend implies that the most probable drop radius to be observed becomes smaller as the Weber number decreases. Keller, Rubinow, and Tu (1973) showed that the eigenvalues of temporal and spatial-temporal waves are related asymptotically as $We \rightarrow \infty$ by

$$-k_i = \omega_i + O(We^{-2}),\qquad(7.22)$$

$$-k_r = \omega_r + O(We^{-2}).$$

Relationship (7.22) is consistent with the Gaster theorem (1962), which relates

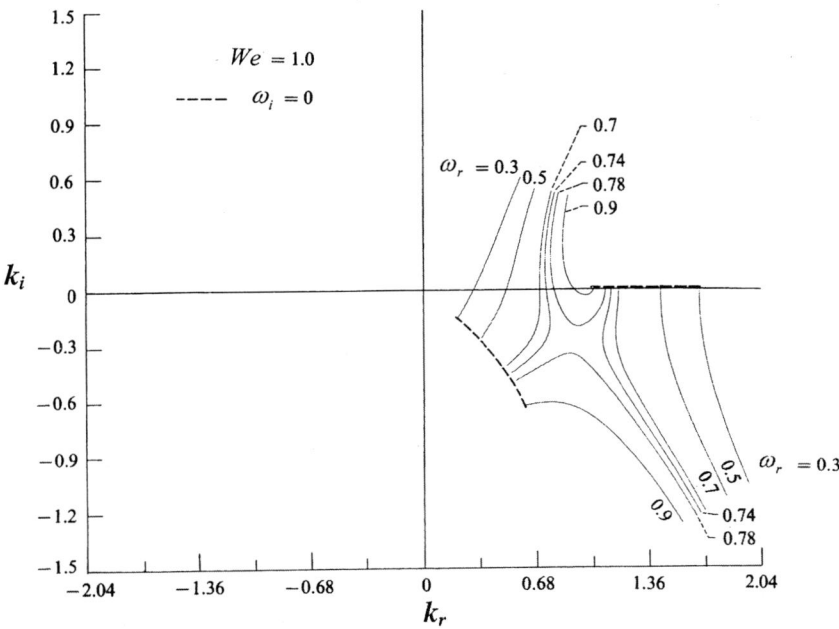

Figure 7.3. Saddle point in k-plane. (From Leib and Goldstein, 1986b).

the spatial growth rate with the temporal growth rate by

$$-k_i = (d\omega_r/dk_r)\,\omega_i.$$

However, the Gaster theorem is valid near the neutral condition, that is, when $k_i \to 0$. Near the maximum growth rate (7.21) does not apply (Lin and Kang, 1987).

When the Weber number is sufficiently small a saddle point singularity appears in the lower half k-plane, signaling absolute instability. An example is given in Figure 7.3. The transition Weber number above which the jet is convectively unstable and below which the jet is absolutely unstable is found to be π (Leib and Goldstein, 1986a). The consequence of absolute instability will be discussed with experimental evidence after the effects of viscosity and density of both liquid and gas are taken into account.

7.4. Effects of Velocity Relaxation

In the previous two sections, the basic flow had a uniform velocity distribution across any section along the jet axis. In practice the fluid inside the nozzle sticks to the wall and has a maximum velocity along the axis. As the fluid leaves the nozzle, it experiences a stick-slip situation. The velocity distribution relaxes and attains an idealized uniform distribution after some distance downstream of the nozzle tip (Scriven and Pigford, 1959). The effect of this velocity relaxation on the jet breakup can be revealed, at least qualitatively, by considering the one-parameter family of basic flow profiles in an inviscid jet,

$$U(r) = \frac{1 - br^2}{1 - \frac{1}{2}b}, \qquad (7.23)$$

which produces progressively flatter profiles as the parameter b decreases from 1 (Hagen–Poiseuille profile) to 0. By using of the Wienner–Hopf technique (Noble, 1958), Leib and Goldstein (1986b) obtained the solution to the stability problem in which b appears only as a parameter in the boundary conditions. Stability analysis reveals that if the Weber number at the nozzle exit is smaller than 3.3, the convectively unstable jet near the nozzle tip becomes absolutely unstable before the jet velocity profile relaxes to a top hat distribution. The Weber number here is based on the mass averaged mean velocity. On the other hand, if the Weber number at the nozzle exit is greater than 3.3 then the jet remains convectively unstable downstream of the nozzle as the jet relaxes to a Rayleigh jet. The velocity relaxation appears to prevent the disturbance from penetrating upstream.

7.5. Effects of Surrounding Gas

We assume that both the jet liquid and the surrounding gas are incompressible. Then $M_1 = M_2 = 0$, but $Q \neq 0$ in (7.16) (cf. Ponstein, 1959, p. 437). Figure 7.4 gives a typical spatial amplification rate curve at the onset of instability $\omega_i = 0$, for the case of quiescent gas where, $W_2 = 0$. The values of $We = 10^3$ and $Q = 0.0013$ in this figure correspond to the case of a water jet emanated at a speed of 201 m/s from a nozzle of radius 0.18 mm into room temperature air under 1 atm pressure. The curve with $Q = 0.013$ is for the same jet at 10 atm pressure. These values are encountered in many common applications. Figure 7.2 showed that the cut-off wave number of the spatial-temporal disturbance approaches that of the temporal one as We becomes very large, which is the case in this example. In the absence of ambient gas, the disturbance with $k > 1$ cannot be amplified. Here when the ambient air is introduced the cut-off wave number is increased to over 1.6. Thus the effect of the ambient gas is to produce a broad distribution of drop sizes, with the smallest drop radius equaling 60% that of the smallest drop produced in vacuum. The linear dependence of k_i on Q in Figure 7.4 clearly indicates that the enhancement of the growth rate in the presence of gas is associated with the gas inertial force.

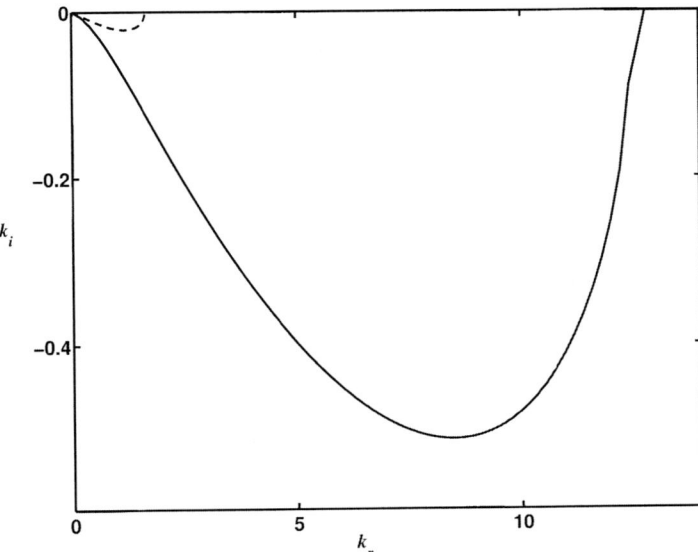

Figure 7.4. The effects of surrounding gas are to increase the production rate and to reduce the size of the drops. $We = 1000.0$, $W_2/W_1 = 0.0$. - - -, $Q = 0.0013$; ———, $Q = 0.013$.

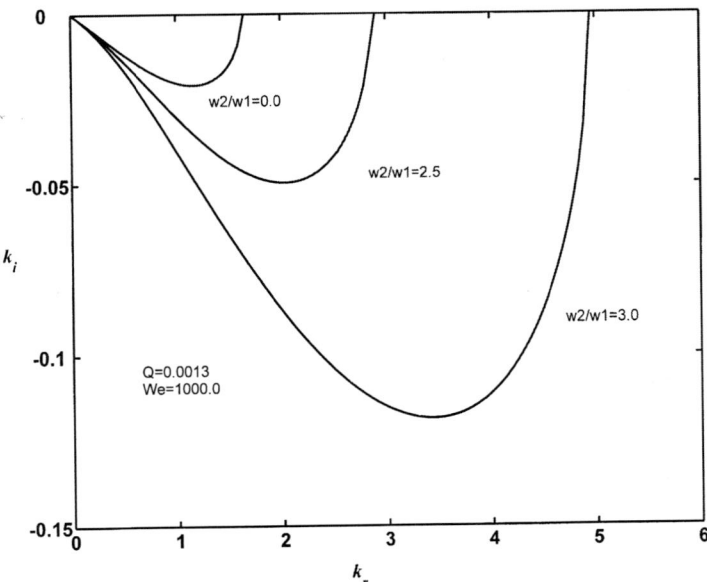

Figure 7.5. Effect of slip velocity is to enhance the jet breakup. $We = 1000$, $Q = 0.0013$.

7.6. Effects of Gas Velocity

The inertial force in the gas phase can also be increased by increasing the gas velocity relative to the liquid jet velocity. Gas velocity is frequently increased to enhance the jet breakup as in the twine-fluid atomizer. Assuming that both liquid and gas are incompressible, then $M_1 = M_2 = 0$ in (7.16). As one increases the gas velocity relative to the liquid jet velocity, both the growth rate and the cut-off wave number increase as shown in Figure 7.5. This is a manifestation of the instability pointed out by Helmholtz (1868) and analyzed by Kelvin (1871). The physical mechanism of the Kelvin–Helmholtz instability was expanded by Batchelor (1967) in terms of vorticity dynamics.

7.7. Effects of Compressibility

In the previous six sections, we assumed that the fluids involved were incompressible. We expect the assumption to be approximately valid for relatively slow fluid motions. Consider a water jet with an average velocity of 26.8 m/s issued into a quiescent air. Then $M_1 = 0.0192$ and $M = M_2(W_1/W_2) = 0.079$ in (7.16), and the numerical solution of (7.16) yields the amplification rate curve given as a solid line in Figure 7.6. $M = W_1/c_2$ is the Mach number of the gas with respect to the liquid jet with a velocity W_1. $We = 1000$

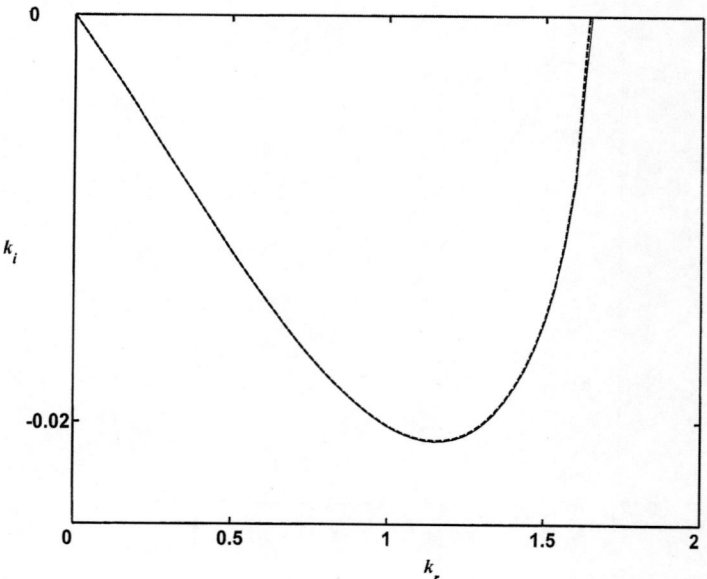

Figure 7.6. Effect of compressibility is negligible at small Mach numbers. $Q = 0.0013$, $We = 1000$. ———, $M_1 = 0.0192$, $M = 0.079$; - - - -, $M_1 = M = 0$.

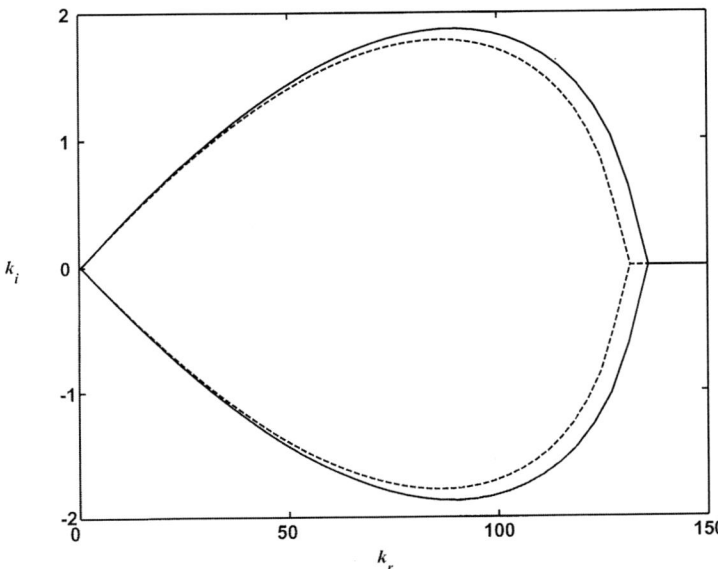

Figure 7.7. Effect of compressibility. $Q = 0.0013$, $We = 10^5$, $W_2/W_1 = 0.$ _ _ _, $M_1 = M = 0;$ ____, $M_1 = 0.061$, $M = 0.251$.

corresponds to a water jet emanating from a nozzle of a diameter 0.02 cm. If we assume the fluids to be incompressible, then the amplification curve is given as a dashed line, which almost falls on top of the solid line. As the water jet velocity is raised to 84.85 m/s, M_1 becomes 0.061 and $M = 0.251$. A water jet issued into quiescent air at this velocity from a nozzle of 0.1 cm radius has a growth rate given as the solid line in Figure 7.7. If we assume the fluid to be incompressible, we obtain the corresponding amplification curve shown in a dashed line. The effect of the fluid compressibility is to increase the amplification rate and the cut-off wave number slightly for short waves (Zhou and Lin, 1992a).

We have shown with examples that the effect of compressibility is negligible for cases of sufficiently small jet velocity. However, for sufficiently large subsonic jet velocity the effect of compressibility cannot be neglected. Consider an axisymmetric water jet issued from a nozzle of diameter 0.036 cm at a speed of 201 m/s in quiescent air under 1 atm pressure. Then the relevant parameter values are $n = 0$, $M_1 = 0.135$, $M = 0.59$, $Q = 0.0013$, and $We = 10^5$. The amplification curves calculated from (7.16) are given in Figure 7.8 There are three branches of $\omega_i = 0$ and a saddle point signaling absolute instability. If we neglect the effect of compressibility $M_1 = M = 0$ with the rest of parameters remaining the same, we obtain the amplification curve

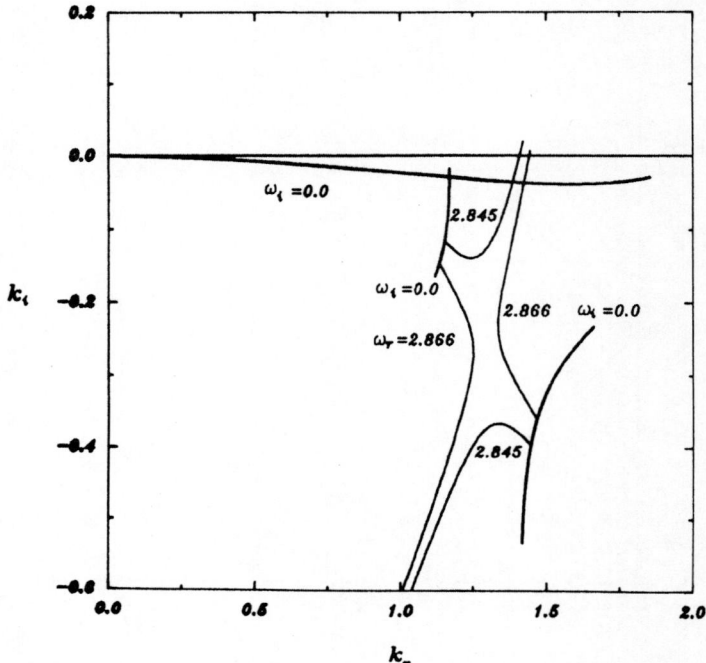

Figure 7.8. Absolute instability of subsonic jet at $Q = 0.0013$, $We = 10^5$, $M_1 = 0.135$, $M = 059$. (From Zhou and Lin, 1992b).

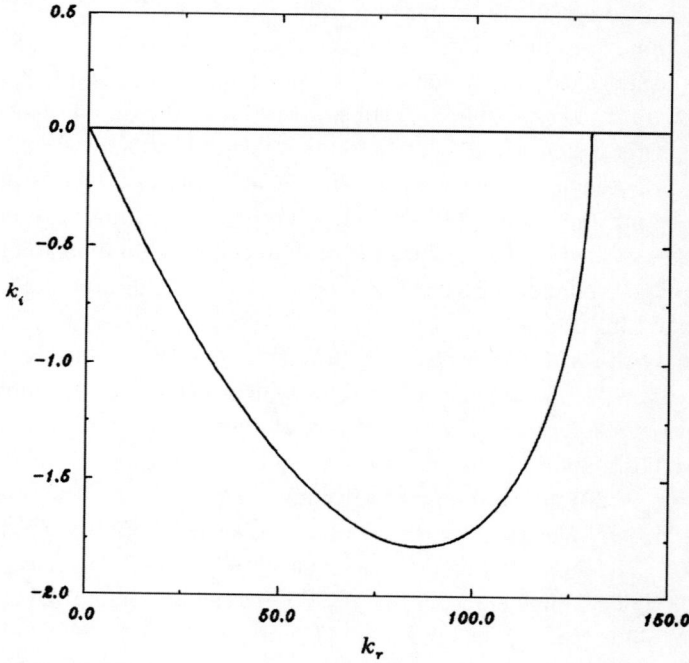

Figure 7.9. Convective instability of incompressible jet. $Q = 0.0013$, $We = 10^5$, $M_1 = M = 0$. (From Zhou and Lin, 1992b).

for the convective instability shown in Figure 7.9. Hence the neglect of compressibility in this case leads to a quantitatively different prediction of the instability characteristics. Absolute instability may occur at $k_r > 1$ or $k_r < 1$. The mechanisms of absolute instability are different, however, for the two cases. The pressure difference $p_1 - p_2$ and the interfacial displacement f obtained from the eigenvector solution are plotted in Figures 7.10 and 7.11 respectively for the cases of $k_r < 1$ and $k_r > 1$. The location of the saddle point with the corresponding frequencies and the relevant parameters are given in the figures. The pressure distribution is nearly 180° out of phase with the displacement in Figure 7.10. Thus $(p_1 - p_2)$ stabilizes the jet by acting against necking and bulging. Hence the surface tension remains the source of instability as in the Rayleigh jet. In Figure 7.11, $(p_1 - p_2)$ is almost 90° out of phase. Hence the pressure difference tends to amplify the interfacial displacement. On the other hand the stabilizing $-k^2 We$ term in (7.16) dominates over the destabilizing We term when $k_r > 1$. Therefore absolute instability

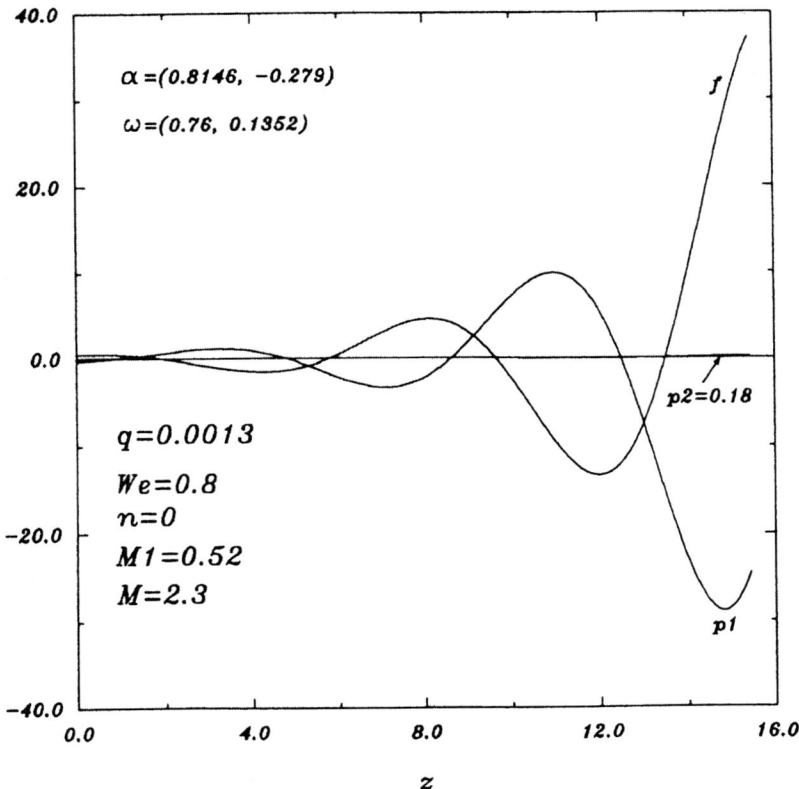

Figure 7.10. Absolute instability at $k_r < 1$. (From Zhou and Lin, 1992a).

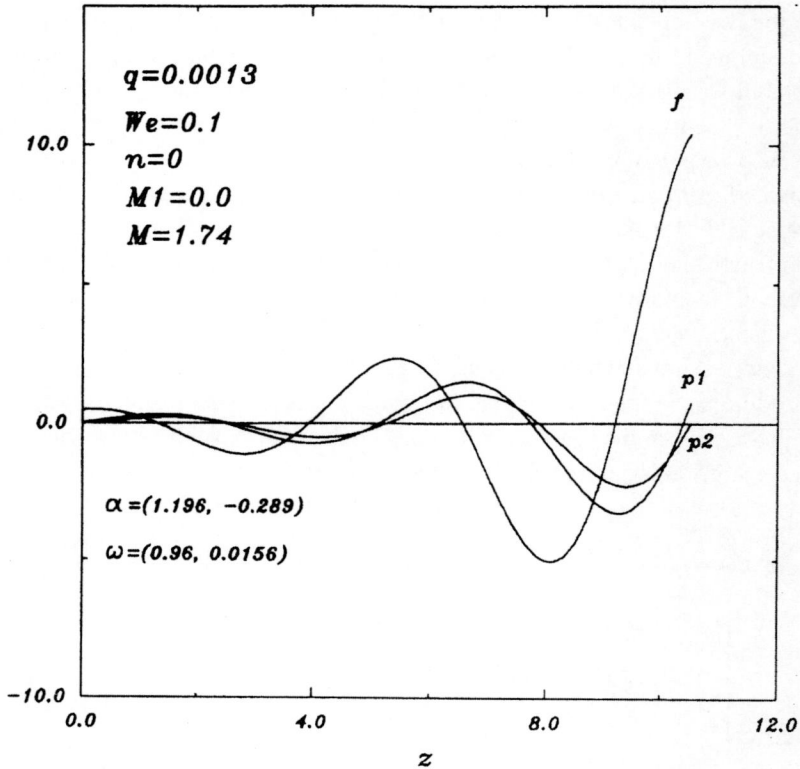

Figure 7.11. Absolute instability at $k_r > 1$. (From Zhou and Lin, 1992a).

for this case is due to the pressure fluctuation not the capillary force (Zhou and Lin, 1992a).

A complete separation of absolute and convective instability can only be achieved in a five-dimensional parameter space involving We, Q, M_1, M_2, and W_2/W_1. This is a tedious task. When the surrounding gas flow relative to the moving jet becomes supersonic the growth rates of the convectively unstable disturbances tend to peak near the relatively large cut-off wave number favoring atomization of the jet into narrowly distributed droplet size. A typical example is given in Figure 7.12 (Zhou and Lin, 1992b).

Li and Kelly (1992) found that when the slip velocity $\Delta U = W_1 - W_2$ is sufficiently large, the sinuous mode with $n = 1$ can actually become more unstable than the symmetric mode with $n = 0$. The amplification curves for these two modes are given in Figure 7.13 for the case of $(W_2/W_1) = 0$, $M\omega_i = 0.01$, and $M^2/We = 6.5 \times 10^{-7}$. The finding is qualitatively consistent with experimental observations of Hoyt and Taylor (1977), Arai and Hashimoto (1986),

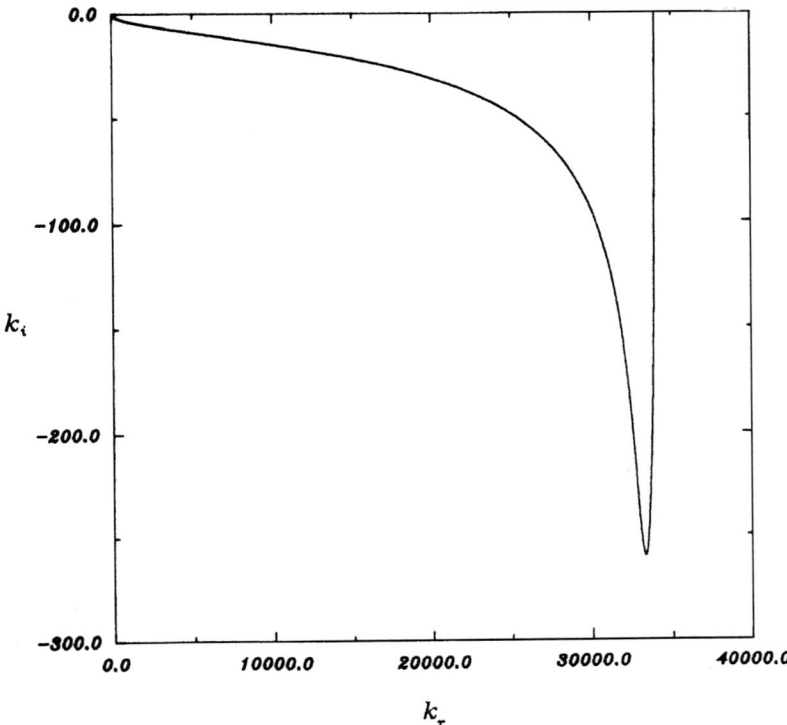

Figure 7.12. Convective instability of subsonic jet. $Q = 0.0013, We = 10^5, M_1 = 0.52$, $M = 2.3$. (From Zhou and Lin, 1992b).

and Chigier and Reitz (1996). Li and Kelly also found that the amplification rates reach a maximum near $M_2 = 1$ for both axisymmetric and the first non-axisymmetric modes. Representative results are given in Figure 7.14 for the flow parameter shown in the figure. The figure also shows that for the subsonic air flow relative to the liquid jet the compressibility tends to enhance the amplification rate, which is consistent with Figure 7.7. On the other hand for the supersonic air flow relative to the liquid, the compressibility tends to reduce the amplification rate, since $-k_i$ decreases as M increases. Making comparisons with earlier related works (Nayfeh and Saric, 1973; Chang and Russel, 1965; Chawala, 1975) Li and Kelly pointed out that the neglected nonlinearity may be more important near the transonic regime.

When the surrounding gas flow relative to the moving jet becomes supersonic, the growth of convectively unstable disturbances tends to peak near a relatively large cut-off wave number favoring atomization into narrowly distributed droplet sizes. From the fairly extensive numerical results of Zhou

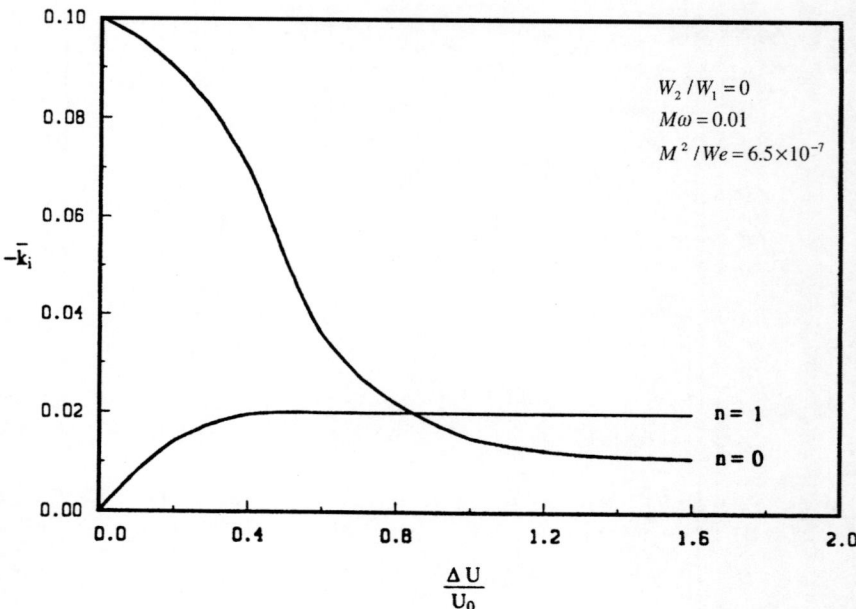

Figure 7.13. Spatial growth rate as a function of velocity difference across the interface. (From Li and Kelly, 1992).

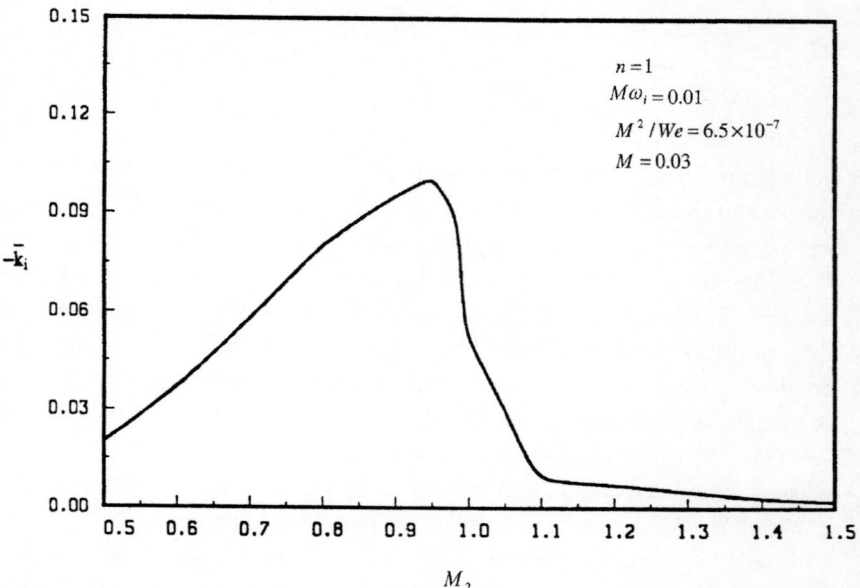

Figure 7.14. Spatial growth rate of nonaxisymmetric mode as a function of M. (From Li and Kelly, 1992).

and Lin (1992a,b) for the cases of $M > 1$, we find that the wave number of the convectively unstable disturbances are proportional to We/M^2. Therefore the corresponding wavelength scales with

$$2\pi S/\rho_1 W_1 c_2.$$

Hence one obtains smaller droplets in a less compressible gas, since the compressibility is inversely proportional to the speed of sound c_2.

On the other hand, when compressibility is negligibly small in the subsonic jet depicted in Figures 7.4 to 7.7, the cut-off wave number for convective instability is proportional to We/M, that is,

$$\frac{2\pi R_0}{\lambda} = k_r \sim WeQ.$$

It follows that the unstable wavelength scales as

$$\lambda/R_0 \sim 2\pi S/\rho_2 W_1^2 R_0 = 2\pi/(WeQ),$$

reflecting the dominant effect of Q and We in determining the outcome of jet breakup as the large Rayleigh drops or the small Taylor droplets in a spray. The breakup phenomena of small droplets have been observed by Joseph, Belanger, and Beavers (1999) in a supersonic flow in a shock tube.

7.8. A Jet with a Swirl

Consider an incompressible jet with a swirl. The circulation Γ in the jet is constant. The surrounding gas having a constant pressure $\bar{P}_2$ is quiescent and assumed incompressible. The potential basic flow in the absence of gravity is given by

$$\mathbf{V}_1 = \mathbf{i}_x W_1 + \mathbf{i}_\theta \left(\Gamma/R\right), \qquad \mathbf{V}_2 = 0, \ H = R_0,$$

$$\bar{P}_1 - \bar{P}_2 = \left(\rho_1 \Gamma^2/2\right) \left(R_0^{-2} - R^{-2}\right) + S/R_0,$$

where $\mathbf{i}$ is the unit vector whose components in the cylindrical coordinates (R, θ, Z) are denoted by subscripts. The rest of the variables are defined as before.

The characteristic equation of the disturbances is given by (cf. Ponstein, 1959, p. 434)

$$(\omega - n\gamma - k)^2 + WeQa_1\omega^2 + k\left[(n^2 - 1 + k^2)We^{-1} - \gamma^2\right]a_2 = 0,$$

$$a_1 = -I_n'(k)\,K_n(k)/I_n(k)\,K_n'(k),$$

$$a_2 = -I_n'(k)/I_n(k),$$

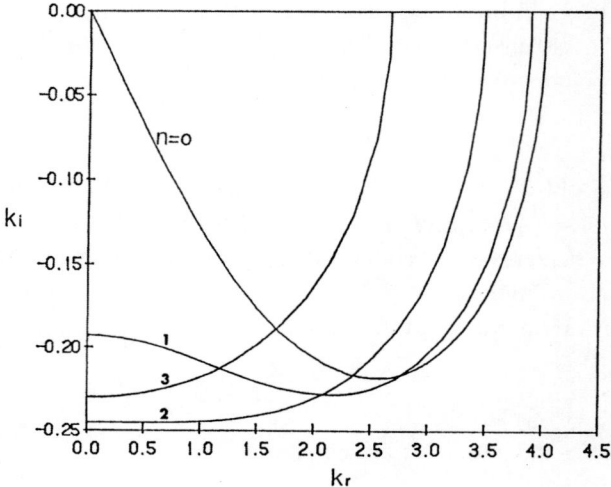

Figure 7.15. Amplification rates of nonaxisymmetric convectively unstable disturbances. $We = 400$, $We\gamma^2 = 15$, $Q = 0$. (From Kang and Lin, 1989).

where $\gamma = \Gamma/R_0 W_1$ is the swirl number. When the Rayleigh jet is imparted
with a constant circulation, the nonaxisymmetric disturbances become more
unstable than the symmetric ones, as shown in Figure 7.15. Moreover, the
swirl enhances the amplification rate and reduces the drop size. The larger
the swirl the larger these effects are, as demonstrated in Figure 7.16. These

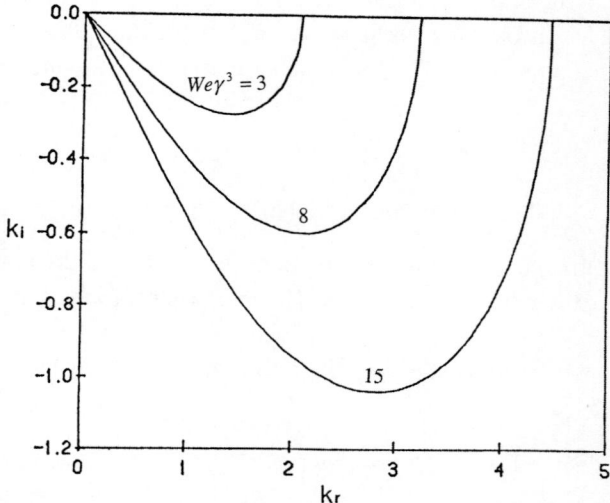

Figure 7.16. Effects of swirl on amplification rates. $We = 25$, $Q = 0$, $n = 0$. (From
Kang and Lin, 1989).

effects amplify in the presence of the surrounding gas with increasing gas density as in the case of nonswirling jets (Kang and Lin, 1989). The swirl appears to have little effect on the onset of absolute instability.

7.9. Initial Instability

So far we have considered only the large time asymptotic behavior of disturbances, although the stability problem has been formulated as an initial-boundary value problem. To know the detail of the initial transience, we must completely evaluate the Laplace–Fourier integrals. Such a task was carried out by Berger (1988) for the temporal instability problem of an infinite inviscid jet in vacuum. He showed that the disturbance grows initially as fast as the square of time but approaches asymptotically at large time to the Rayleigh's normal mode results. The cuto-ff wave number of stability $k_r = 1$ remains unchanged. However, the initial growth rate is larger than the large time asymptotic growth rate. This fact may affect the manner in which the linearly unstable disturbance evolves into finite amplitude disturbances. Algebraic initial growth of disturbances has been encountered in other stability problems. It occurs when the eigenvalues are not discrete. Even if they are discrete they can sometimes be so close together that numerically or experimentally, because of the limitation in accuracy control, the disturbance can grow algebraically. The initial value approach is of more physical significance when it leads to instability prediction while the large time asymptotic behavior is stable. The initial value considerations have been given in other stability problems (Case, 1960a,b; Prosperetti, 1976, 1980, 1981; Menickoff et al., 1978).

Exercises

7.1. Show that in a Rayleigh jet, the temporally unstable disturbances are nondispersive and their phase and group velocities are equal to the jet velocity. For spatially growing disturbances this is true only when $We^{-1} \to 0$.

7.2. Consider an inviscid swirling liquid jet with a constant circulation Γ within the jet radius. Thus the basic flow velocity distributions are given respectively by

$$\mathbf{V}_1 = \mathbf{i}_z W_0 + \mathbf{i}_\theta \left(\Gamma / R \right), \qquad \mathbf{V}_2 = 0,$$

and

$$\bar{P}_1 - \bar{P}_2 = \left(\rho_1 \Gamma^2 / 2 \right) \left(R_0^{-2} - R^{-2} \right) + S / R_0,$$

where $(\mathbf{i}_R, \mathbf{i}_z, \mathbf{i}_\theta)$ is the unit vector in the cylindrical coordinates (R, Z, θ). Show that the dispersion equation for the dynamic response of the jet is given by

$$(\omega - n\gamma - k)^2 - k\left[(n^2 + k^2 - 1)\,We^{-1} - \gamma^2\right] I_n'(k)/I_n(k) = 0,$$

where $\gamma = (\Gamma/R_0 W_0)$ is the swirl number. Solve this equation for the spatial amplification rate $-k_i$ as a function of the wave number k_r at the onset of instability $\omega_i = 0$. Show that introducing the swirl into the jet forces the jet to become more unstable relative to nonaxisymmetric disturbances as shown in Figure 7.15 (Kang and Lin, 1989).

References

Arai, T., and Hashimoto, H. 1986. Surface instability of cylindrical liquid jet in concurrent gas stream. *Bull. Jpn. Soc. Mech. Eng.* **29**, 77.

Batchelor, G. K. 1967. *An Introduction to Fluid Dynamics*. Cambridge University Press. p. 516.

Berger, S. A. 1988. Initial value stability of a liquid jet. *J. Appl. Math.* **48**, 973–991.

Case, K. M. 1960a. Stability of an idealized atmosphere. I. Discussion of results. *Phys. Fluids* **3**, 139–154.

Case, K. M. 1960b. Taylor instability of an inverted atmosphere. *Phys. Fluids* **3**, 366–368.

Case, K. M. 1962. Hydrodynamic stability and initial value problem. In *Proc. Symposia in Applied Mathematics*, Vol. 13, Hydrodynamic Instability. American Mathematical Society, 25–33.

Chang, I. D., and Russel, P. E. 1965. Stability of a liquid layer adjacent to a high speed gas stream. *Phys. Fluids*. **8**, 1018.

Chawala, T. C. 1975. The Kelvin–Helmholtz instability of the gas-liquid interface of a sonic gas jet submerged in a liquid. *J. Fluid Mech.* **67**, 513.

Chigier, N., and Reitz, R. D. 1996. Regimes of jet breakup and breakup mechanisms (physical aspects). In *Recent Advances in Spray Combustion: Spray Atomization and Drop Burning Phenomena* (Ed. K. K. Kuo), Vol. 1, pp. 109–136.

Gaster, M. 1962. *J. Fluid Mech.* **14**, 222. AIAA.

Helmholtz, H. von. 1868. Über discontinuriche Flüssigkeitsbewogungen. *Monato Königl. Preuss. Akad. Wirs. Berlin* **23**, 215.

Hoyt, J. W., and Taylor, J. J. 1977. Waves on water jets. *J. Fluid Mech.* **83**, 119.

Joseph, D. D., Belanger, J., and Beavers, G. S. 1999. *J. Multiphase Flow*, **25**, 1263.

Kang, D. J., and Lin, S. P. 1989. Breakup of a swirling liquid jet. *Int. J. Eng. Fluid Mech.* **2**, 47–62.

Keller, J. B., Rubinow, S. I., and Tu, Y. O. 1973. Spatial instability of a jet. *Phys. Fluids* **16**, 2052–2055.

Kelvin, L. 1871. Hydrokinetic solutions and observations. *Phil. Mag.* **42**, 362.

Leib, S. J., and Goldstein, M. E. 1986a. The generation of capillary instability on a liquid jet. *J. Fluid Mech.* **168**, 479–500.

Leib, S. J., and Goldstein, M. E. 1986b. Convective and absolute instability of a viscous liquid jet. *Phys. Fluids* **29**, 952–954.

Li, H. S., and Kelly, R. E. 1992. The instability of a liquid jet in a compressible air stream. *Phys. Fluids* A **4**, 2162–2168.

Lin, S. P., and Kang, D. J. 1987. Atomization of a liquid jet. *Phys. Fluids* **30**, 2000–2006.

Menickoff, R., Mjolsness, R. C., Sharp, D. H., and Zemach, C. 1978. Initial value problem for Rayleigh–Taylor instability of viscous fluids. *Phys. Fluids*, **21**, 1674–1687.

Nayfeh, A. H., and Saric, W. S. 1973. Nonlinear stability of a liquid film adjacent to a supersonic stream. *J. Fluid Mech.* **58**, 39.

Noble, B. 1958. *Methods Based on Wiener-Hopf Technique for the Solution of Partial Differential Equations*. Pergamon.

Ponstein, J. 1959. *Appl. Sci. Res.* A **8**, 424.

Prosperetti, A. 1976. Viscous effects on small amplitude waves. *Phys. Fluids*, **19**, 195–203.

Prosperetti, A. 1980. Free oscillation of drops and bubbles: The initial value problem, *J. Fluid Mech.* **100**, 333–347.

Prosperetti, A. 1981. Motion of two superposed fluids. *Phys. Fluids*, **24**, 1217–1223.

Rayleigh, L. 1878. On the instability of jets. *Proc. Lond. Math. Soc.* **10**, 4–13.

Schetz, J. A., Kush, E. A., and Joshi, P. B. 1980. Wave phenomena in liquid jet breakup in a supersonic crossflow. *AIAA J.* **18**, 774.

Scriven, L. E., and Pigford, R. L. 1959. Fluid dynamics and diffusion calculation for laminar liquid jets. *AIChE J.* **5**, 397–402.

Sherman, A., and Schetz, J. A. 1971. Breakup of liquid sheets and jets in a supersonic gas stream. *AIAA J.* **9**, 666.

Tam, C. K. W., and Hu, Q. 1989. On the three families of instability waves of high speed jets. *J. Fluid Mech.* **201**, 447.

Zhou, Z. W., and Lin, S. P. 1992a. Absolute and convective instability of a compressible jet. *Phys. Fluids* A **4**, 227.

Zhou, Z. W., and Lin, S. P. 1992b. Effects of compressibility on the atomization of liquid jets. *J. Propul. Power* **8**, 736.

8

A Viscous Jet

The effects of surface tension, inertial force, fluid compressibility, swirl, and jet velocity relaxation on jet breakup were analyzed in Chapter 7 while neglecting the effects of fluid viscosity. This chapter investigates the effect of liquid viscosity and discusses the practical implications of the results. A fairly general dispersion equation that includes many known special cases is obtained in Section 8.1. Individual as well as coupled effects of various physical factors are then discussed in subsequent sections.

8.1. Onset of Instability

Consider an incompressible viscous liquid jet emanating from a circular cylindrical nozzle into an unbounded inviscid swirling gas. The basic flow velocity $\bar{v}_i$ and pressure fields $\bar{p}_i$ are given respectively in the cylindrical coordinates (r, θ, z) by

$$(\bar{u}_1, \bar{v}_1, \bar{w}_1) = (0, 0, 1), \qquad (\bar{u}_2, \bar{v}_2, \bar{w}_2) = (0, E/r, W_r), \qquad (8.1)$$

and

$$\bar{p}_1 = \text{constant}, \qquad \bar{p}_2 = \bar{p}_1 - We^{-1} + \frac{1}{2}QE^2(1 - r),$$

where the velocity and pressure are normalized respectively with the uniform jet velocity W_1 and $\rho_1 W_1^2$. The parameters We, Q, and W_r in (8.1) are the familiar Weber number, the gas to liquid density ratio, and the gas to liquid velocity ratio respectively. E in (8.1) is the swirl number $\Gamma/R_0 W_1$ arising from the potential swirl with a constant circulation $2\pi\Gamma$ in the gas, A being a constant representing the strength of the circulation. Equation (8.1) satisfies exactly the Navier–Stokes equations for incompressible fluids.

We impulsively introduce a disturbance to the basic flow. Any Fourier component of the dynamic response can be written as

$$(\mathbf{v}_i, p_i) = (\hat{u}_i, \hat{v}_i, \hat{w}_i, \hat{p}_i) \exp[i(kz + n\theta - \omega\tau)],$$

where the perturbation variables without upper bars correspond to their counterparts in the basic flow, and k, n, τ, and ω are as defined in the previous chapter. The linearized governing equations for the perturbed field are

$$\partial_\tau + (W_i/W_1)\,\mathbf{v}_i = -(\rho_i/\rho_1)\nabla p_i + Re^{-1}\delta_{i1}\nabla^2 \mathbf{v}_i, \qquad \nabla \cdot \mathbf{v}_i = 0, \quad (8.2)$$

where $\delta_{i1} = 1$ or 0 depending on whether $i = 1$ or 0, z is positive in the flow direction, and Re is the Reynolds number $Re = W_1 R_0 / \nu_1$.

The bounded normal mode solution of (8.2) is given by

$$\hat{u}_1 = -i\left(\frac{Bk I_n'(kr)}{\omega - k} + \frac{Dk I_n'(lr)}{l} + \frac{F n I_n(lr)}{(lr)}\right),$$
$$\hat{v}_1 = Bn I_n(kr)/(\omega - k) + Dnk I_n(lr)/(l^2 r) + F I_n'(lr),$$
$$\hat{w}_1 = Bk I_n(kr)/(\omega - k) + D I_n(lr),$$
$$l^2 = k^2 - iRe(\omega - k), \qquad (8.3)$$
$$\hat{u}_2 = -iC K_n(kr),$$
$$\hat{v}_2 = C n K_n(kr)/(kr),$$
$$\hat{w}_2 = C K_n(kr),$$
$$\hat{p}_2 = C Q(\omega - kW_r - nE)K_n(kr)/k,$$

where primes on the Bessel functions denote differentiation with their arguments, and B, C, D, and F are integration constants to be determined to satisfy the boundary conditions.

There are five boundary conditions at the perturbed interface

$$r = 1 + \zeta(z, \theta, \tau) = 1 + f \exp[i(kz + n\theta - \omega\tau)].$$

The additional unknown interfacial displacement ζ is related to the radial component of velocity by

$$\hat{u}_1 = -i(\omega - k)f, \qquad (8.4)$$
$$\hat{u}_2 = i(kW_r - \omega + nE)f. \qquad (8.5)$$

The balance of normal force per unit area of the interface requires

$$\hat{p}_1 - \hat{p}_2 - QE^2 f - 2Re^{-1}u_1' - We^{-1}(k^2 + n^2 - 1)f = 0, \qquad (8.6)$$

where the prime denotes differentiation with respect to r. The balances of the tangential forces at the interface in the axial and azimuthal directions give, respectively,

$$ik\hat{u}_1 + \hat{w}_1' = 0, \qquad (8.7)$$

and

$$\hat{v}_1' - \hat{u}_1 + in\hat{u}_1 = 0. \qquad (8.8)$$

Substituting of the solution (8.3) into the boundary conditions (8.4)-(8.8) and requiring the solution for B, C, D, and F to be nontrivial, we arrive at the dispersion equation

$$a_1 \left(\frac{I_n'(k)}{I_n(k)} \right) \left(\frac{I_n'(\lambda)}{I_n(\lambda)} \right)^2 + a_2 \left(\frac{I_n'(\lambda)}{I_n(\lambda)} \right)^2 + a_3 \left(\frac{I_n'(k) I_n'(\lambda)}{I_n(k) I_n(\lambda)} \right)$$

$$+ a_4 \left(\frac{I_n'(k)}{I_n(k)} \right) + a_5 \left(\frac{I_n'(\lambda)}{I_n(\lambda)} \right) + a_6 = 0, \qquad (8.9)$$

where

$$a_1 = 2k \, (\lambda^2 - k^2)[2(1 - n^2) - iRe \, h],$$

$$a_2 = -2 \, (\lambda^2 + k^2)^2,$$

$$a_3 = 2k\lambda \, (5k^2 - \lambda^2) + ik\lambda Re \, (\lambda^2 - k^2) \, h,$$

$$a_4 = 2k^3 \{ 2 \lfloor n^2 - (n^2 + \lambda^2)^2 \rfloor - in^2 Re \, h \} / \lambda^2, \qquad (8.10)$$

$$a_5 = \lambda \left[(2n^2 + k^2 + \lambda^2)^2 - 4n^2 + 2in^2 Re \, h \right],$$

$$a_6 = 2n^2 k^2 \, (k^2 - \lambda^2)/\lambda^2,$$

$$\lambda^2 = k^2 - iRe(\omega - k),$$

$$h = \left[(Q/k) \, (kWr - \omega - nE)^2 \, K_n(k)/K_n'(k) \right. \\ \left. + We^{-1} \, (n^2 + k^2 - 1) + QE^2 \right].$$

8.2. Viscous Jets in a Vacuum

For a viscous liquid jet issued without swirl into a vacuum under the influence of axisymmetric disturbances, $E = Q = n = 0$ and (8.9) is reduced to that obtained by Ponstein (1959) and Chandrasekhar (1961) written in a different form (cf. Levich, 1962; Weber, 1931; Rayleigh, 1892)

$$(k - \omega)^2 - \frac{2}{Re} ik^2 \left[1 + \frac{I_1'(k)}{I_0(k)} \right] (k - \omega)$$

$$- \frac{4k^3}{Re^2} \left[k \frac{I_1'(k)}{I_0(k)} - l \frac{I_1'(l) \, I_1(k)}{I_1(l) \, I_0(k)} \right] + \frac{k \, (1 - k^2) \, I_1(k)}{We \quad I_0(k)} = 0. \quad (8.11)$$

The spatial growth rate $-k_i$ at the onset of convective instability $\omega_i = 0$ is plotted against ω_r in Figure 8.1 for various values of We, and $Re = 50$. ω_r is approximately equal to k_r. If the value of We is reduced below the smallest value of We in this figure, the flow may become absolutely unstable. The transition Weber number below which the jet is absolutely unstable is plotted as a function of $2Re$ in Figure 8.2. A viscous jet suffers from absolute

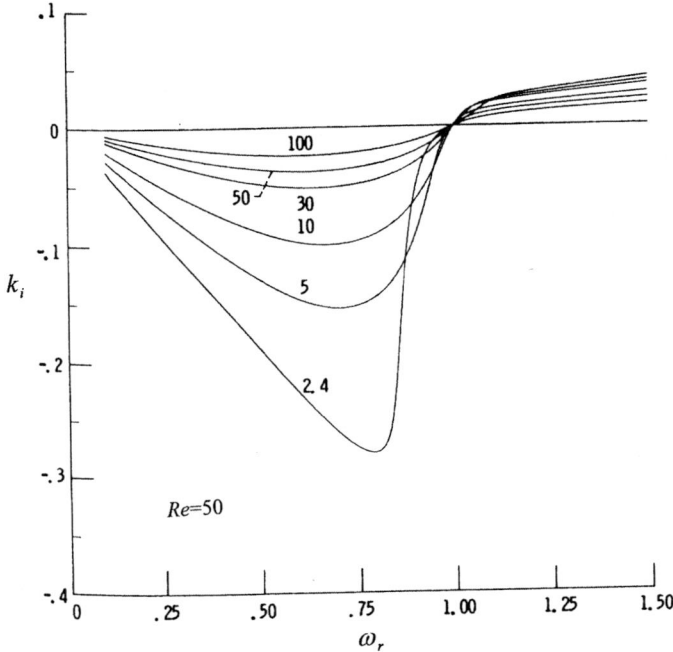

Figure 8.1. Growth rates for unstable waves. (From Leib and Goldstein, 1986).

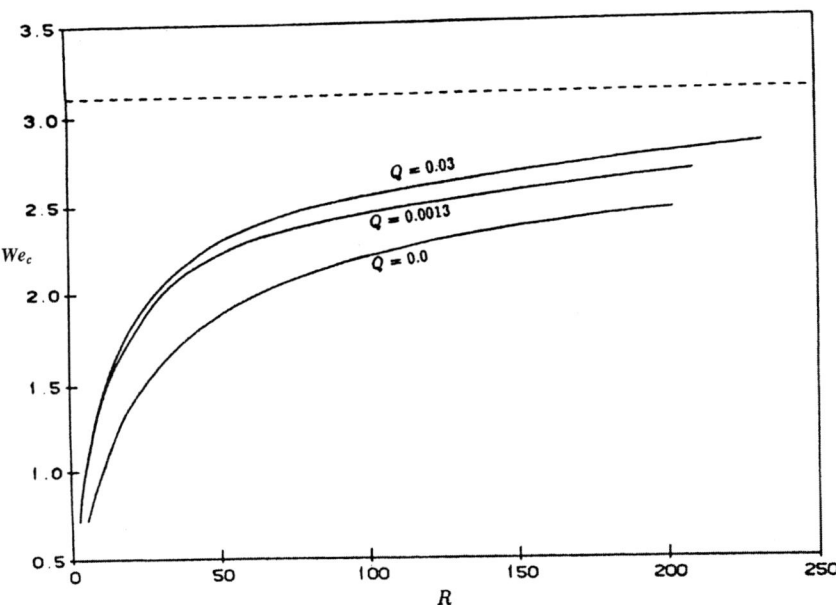

Figure 8.2. Critical Weber number versus Reynolds number. (From Lin and Lian, 1989).

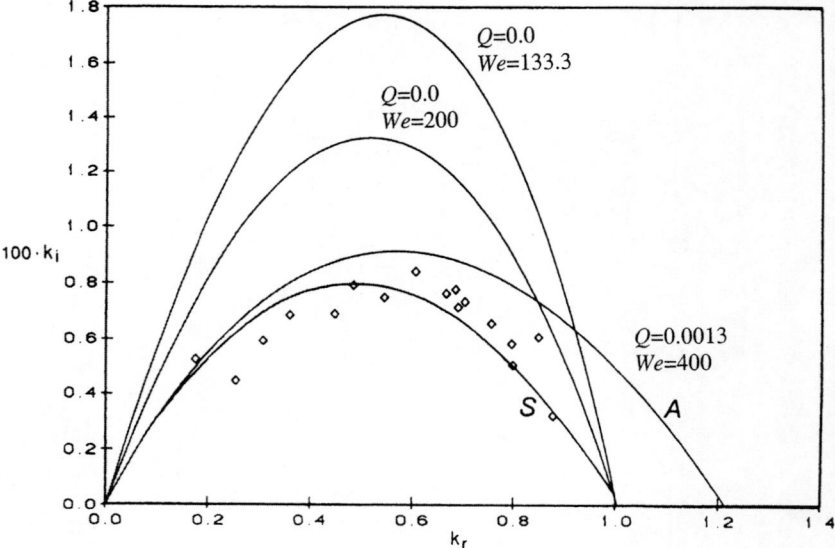

Figure 8.3. Ambient gas and surface tension are destabilizing in the viscous Rayleigh jet. $R = 34.5$. $\diamond$, experiments of Goedde and Yuen. (From Lin and Lian, 1990).

instability at a jet velocity considerably lower than that of an inviscid jet. The transition Weber number for an inviscid jet is given as a dashed line in the figure. The effect of viscosity in a convectively unstable jet is to change neutral stability in an inviscid jet to stability for $k_r > 1$ and to reduce the amplification rate for $k_r < 1$, as can be seen by comparing Figure 8.1 with Figure 7.2. The amplification rates increase as the Weber number is decreased in Figure 8.1. Therefore the origin of the breakup of the Rayleigh jet remains that of surface tension even when the effect of the liquid is included. However, the viscosity tends to reduce the most amplified wave number from 0.697 for the case of inviscid jets, as illustrated in Figure 8.3 for the case of $Re = 34.5$.

8.3. Effects of Surrounding Gas

For an axisymmetrically perturbed jet without swirl $n = 0$, $E = 0$, and (8.9) is reduced to that obtained by Li and Shen (1996)

$$(k - \omega)^2 + Q\,(kW_r - \omega)^2\,\frac{K_0\,(k)\,I_1\,(k)}{K_1\,(k)\,I_0\,(k)} - i\,\frac{2k^2}{Re}\left[1 + \frac{I_1'\,(k)}{I_0\,(k)}\right](k - \omega)$$

$$- \frac{4k^3}{Re^2}\left[k\frac{I_1'\,(k)}{I_0\,(k)} - \lambda\frac{I_1'\,(\lambda)\,I_1\,(k)}{I_1\,(\lambda)\,I_0\,(k)}\right] + \frac{k\,(1 - k^2)}{We}\frac{I_1\,(k)}{I_0\,(k)} = 0. \quad (8.12)$$

When the surrounding gas is quiescent, $W_r = 0$, and (8.12) is reduced to

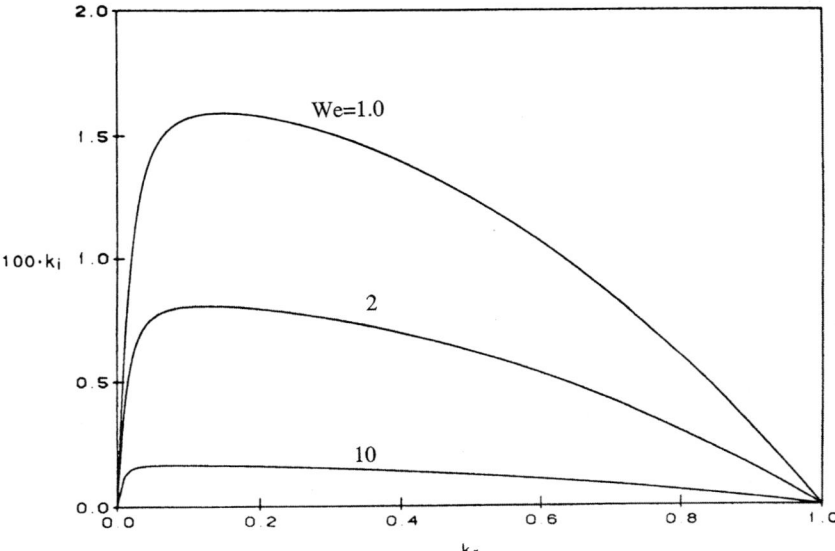

Figure 8.4. Surface tension is destabilizing in a very viscous thread. $Re = 0.1$, $Q = 0.001$.

that obtained by Sterling and Sleicher (1975). Based on a physical argument, they multiplied a term associated with Q by a constant factor chosen so that their solution for temporal disturbance would find a better agreement with their experiments. They succeeded partially in improving the solution of Weber (1931) who considered the case of $k < 1$. It can be shown (Lin and Kang, 1987) that when the jet radius approaches infinity (8.10) reduces to the characteristic equation obtained by Taylor (1963) in his study of the generation of ripples by wind over a viscous fluid of infinite depth.

The effects of the liquid viscosity and the gas density on the spatial amplification of disturbances have been examined by Lin and Lian (1990). Figure 8.4 shows that for a very viscous liquid jet with $Re = 0.1$ in air under slightly lower than atmospheric pressure, the most amplified disturbance wavelength becomes less than 1/7 of that for an inviscid jet in vacuum, and less than 1/6 of that for the jet in Figure 8.3 with $Re = 34.5$ and $Q = 0.0013$. Thus the most obvious manifestation of the effects of viscosity is to produce drops of size larger than that predicted by the Rayleigh theory as observed in the initial stage of the jet breakup regime experiments in Chapter 6.

As the Weber number increases to two orders of magnitude larger than $Q = 0.0013$, the cut-off wave number also increases by two orders of magnitude as in the case of inviscid jets in gases. The example given in Figure 8.5 also shows that liquid viscosity tends to reduce the spatial amplification

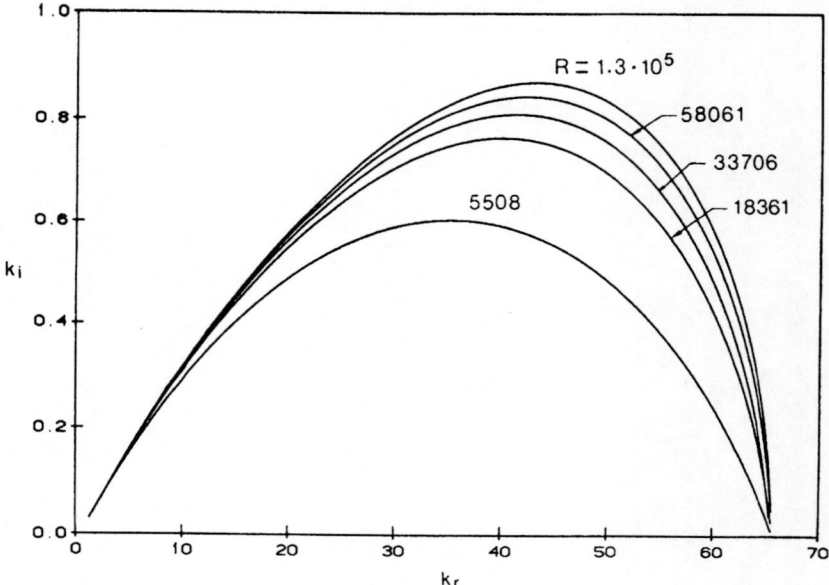

Figure 8.5. Viscous damping of the Taylor mode. $We = 1.964 \times 10^{-5}$, $Q = 0.0013$.

rate of the convectively unstable disturbances, which produce small droplets belonging to the Taylor mode.

In the experiment of Hoyt and Taylor (1977a,b) $R_0 = 3.175$ mm, $U_0 = 21$ m/s, and the water jet is issued into the atmosphere. Thus $Re = 6.678 \times 10^4$ and $We = 1.95 \times 10^4$. This Weber number is of the same order of magnitude as that in Figure 8.5. The amplification curve corresponding to the experimental value of Re lies between the top two curves in Figure 8.5. Hence the theoretical wave number corresponding to the most amplified disturbance lies between 43 and 45. This predicted wave number compares quite well with the measured value of $k = 43.35$ corresponding to the measured wavelength $\lambda = R_0/7.4$. However, for a closer comparison between theory and Hoyt and Taylor's experiments, one must obtain an amplification curve for $Re = 6.678 \times 10^4$, $We = 1.95 \times 10^4$, and $Q = 0.0013$ (Exercise 8.2).

The results presented in Figures 8.3 and 8.5 will be used to show that the nonvanishing of Q is essential for the Taylor mode of atomization. In the limiting case of $Q \to 0$, after using (8.10), (8.11) is reduced to

$$2k^2(k^2 + \lambda^2)\frac{I'(k)}{I_0(k)}\left[1 - \frac{2k\lambda}{k^2 + \lambda^2} \cdot \frac{I_1(k)I_1'(\lambda)}{I_1'(k)I_1(\lambda)}\right]$$
$$- (k^2 - \lambda^2) - J(1 - k^2)k\frac{I_1(k)}{I_0(k)} = 0, \qquad (8.13)$$

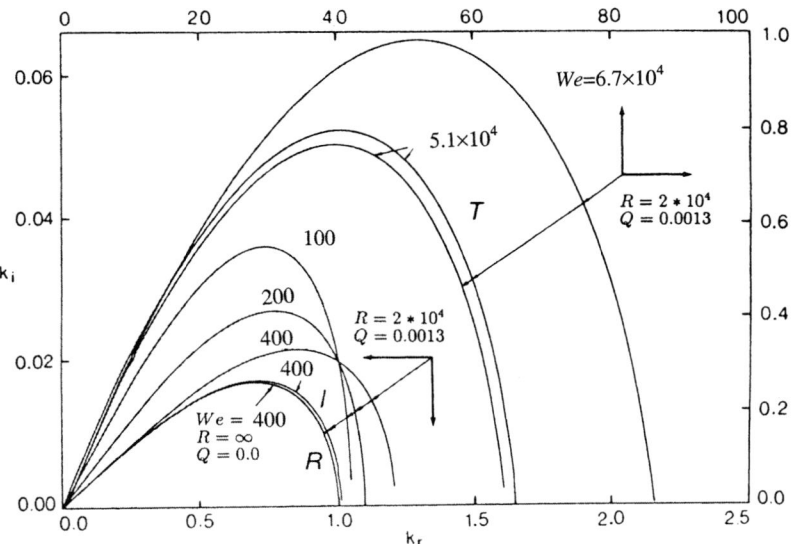

Figure 8.6. The Rayleigh and Taylor modes.

where $J = Re^2/We$. Thus the only relevant parameter is J, which is called the Ohnesorge number. The curve S in Figure 8.3 is obtained from (8.13) with $Re = 34.5$, $We = 400$, or $J = 2.976$. Hence this curve can be used for any Re and We as long as J remains 2.976. Consider the case of $Q = 0.0013$, $We = 1.34409 \times 10^8$, and $R = 2 \times 10^4$ and thus $J = 2.976$. The corresponding amplification rate curve lies far above the highest amplification curve in Figure 8.6. The cut-off wave number is of order $We \cdot Q \sim 10^5$. Thus the jet breakup takes place in the Taylor atomization regime in which the unstable disturbance wavelength is much smaller than the jet radius (see Section 7.7). Retaining the same values of We and Re but putting $Q = 0$, we obtain an amplification curve that is exactly the curve S given in Figure 8.3. Thus the curve in the Taylor regime is brought down to the Rayleigh regime simply by reducing the value of Q from 0.0013 to 0. Hence atomization cannot occur without the surrounding gas, even when the surface tension is small or the jet velocity is so large that $We = 1.344 \times 10^8$.

Figure 8.6 summarizes the convective instability results for an incompressible viscous liquid jet in an incompressible inviscid quiescent gas. The amplification curve R is obtained in the Rayleigh limit of $Q = 0$ and $Re \to \infty$. Curve T is obtained in the Taylor limit of $r_0 \to \infty$. The limiting case of Chandrasekhar, that is, $Q = 0$, $Re/We \to 0$, as well as the case of $Q \neq 0$, $Re \to 0$ depicted in Figures 8.3 and 8.4 are not included in this figure because of their relatively small amplification rates. The effect of gas density on absolute instability is shown in Figure 8.2.

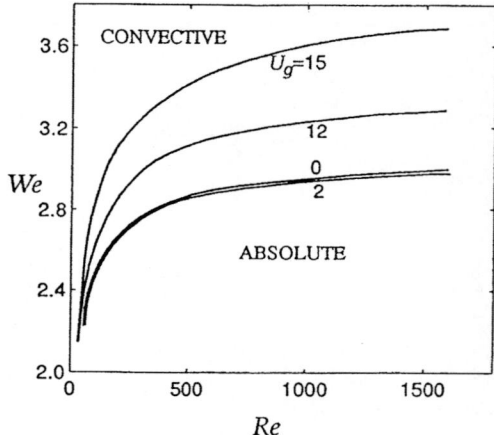

Figure 8.7. Dependence of critical Weber number on Reynolds number for different gas velocities and $Q = 0.0013$. (From X. Li and S. P. Shen, private communication).

The effect of slip velocity W_r on the convective and absolute instability of an axisymmetric jet can be analyzed using (8.12). With the slip velocity included, the effects of parameters We, Q, and Re remain qualitatively the same except they are amplified (X. Li and S. P. Shen, private communication). In particular the region of absolute instability is increased as W_r is increased as illustrated in Figure 8.7.

8.4. Spray Angles and Intact Length

A spray angle θ is the angle sustained by the tangent planes to the spray boundary at a location where the spray appears to start to form as depicted in Figure 8.8 for an axisymmetric jet. Angle θ was estimated theoretically, by earlier workers (Ranze, 1958; Ranze and Dreier, 1964; Reitz and Bracco, 1982) using temporal instability theories. The method is based on two assumptions. The first is that the temporally growing disturbances are related to the spatially growing disturbances by the Gaster theorem. The second is that the group velocity corresponding to the most temporally amplified disturbance is equal to the jet velocity. Then the tangent of the half spray angle is equal to the disturbance amplitude increase over time divided by the distance traveled by the disturbance over the same time period, that is,

$$tan\left(\frac{\theta}{2}\right) = \frac{d\,|H_m|}{dt}\frac{t}{W_1 t}, \qquad (8.14)$$

where $|H_m|$ is the amplitude of the interfacial displacement corresponding to

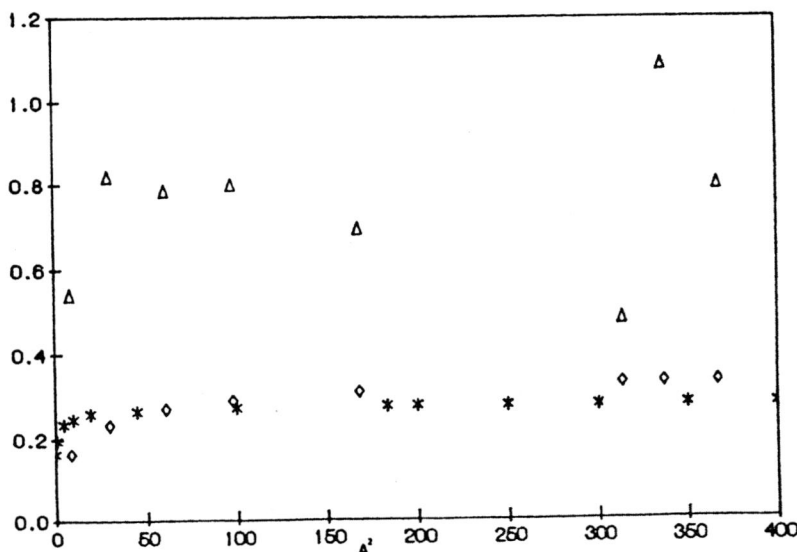

Figure 8.8. Variation of spray angles with A^2 and Q. Δ, experiments $\tan\beta/\sqrt{Q}$; $\diamond$, spatial theory $-k_i/\sqrt{Q}$; *, temporal theory $(Sk_r)_m/A$. (From Lin and Kang, 1987).

the most amplified temporally growing Fourier component of the disturbance. Hence

$$H = \frac{2\pi}{M}r_0\, exp\,[i\,(K_m X - \Omega_m t)]\,, \qquad (8.15)$$

where M is an arbitrary constant arising from the linear theory, and K_m and Ω_m are respectively the dimensional wave number and frequency corresponding to the most amplified disturbance. Earlier workers used the numerical values of Ω_m obtained by Taylor (1963). He expressed Ω as

$$\Omega = \left(\frac{2}{r_0}\right) W_1 Q^{1/2}\left(Sk_m^2/A\right), \qquad (8.16)$$

where $A = Re/(WeQ^{1/2})$ and calculated numerically $k_m = K_m/(S/\rho_2 W_1^2)$ from (8.12) in the limit of $r_0 \to \infty$, with the approximation that the real part of Ω_m is negligibly small (see Exercise 8.1). Without this approximation, we calculated θ from

$$\frac{\tan(\theta/2)}{Q^{1/2}} = \frac{4\pi}{M}W_1\left(\frac{Sk_m^2}{A}\right),$$

which is obtained by substituting (8.15) and (8.16) into (8.14). The results are given in Figure 8.8 together with the experimental results of Reitz and Bracco, and the results obtained with spatial theory, which is described below.

The tangent of the half spray angle is given by

$$tan\left(\frac{\theta}{2}\right) = \frac{d}{dy}\text{(envelope of }\zeta) = -C_0 k_i \, exp\,(-k_i z).$$

where C_0 is an arbitrary constant. For small spatial growth rate k_i, we have approximately

$$tan\left(\frac{\theta}{2}\right) = -C_0 k_i \, (1 - k_i z + \cdots). \tag{8.17}$$

Hence, unless $C_0 k_i$ is sufficiently large, a discernable θ may not appear until some distance downstream in the flow direction. This is the origin of the so-called intact length over which the jet does not appear to diverge. Based on Taylor's dispersion equation (see Exercise 8.1), treating both k and ω as complex, we calculate $\theta/2$ from (8.15) at $z = 0$ with $C_0 = 1$. The results are given in Figure 8.8 and Table 8.1.

As seen in Figure 8.8, spatial theory predicts larger spray angles than does temporal theory when $A^2 \geq 100$, but the converse is true when $A^2 \leq 100$. The curve passing through theoretical points becomes almost horizontal when $A^2 \geq 100$. Hence we have

$$tan\left(\frac{\theta}{2}\right) = C_0 Q^{1/2}, \qquad (Re/We)^2 \, Q > 100.$$

On the other hand when $A^2 \ll 100$, A^2 increases almost parabolically. Hence

$$tan\left(\frac{\theta}{2}\right) = C_0 \, (Re/We), \qquad A^2 \ll 100.$$

However, the theoretical points in Figure 8.9 are all below the experimental points. If the value of the arbitrary constant C_0 is raised from 1 to the values given in Table 8.1, then the products of the values in the C_0 and k_i columns give the experimentally observed half spray angle β. This adjustment amounts to assigning different values to the height of the first wave crest at the nozzle exit. Although the dimensional wave amplitudes $C_0 a$ of these initial waves are all smaller than 10^{-4} cm, which is two orders of magnitude smaller than the nozzle diameter $d_0 = 0.034$ cm, and because different initial wave amplitudes for the same nozzle may be actually encountered in experiments, we are not really entitled to speak of any finite amplitude in the linear theory. However, C_0 values in a different series of tests in Table 8.1 are all very close, indicating that the linear theory predicts the phenomena adequately up to an arbitrary constant.

When the viscosity of the jet liquid increases while all other physical quantities remain fixed, A decreases. It follows from the results in Figure 8.8

Table 8.1. *Experimental and Theoretical Data*

Series	$\rho^2 \times 10^3$	A^2	$\tan \beta$	x_1/d_0	C_0	$a_t \times 10^4$ cm	k_i	k_{rm}	ω_i	$a \times 10^4$ cm
22	1.3	337	0.0393	2.0	3.29	15.2	0.01194	0.60	305.1346	4.628
23	1.3	314	0.0175	20.0	1.47	6.3	0.01188	0.59	289.6332	4.309
24	7.7	168	0.0612	15.0	2.26	5.2	0.02713	0.56	82.18053	2.308
25	8.6	367	0.0743	1.5	2.40	1.2	0.03075	0.59	120.9591	5.036
26	12.9	61	0.0892	0.2	2.93	2.5	0.03045	0.50	34.03625	8.393
28	17.2	98	0.1051	0	2.79	0	0.03766	0.52	38.71045	1.350
30	25.8	30	0.1317	0	3.54	0	0.03716	0.43	14.38934	0.410
34	20.4	2.2×10^{-4}	0	>40	0	0	0.00097	0.07	0.00726	0.613
35	51.5	8.5	0.1228	0	3.33	0	0.03684	0.030359	3.76927	0.117

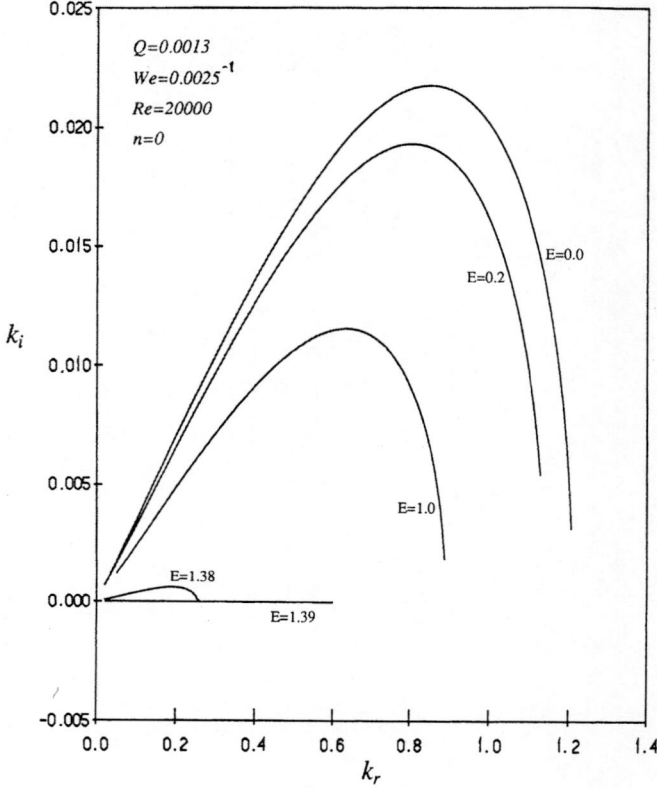

Figure 8.9. An efficient reduction of the amplification rate with gas swirl in the Rayleigh mode. (From Lin and Lian, 1990).

that the amplification rate and the spray angle decrease as the viscosity increases. The effect is very dramatic when A is very small. The series number 34 test in Table 8.1 is a case in point. In this test the liquid is glycerol. Because of its high viscosity, the amplification rate k_i is only 0.00097. It follows from (8.17) that even if the largest initial amplitude coefficient of $C_0 = 3.54$ in Table 8.1 is used, the distance required to amplify the spray angle to a mere $1°$ is $Z/a = 10022.9$. The dimensional distance is therefore $Z = 0.6144$ cm, which is 18 nozzle diameters. This length is of the same order of magnitude as the intact length of 40 diameters reported by Reitz and Bracco. Additional numerical results for the cases of finite jet radius obtained from (8.12) with $W_r = 0$ are given in Table 8.2. Because the jet is convectively unstable, an impulse-created disturbance at a fixed point in space near the nozzle will decay in time but amplify as it is convected downstream. Hence both the

Table 8.2. *Spray Angles of Atomizing Jets*

Series	$Q \times 10^3$	$We \times 10^5$	r_o/a	$tan(\theta/2) \times 10^2$	C_0	$k_i \times 10^2$	$a \times 10^5$
64	1.3	2.754	944	1.66	0.6	2.77	36.02
65	7.7	2.754	559	8.31	1.43	5.81	6.08
66	25.8	2.504	2061	16.46	2.13	7.73	1.65
67	51.5	2.527	4096	14.95	2.60	5.75	0.83

intact length and the spray angle may be viewed as the signature of convective instability.

8.5. Effects of Swirl in Gas

The effects of swirl with a constant circulation $2\pi\Gamma$ in the surrounding gas on the convective instability of the Rayleigh and Taylor modes are given

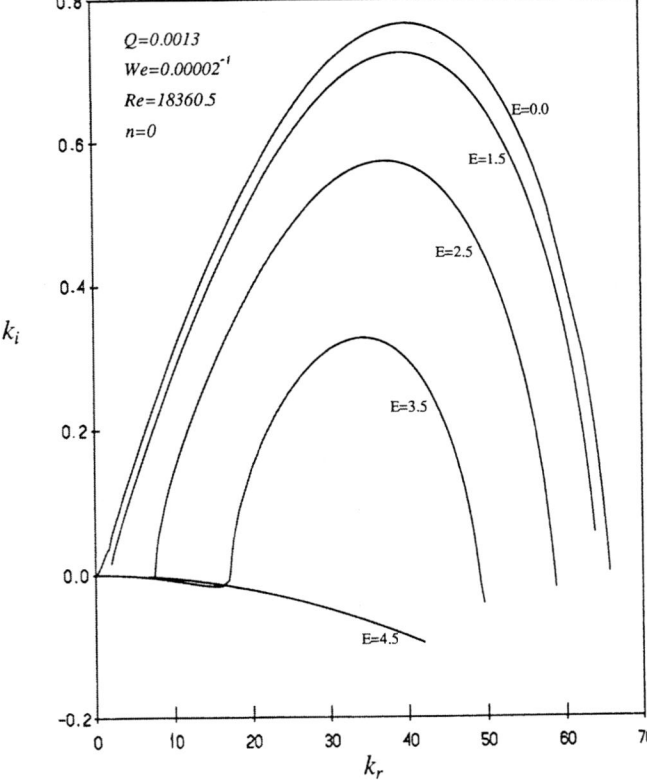

Figure 8.10. Slight reduction of the amplification rate with gas swirl in the Taylor mode. (From Lin and Lian, 1990).

respectively in Figure 8.9 and 8.10 (Lin and Lian, 1990). In contrast to swirl in the liquid jet, swirl in the surrounding gas is seen to stabilize both the Rayleigh and Taylor modes of instability. Instability is completely suppressed at critical swirl numbers $E_c = 1.39$ and $E_c = 4.5$ respectively for the Rayleigh and Taylor modes. Hence a potential swirl may be applied to enhance jet cutting processes.

Exercises

8.1. Consider the limiting case of $r_0 \to \infty$ of (8.12). Rescaling the resulting equation with characteristic length and time given respectively by $a = S/\rho_2 W_1^2$ and a^2/υ_1 and neglecting the real part of the frequency, obtain the dispersion equation of Taylor in an inertial reference frame.

$$\left[-i \left(\frac{kA}{Q^{1/2}} + \omega_r \right) + 2k^2 \right]^2 + A^2 k^3 - Q\omega_r^2$$

$$- 4k^3 \left[k^2 - i \left(\frac{kA}{Q^{1/2}} + \omega_r \right) \right]^{1/2} = 0,$$

where $A = (S/\rho_1 \upsilon_1 W_1) Q^{-1/2}$ is the Taylor parameter (1963).

8.2. Obtain the spatial amplification curve for $We = 1.95 \times 10^4$, $Re = 6.678 \times 10^4$, $Q = 0.0013$, and compare the wavelength of the most amplified disturbance with the measured wavelength $\lambda = R_0/7.4$ (Hoyt and Taylor, 1977a,b).

References

Chandrasekhar, S. 1961. *Hydrodynamic and Hydromagnetic Stability*. Clarendon Press.

Gasler, M. 1962. *J. Fluid Mech.* **14**, 222.

Hoyt, J. W., and Taylor, J. J. 1977a. *J. Fluid Mech.* **83**, 119.

Hoyt, J. W., and Taylor, J. J. 1977b. *Phys. Fluids* **20**, 5253.

Leib, S. J., and Goldstein, M. E. 1986. Convective and absolute instability of a viscous liquid jet. *Phys. Fluids* A **29**, 952.

Levich, V. G. 1962. *Physicochemical Hydrodynamics*, p. 631. Prentice-Hall.

Lian, Z. W., and Lin, S. P. 1990. Breakup of a liquid jet in a swirling gas. *Phys. Fluids* **A2**, 2134.

Lin, S. P., and Kang, D. J. 1987. Atomization of a liquid jet. *Phys. Fluids*. **30**, 2000.

Lin, S. P., and Lian, Z. W. 1990. Mechanism of the breakup of liquid jets. *AIAA J.* **28**, 120.

Lin, S. P., and Lian, Z. W. 1989. Absolute instability of a liquid jet in a gas. *Phys. Fluids* **A1**, 490.

Ponstein, J. 1959. *Appl. Sci. Res.* A **8**, 425.

Ranze, W. E. 1958. *J. Chem. Eng.* **36**, 175.

Ranze, W. E., and Dreier, W. M. 1964. *Lnd. Eng. Chem. Fourd.* **3**, 53.

Rayleigh, L. 1892. *J. Sci.* 5th series **34**, 145.

Reitz, R. D., and Bracco, F. V. 1982. *Phys. Fluids* **25**, 1730.

Sterling, A. M., and Sleicher, C. A. 1975. *J. Fluid Mech.* **68**, 477.

Taylor, G. I. 1963. In *The Scientific Papers of G. I. Taylor* (Ed. G. K. Batchelor) Vol. 3, 244–54. Cambridge University Press.

Tennekes, H., and Lumley, J. L. 1975. *A First Course in Turbulence.* MIT Press, pp. 248–286.

Weber, C. E. 1931. *Angew. Math. Mech.* **11**, 136.

Wu, P. K., and Faeth, G. M. 1993. *Atomiz. Sprays.* **3**, 265.

Wu, P. K., Tseng, L. K., and Faeth, G. M. 1992. *Atomiz. Sprays.* **2**, 295.

9

Roles Played by Interfacial Shear

This chapter elucidates the role of interfacial shear on the onset of instability of a cylindrical viscous liquid jet in a viscous gas surrounded by a coaxial circular pipe by using an energy budget associated with the disturbance. It is shown that the shear force at the liquid-gas interface retards the Rayleigh mode instability, which leads to the breakup of the liquid jet into drops of diameter comparable to the jet diameter because of capillary force. On the other hand the interfacial shear and pressure work in concert to cause the Taylor mode instability, which leads the jet to breakup into droplets of diameter much smaller than the jet diameter. While the interfacial pressure plays a slightly more important role than the interfacial shear in amplifying the longer wave spectrum in the Taylor mode, shear stress plays the main role of generating shorter wavelength disturbances.

9.1. Basic Flow

Consider the instability of an incompressible Newtonian liquid jet of radius R_1. The jet is surrounded by a viscous gas enclosed in a vertical pipe of radius R_2, which is concentric with the jet. For the jet to maintain a constant radius, the dynamic pressure gradients in the steady liquid and gas flows must maintain the same constant. This will allow the pressure force difference across the liquid-gas interface to be exactly balanced by the surface tension force as required. Such coaxial flows, which satisfy exactly the Navier–Stokes equations, are given by (Lin and Ibrahim, 1990).

$$W_1(r) = -1 + \frac{Nr^2}{[N-(1-l^2)]}\left\{1 - \frac{(1-Q)}{4N}R[2\ln l + (1-l^2)]\right\},$$

$$W_2(r) = -\frac{(l^2-r^2)}{[N-(1-l^2)]}\left\{1 - \frac{(1-Q)}{4N}R[2\ln l + (1-l^2)]\right\}$$

$$+ \frac{(1-Q)}{4N}R\left[l^2 - r^2 - 2\ln\left(\frac{l}{r}\right)\right], \tag{9.1}$$

146

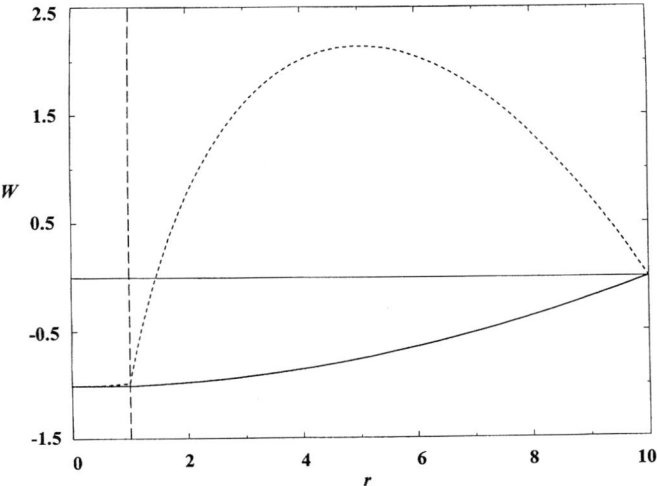

Figure 9.1. Basic flow velocity distribution, $N = 0.018$, $l = 10$, $Re = 1000$. ———,
$1/Fr = 0$, $We = 400$, $Q = 0.0013$; ----, $1/Fr = 0.0001$, $We = 4761.9$, $Q = 0.013$;
— — —, liquid-gas interface.

in which

$$N = \frac{\mu_2}{\mu_1}, \qquad l = \frac{R_2}{R_1}, \qquad Q = \frac{\rho_2}{\rho_1},$$

$$Re = \rho_1 W_0 R_1 / \mu_1, \qquad Fr = W_0^2 / g R_1,$$

$$R = Re/Fr,$$

where the subscript 1 and 2 stand for the liquid and gas phase respectively, W_0 is the magnitude of the jet velocity in the z-axis, r is the radial distance normalized with R_1, $W(r)$ is the axial velocity distribution normalized with W_0, μ is the dynamic viscosity, ρ is the density, and g is the gravitational acceleration in the negative z-direction. Two different basic flow velocity distributions are given in Figure 9.1. They were obtained in two different regions of the parameter space.

9.2. Equations Governing Disturbances

The onset of instability of the basic flow described by (9.1) with respect to infinitesimal disturbances is governed by the linearized Navier–Stokes equations

$$\bar{Q}_\alpha \partial_t \mathbf{v}_\alpha + \mathbf{W}_\alpha \cdot \nabla \mathbf{v}_\alpha + \mathbf{v}_\alpha \cdot \nabla \mathbf{W}_\alpha = -\nabla p_\alpha + Re^{-1}\bar{N}_\alpha \nabla \cdot \tau_\alpha, \quad (9.2)$$

$$\nabla \cdot \mathbf{v}_\alpha = 0, \qquad (\alpha = 1, 2), \tag{9.3}$$

$$\bar{Q}_\alpha = \rho_\alpha/\rho_1, \qquad \bar{N}_\alpha = \mu_\alpha/\mu_1,$$

where $\mathbf{W}_\alpha$ is the basic flow velocity vector, the magnitude of which is given by (9.1), $\mathbf{v}_\alpha$, p_α, and t are dimensionless velocity perturbation, pressure disturbance, and time normalized respectively with W_0, $\rho_1 W_0^2$, and R_1/W_0. For the Newtonian fluids considered, the rate of strain τ_α in (9.2) is given by

$$\tau_\alpha = \left[\nabla \mathbf{v}_\alpha + (\nabla \mathbf{v}_\alpha)^T\right],$$

where the superscript T stands for transposition.

The boundary conditions at the perturbed liquid-gas interface, which is at the radial location $r = 1 + f$ when measured in the unit of R_1, can be linearlized by use of the Taylor series expansions of all variables involved about $r = 1$, and retaining only terms of the first order in perturbations. Hence the interfacial conditions are evaluated at $r = 1$ with f as an additional unknown. Because the interface is a material surface, f must satisfy at $r = 1$ the kinematic condition

$$f_{,t} + W_1 f_{,z} = u_1, \tag{9.4}$$

where u is the radial velocity component in the cylindrical coordinate (r, θ, z), and the commas followed by the subscripts denote partial differentiations with respect to the independent variables designated by the subscripts. Other interfacial kinematic conditions are the continuity of the radial and axial components of the velocity across the interface respectively given by

$$[u_\alpha]_2^1 \equiv u_1 - u_2 = 0, \tag{9.5}$$

and

$$[W_{\alpha,r} f + w_{,\alpha}]_2^1 = 0, \tag{9.6}$$

where w is the axial component of the perturbation velocity (u, v, w) in the cylindrical coordinate system. Note that the disturbance is considered to be axisymmetric and thus $v = 0$. The balancing of forces per unit area of interface in the tangential and normal directions leads respectively to the dynamic conditions at $r = 1$,

$$[\bar{N} \{f W_{\alpha,rr} + w_{\alpha,r} + u_{\alpha,z}\}]_2^1 = 0, \tag{9.7}$$

$$[p_\alpha - (2/Re) \bar{N}_\alpha u_{\alpha,r}]_2^1 + (f + f_{,zz})/We = 0. \tag{9.8}$$

The boundary condition at the pipe wall is the no-slip condition at $r = R_2/R_1 = l$.

9.3. Energy Budget

To trace the energy sources of instability, we will balance the energy budget in a disturbed liquid jet. A history of the use of energy budgets can be found in Joseph and Renardy (1992). Consider a control volume of the liquid over one wavelength λ of the disturbance. By forming the dot product of (9.2) for liquid ($\alpha = 1$) with $\mathbf{v}_1$, integrating over the control volume, using (9.3) and the Gauss theorem to reduce some of the volume integrals to surface integrals, and averaging over one wavelength λ and one wave period $T = 2\pi/\omega_i$, ω_i being the wave frequency, we arrive at the energy equation

$$\frac{1}{T\lambda} \int_0^T \int_V (\partial_t + \mathbf{W}_1 \cdot \nabla) e \, dV \, dt$$

$$= \frac{1}{T\lambda} \int_0^T \int_V \rho_1 \mathbf{v}_1 \cdot (\mathbf{v}_1 \cdot \nabla) \mathbf{W}_1 \, dV \, dt - \frac{1}{T\lambda} \int_0^T \int_A p_1 \mathbf{v}_1 \cdot \mathbf{n} \, dA \, dt$$

$$+ \frac{1}{T\lambda Re} \int_0^T \int_V (\mathbf{v}_1 \cdot \tau_1) \cdot \mathbf{n} \, dA \, dt - \frac{1}{2ReT\lambda} \int_0^T \int_T \tau_1 \cdot \tau_1 dV \, dt,$$

$$(9.9)$$

where $e = \mathbf{v} \cdot \mathbf{v}/2$ is the disturbance kinetic energy, and V and A stand respectively for the control volume and surface area. Thus the left side of (9.9) represents the time rate of change of the disturbance kinetic energy in the control volume. The first term on the right side of (9.9) gives the time rate of mechanical energy transfer between the disturbance and the basic flow through the Reynolds stress, and the last term gives the mechanical energy dissipation through liquid viscosity. The second and third integrals on the right side of (9.9) give respectively the rates of work done by the pressure and the viscous stress of the liquid. These two integrals will be transformed by use of the boundary conditions (9.7) and (9.8) to better explain the physics involved. Applying (9.7) to the evaluation of the surface integrals involving p_1, and applying (9.8) to the evaluation of the surface integral involving the shear stress allows us to write the energy budget (9.9) as

$$KE = PRG + SUT + SHG + SHB + NVG + SHL$$
$$+ NVL + PRL + REY + DIS,$$
$$(9.10)$$

where

$$KE = \left(2\pi^2/\omega_i\lambda\right) \int_0^T \int_0^1 \int_{-\lambda}^0 \left(\partial_t + W_1\partial_z\right)\left(u_1^2 + w_1^2\right)\pi r\,dz\,dr\,dt,$$

$$PRG = -\left(4\pi^2/\omega_i\lambda\right) \int_{-\lambda}^0 \left(u_1 p_2\right)_{r=1} dz,$$

$$SUT = \left(4\pi^2/\lambda\omega_i\,We\right) \int_0^T \int_{-\lambda}^0 \left[u_1\left(d + d_{,zz}\right)\right]_{r=1} dz\,dt,$$

$$SHG = \left(4\pi^2/\lambda\omega_i\,Re\right) \int_0^T \int_{-\lambda}^0 N w_1\left(w_{2,r} + u_{2,z}\right)_{r=1} dz\,dt,$$

$$SHB = \left(4\pi^2/\lambda\omega_i\,Re\right) \int_0^T \int_{-\lambda}^0 w_1 d\left(N w_{2,rr} - w_{1,rr}\right)_{r=1} dz\,dt,$$

$$NVG = \left(4\pi^2 N/\lambda\omega_i\,Re\right) \int_0^T \int_{-\lambda}^0 \left(2u_1 u_{2,r}\right)_{r=1} dz\,dt,$$

$$SHL = \left(4\pi^2/\lambda\omega_i\,Re\right) \int_0^T \int_0^1 \left[u_1\left(w_{1,r} + u_{1,z}\right)\right]_{z=-\lambda}^{z=0} r\,dr\,dt,$$

$$NVL = \left(4\pi^2/\lambda\omega_i\,Re\right) \int_0^T \int_0^1 2\left[w_1 w_{1,z}\right]_{z=-\lambda}^{z=0} dr\,dt, \qquad (9.11)$$

$$PRL = -\left(4\pi^2/\lambda\omega_i\right) \int_0^T \int_0^1 \left[w_1 p_1\right]_{z=-\lambda}^{z=0} r\,dr\,dt,$$

$$REY = -\left(4\pi^2/\lambda\omega_i\right) \int_0^T \int_0^1 \int_{-\lambda}^0 u_1 w_1 W_{1,r} r\,dz\,dr\,dt,$$

$$DIS = -\left(4\pi^2/\lambda\omega_i\,Re\right) \int_0^T \int_0^1 \int_{-\lambda}^0 \left[2\left(u_{1,r}\right)^2 + 2\left(w_{1,r}\right)^2\right.$$
$$\left. + \left(w_{1,r} + u_{1,z}\right)^2\right] r\,dz\,dr\,dt.$$

Each term in (9.10) represents the phase-averaged time rate of change of a physically distinctive factor per unit length of the liquid control volume enclosed by the interface of length λ on the side and by the two circular lids at $z = 0$ and $-\lambda$. In (9.10) KE is the time rate of change of the disturbance kinetic energy. The last term DIS is the rate of mechanical energy dissipation through viscosity in the volume, which tends to reduce KE as it is always negative. The energy transfer between the disturbance and the basic flow through the Reynolds stress is represented by REY, the sign of which depends on the flow parameters. The rest of the surface integrals in (9.10) represent various rates of work done on the control surface. PRG represents the rate of work done by the gas pressure fluctuation on the liquid jet, if it is positive. If it is negative, the work is done by the liquid jet on the surrounding gas at the expense of the disturbance kinetic energy. The same sign convention is followed by the rest of the work terms. *SUT* is the rate of work done by the surface tension. *SHG* is the rate of work done by the shear stress exerted by the fluctuating gas at the interface. *SHB* is the rate of work done by the shear stress associated with the basic flow distortion caused by the interfacial displacement. *NVG* represents the rate of work done by normal viscous stress exerted by the fluctuating gas at the interface. *SHL* and *NVL* represent respectively the rates of work done by the normal and tangential components of the viscous stress at the top and bottom ends of the control volume. *PRL* gives the rate of the pressure work at the top and bottom ends of the control volume.

Each integral on the right side of (9.10) represents a different physical factor affecting the instability of the liquid jet. Therefore the relative magnitude as well as the sign of each term must be evaluated. To achieve this, we must carry out a stability analysis that provides the functions appearing in the integrands of (9.10). An accurate eigenvector solution is obtained by use of the Chebyshev collocation method (Boyd, 1989) which is described in the next section.

9.4. Stability Analysis

Let us now consider the onset of instability of the basic flow with respect to axisymmetric disturbances. With (9.3), we can express the radial and axial components of the velocity disturbance in terms of the Stokes stream function ψ_α as

$$u_\alpha = \frac{1}{r}\psi_{\alpha,z} \quad \text{and} \quad w_\alpha = -\frac{1}{r}\psi_{\alpha,r}. \tag{9.12}$$

Arbitrary Fourier components of ψ_α and the corresponding pressure and interfacial displacement d can be written as

$$[\psi_\alpha(r,z,t), p_\alpha(r,z,t), f(z,t)] = [\phi_\alpha(r), \zeta_\alpha(r), \xi]e^{ikz+\omega t}, \tag{9.13}$$

where $[\phi_\alpha, \zeta_\alpha, \xi]$ is the perturbation amplitude corresponding to the complex wave number $k = k_r + ik_i$ and the complex wave frequency $\omega = \omega_r + i\omega_i$. Substitution of the Fourier mode solution into the curle of the linearized Navier–Stokes equation results in the Orr–Sommerfeld equation (Drazin and Reid, 1981),

$$\left(\omega - \frac{v_\alpha}{v_1}\frac{1}{Re}E^2\right)E^2\phi_\alpha(r) + ikW_\alpha(r)E^2\phi_\alpha(r)$$

$$- ikr\frac{d}{dr}\left[\frac{1}{r}\frac{dW_\alpha}{dr}\right]\phi_\alpha(r) = 0, \tag{9.14}$$

where

$$E^2 \equiv \frac{d^2}{dr^2} - \frac{1}{r}\frac{d}{dr} - k^2. \tag{9.15}$$

The boundary condition at the vertical pipe wall $r = l$ is the no-slip condition,

$$\phi_2(l) = 0, \tag{9.16}$$

$$D\phi_2(l) = 0, \qquad D = \frac{d}{dr}. \tag{9.17}$$

Using (9.12) and (9.13), we can write the interface boundary conditions in (9.4) to (9.8) respectively as

$$ik\phi_1 = [\omega + ikW_1(1)]\xi, \tag{9.18}$$

$$[\phi_\alpha]_2^1 \equiv \phi_1(1) - \phi_2(1) = 0, \tag{9.19}$$

$$[D\phi_\alpha - \xi DW_\alpha]_2^1 = 0, \tag{9.20}$$

$$\left[\frac{\mu_\alpha}{\mu_1}(E^2 + 2k^2)\phi_\alpha - \frac{\rho_\alpha}{\rho_1}R\xi\right]_2^1 = 0, \tag{9.21}$$

$$\left[\zeta_\alpha - \frac{2ik}{Re}\frac{\mu_\alpha}{\mu_1}(D\phi_\alpha - \phi_\alpha)\right]_2^1 + \xi(1 - k^2)/We = 0, \tag{9.22}$$

where the pressure amplitude can be obtained from the linearized Navier–Stokes equation and is given by

$$ik\,[\zeta_\alpha]_2^1 = \left[\frac{\rho_\alpha}{\rho_1}(\omega + ikW_\alpha)D\phi_\alpha - ik\left(\frac{\rho_\alpha}{\rho_1}\right)\phi_\alpha DW_\alpha\right.$$
$$\left. - \frac{1}{Re}\left(\frac{\mu_\alpha}{\mu_1}\right)DE^2\phi_\alpha\right]_2^1. \tag{9.23}$$

The disturbance along the axis $r = 0$ must be bounded. It follows from (9.12) that

$$\phi_1(0) = 0, \tag{9.24}$$

$$\frac{d\phi_1}{dr}(0) = 0. \tag{9.25}$$

Equations (9.14) to (9.25) constitute an eigenvalue problem.

9.5. Chebyshev Polynomial Expansion

We will obtain the solution for ϕ_α by using the collocation method and applying the Chebyshev polynomials (Gottlieb and Orszag, 1977; Canuto et al.,

1988; Boyd, 1989) as the cardinal function of Lagrange. First we map the liquid region $r \in [0, 1]$ into the Chebyshev space $y \in [-1, 1]$ by means of the linear transformation

$$r = \frac{1}{2}(y + 1), \tag{9.26}$$

and map the gas region $r \in [1, l]$ into $y \in [1, -1]$ by the transformation

$$r = 1 - \frac{(l - 1)}{2}(y - 1). \tag{9.27}$$

It follows from (9.26) and (9.27) that the transformed Orr–Sommerfeld equation in y remains the same except the p-th derivative in r in (9.14) must be replaced by

$$\frac{d^p}{dr^p} = q_\alpha^p \frac{d^p}{dy^p},$$

where $q_1 = 2$ for the liquid domain, and $q_2 = 2/(1 - l)$ for the gas domain. The same modification must be made in the boundary conditions. The pipe axis and the pipe wall are now both at $y = -1$, and the interface is at $y = 1$. The solution for $\phi_\alpha(y)$ will be expanded as

$$\phi_\alpha(y) = \sum_{j=0}^{N_\alpha} h_{\alpha j}(y)\phi_\alpha(y_{\alpha,j}), \tag{9.28}$$

$$h_{\alpha j}(y) = \frac{(-1)^{j+1}(1 - y^2)}{C_j N_\alpha^2(y - y_{\alpha,j})} T'_{N_\alpha}(y),$$

$$C_0 = C_{N_\alpha} = 2, \qquad C_j = 1 \quad \text{for} \quad 0 < j < N_\alpha,$$

where $h_{\alpha j}$ is the Lagrange cardinal function, T_{N_α} is the Chebyshev polynomial, the upper prime denotes differentiation with y, and $y_{\alpha,j}$ is the Gauss–Lobatto collocation points (Boyd, 1989; Canuto and Hussaini, 1988) given by

$$y_{\alpha,j} = \cos\left(\frac{\pi j}{N_\alpha}\right), \qquad j = 0, 1, \ldots, N_\alpha. \tag{9.29}$$

The method of evaluation of the derivatives of ϕ at the collocation points can be found in Boyd (1989). Substitution of ϕ in (9.28) and its derivatives into the transformed Orr–Sommerfeld equation results in an equation of $(N_1 + N_2 + 2)$ unknown $\phi_{\alpha,j}$. An additional unknown ξ appears in the boundary conditions. Hence there are $(N_1 + N_2 + 3)$ unknowns. The resulting system evaluated at $y_{\alpha,i}(i = 2, 3, \ldots, N_\alpha - 2)$ provides $(N_\alpha - 3)$ equations from the

Orr–Sommerfeld equation for each α, and nine equations from the boundary conditions. The number of equations has been made the same as the number of unknowns by not evaluating the Orr–Sommerfeld equation at $y_{\alpha,i}$ ($i = 0$, $1, \ldots, N_\alpha - 1$, N_α). This is the so-called the Lanczos method (1956). The vanishing of the determinant of coefficient matrix of the eigenvector $\phi_{\alpha,j}$ gives the characteristic equation, which has the form

$$(A_{ij} - \omega B_{ij}) \phi_j = 0, \tag{9.30}$$

where A_{ij} and B_{ij} are the parts of the coefficient matrix arising respectively from the time-independent and time-dependent part of the Orr–Sommerfeld equation and its boundary conditions. The elements of matrices A_{ij} and B_{ij} can be found in Appendix A, which can be applied to the present problem as a special case.

9.6. Numerical Eigenvalue Evaluation

For a given set of flow parameters (Re, Fr, We, N, Q, l) the complex eigenvalues (k, ω) are obtained from the characteristic equation. Assuming the jet to be convectively unstable, we assign a value of (k_r, k_i) such that $k_r > 0$, $k_i < 0$, and obtain ω_r and ω_i with the subroutine DG2LCG in the IMSL library. If $\omega_r > 0$, we keep the same value of k_r but increase the value of k_i and substitute the new value of (k_r, k_i) into the characteristic equation, and then solve for (ω_r, ω_i) again with the same subroutine. We repeat the same computation until a point $(\omega_r = 0, k_i > 0)$ is reached from the domain $\omega_r > 0, k_i < 0$. We repeat the same procedure for increasing values of k_r, and obtain the spatial amplification curve $\omega_r = 0$ over a range of k_r. This procedure allows us to ascertain that the amplification curve in the complex k-plane can be reached from the domain $k_i < 0, \omega_r > 0$ so that the causality condition, that is, the condition that the disturbance does not exist in $t < 0$, can be satisfied. This procedure breaks down in certain parameter space when saddle point and branch point singularities appear in the complex k- and ω-planes. The jet is then absolutely unstable.

To test for possible syntax and computer program errors, the results for the special cases included in the present problem are checked against the known results of axisymmetric Poiseuille flow (Davey and Drazin 1969) and core-annular flow (Preziosi, Chen, and Joseph, 1989). The specific comparison is given in Appendix B. The numbers of terms retained in (9.28) are systematically increased until the desired significant digits are obtained for the eigenvalue, eigenvector, and its derivatives. A typical example of the

Table 9.1. Convergence Test: $Re = 500$, $Fr^{-1} = 0.0005$, $We = 10^5$, $Q = 0.0013$, $N = 0.018$, $l = 10$, $k_r = 5.0$

N_1	N_2	k_i	ω_i	$\phi_1(1)$	$\phi_2^{iv}(1)$
25	60	0.08025	4.6843	−0.00370	−55260
20	70	0.08112	4.6834	−0.00389	−56280
30	70	0.08112	4.6834	−0.00389	−56340
20	80	0.08106	4.6833	−0.00391	−56340
20	90	0.08105	4.6833	−0.00391	−56340
30	90	0.08106	4.6833	−0.00391	−56340
20	100	0.08105	4.6833	−0.00391	−56340

convergence test is given in Table 9.1 in which six, five, and four significant digits are obtained for the eigenvalue, eigenvector, and its fourth derivative. The eigenvectors and their derivatives corresponding to the calculated eigenvalues are obtained by use of the IMSL subroutine G2CCG.

9.7. Convective Instability

Figure 9.2 gives a spatial amplification curve $\omega_r = 0$ for the flow parameters specified in the caption. Q and We are so chosen that $Q \sim We^{-1}$, and thus we

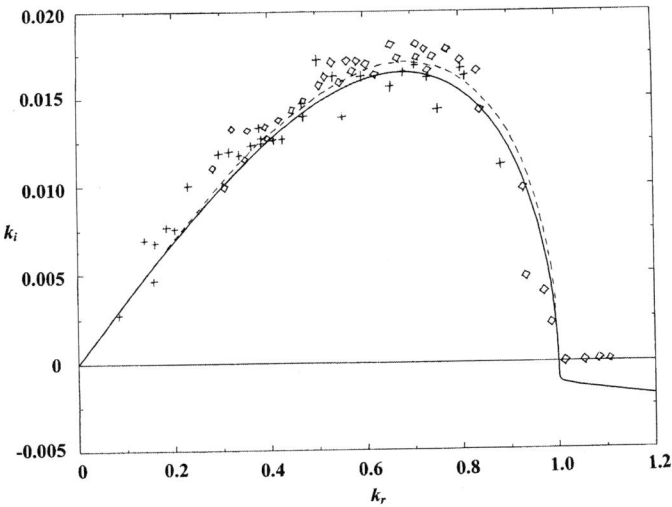

Figure 9.2. Amplification curve for Rayleigh mode. _____, $Re = 1000$, $1/Fr = 0$, $We = 400$, $Q = 0.0013$, $N = 0.018$, $l = 10$; _ _ _, $Re = \infty$, $Q = 0$, $1/Fr = 0$, $We = 400$; +, experiments of Goedde and Yuen (1970); $\diamond$, experiments of Donnelly and Glaberson (1996).

obtain the Rayleigh mode. This curve can be approached from below starting with $\omega_r > 0$, $k_i < 0$ for a given k_r. Hence the liquid jet for the given flow parameters is convectively unstable. The maximum spatial amplification rate k_{im} occurs at the wave number $k_{rm} = 0.684$, which is slightly smaller than the wave number for the maximum temporal growth rate found by Rayleigh for an inviscid jet in a vacuum. The cut-off wave number below which the viscous jet is stable is $k_{rc} = 1$. This value is the same as that found by Rayleigh, except the jet is neutrally stable for the inviscid case. Rayleigh's temporal amplification curve for an inviscid jet in a vacuum is converted to a spatial amplification curve by use of (7.22) and is included in Figure 9.2 as a dashed line. The temporal growth rates measured by Geodde and Yuen (1970) and by Donnelly and Glaberson (1996) are likewise converted to spatial growth rates and given in Figure 9.2 for comparison. They did not record the Reynolds numbers and the Weber numbers corresponding to each experimental point. Agreement of inviscid theory and viscous theory with the given values of Re, We, and Q is remarkable. However, the spatial amplification rate increases with increasing values of We. Thus we cannot expect good agreement with their experiments for all We.

Figure 9.3 demonstrates this point. Another spatial amplification curve for a different set of flow parameters is given in Figure 9.3. Q and We are so chosen that $Q \gg We^{-1}$ to yield the Taylor mode. The jet is again convectively

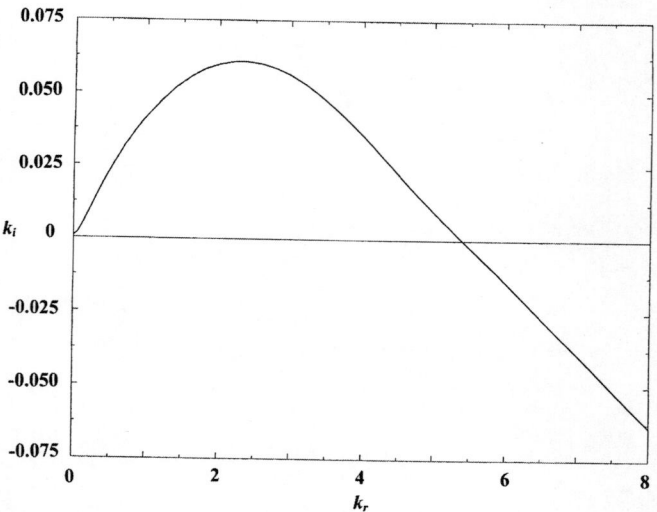

Figure 9.3. Amplification curve for Taylor mode. $Re = 1000$, $1/Fr = 0.0001$, $We = 4761.9$, $Q = 0.013$, $N = 0.019$, $l = 10$.

unstable. However, both k_{rm} and k_{rc} are much larger than those in Figure 9.1. Hence the unstable disturbances with a broader bandwidth of wavelength manifest themselves at the onset of instability, which leads to the formation of smaller droplets. Although the jets specified in Figures 9.2 and 9.3 are both convectively unstable, the physical origins of instability are totally different, as will be explained in the next section.

9.8. Absolute Instability

Figure 9.4 shows the emergence of a saddle point singularity in the $k_r - k_i$ plane. At $We = 2.703$ and the rest of the flow parameters specified in the figure caption, the jet is convectively unstable. The spatial amplification curve is given by the solid line. When the value of We is decreased to 2.632 while the rest of the flow parameters remain the same, a saddle point at $k_r = 1.2162$, $k_i = 0.21658$ appears, and the jet becomes absolutely unstable. The curve $\omega_r = 0$ has two branches shown in dotted lines. ω_i increases with increasing k_r along the upper branch but decreases along the lower branch. Thus for the disturbances characterized by the upper and lower branches the wave packet propagates respectively upstream and downstream as they grow. The jet is absolutely unstable. The physical mechanism of absolute instability is

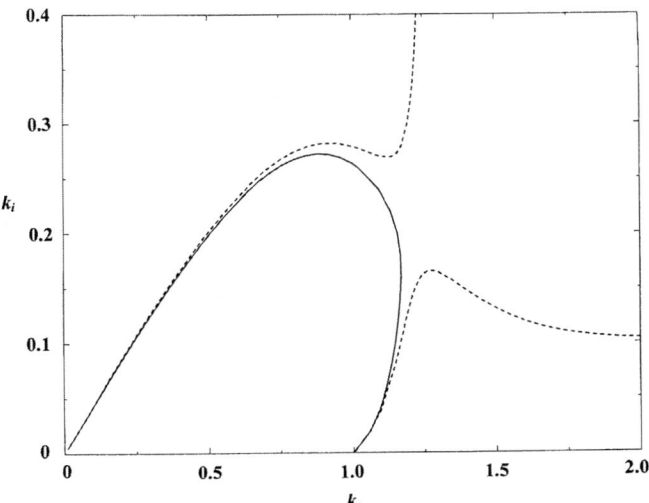

Figure 9.4. Emergence of a saddle point. $Re = 200$, $1/Fr = 0$, $Q = 0.0013$, $N = 0.018$, $l = 10$. ———, $We = 2.703$; - - - -, $We = 2.632$.

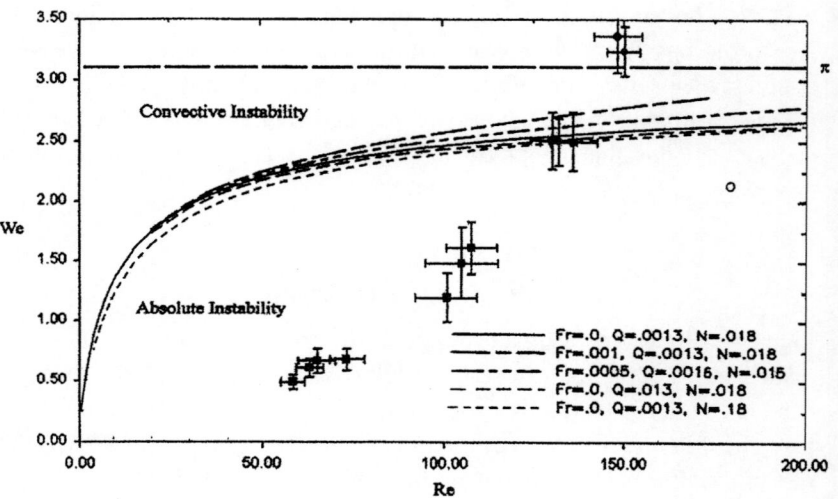

Figure 9.5. Critical Weber number of transition to absolute instability. o, experimental point of Monkewitz (1990).

basically the same as that of the Rayleigh mode convective instability, as will be shown in Section 9.11.

The curves in Figure 9.5 give the transition Weber number as a function of the Reynolds number for various gas to liquid density ratios Q, the Froude number, and the dynamic viscosity ratio N. Below each curve the stability characteristic equation encounters a saddle point singularity in Fourier space and a pinch point singularity in Laplace space. Hence in the region below each curve the jet is absolutely unstable. Above each curve such singularities do not exist, and the disturbance can only propagate in the downstream direction as it grows in time, so the jet eventually breaks up into drops downstream of the nozzle. It is seen from Figure 9.5 that the transition Weber number as a function of Re depends weakly on the density ratio Q and the viscosity ratio N under various microgravity conditions. The condition ($Q = 0.0013$, $N = 0.018$) corresponds to a water jet in atmosphere; ($Q = 0.0016$, $N = 0.015$) corresponds to an alcohol jet in our experiments. The inviscid asymptotic limit $We = \pi$ for the case of $Q = 0$ and $\nu = 0$ in the figure is from Leib and Goldstein (1986a,b). The most important feature is that transition from convective to absolute instability for viscous jets takes place over a wide range of Re. All transition curves pass through the origin in the $We - Re$ space. Thus a dripping jet is certainly a candidate for the manifestation of absolute instability. However, the transition can occur at large Reynolds numbers, according to Figure 9.5.

9.9. Comparisons with Experiments

The experiment was conducted by Vihinen, Honohan, and Lin (1997) and O'Donnel, Chen, and Lin (2001) in the NASA Glenn 2.2 sec drop tower. Only a brief description of the experiment will be given here as the detailed description of the equipment and procedure can be found in the literature. The compressed gas stored in the Whitney sample cylinder was released through the TESCOM pressure regulator to compress the test liquid stored in a bladder housed in a tank. The outlet of the bladder was connected to a stainless steel nozzle of 1.067 mm inner diameter. The test liquid jet ejected from the nozzle was photographed with a Kodak Ecta Pro RO digital camera before it reached a catch pan. The frame rate and exposure settings were respectively 250 frames per second and 48 μsec. The pressure, temperature, and flow rate at the inlet of the nozzle were measured respectively with a Setra model 206 pressure transducer, Omega Type T thermocouple, and an AWCO model 2HM-01 flow meter. A small on-board computer based on an Onset model 4A Tattle Tab data-logger was used to acquire and store the data on pressure, temperature, and flow rate during each drop test. Three DC batteries were used to power the system. The entries system was housed in a drop rig of dimension $41.4 \times 40.6 \times 81.2$ cm, and of weight 155 kg. The rig attained 10^{-4} g in the tower and survived 30 g impacts for more than 200 drops. Alcohol of $v = 1.5 \times 10^{-6}$ m^2/s, $\rho = 789$ kg/m^3, $S = 0.024$ N/m at 20°C was used for the test fluid to attain Reynolds numbers much higher than those attained in the previous experiments in which glycerine, SAE 10 oil, and silicon oil were used as test fluids. The Reynolds numbers attained in the experiments of Vihinen et al. were less than 10, and the transition from convective to absolute instability was observed at Reynolds numbers less than 0.2. The transition observed by O'Donnel et al. took place at Reynolds numbers of $O(100)$. Because of the relatively large Reynolds number, the observation location had to be moved farther downstream, which required a periscope mirror arrangement in the small drop rig for image recording. All of the drop tests were carried out in air under 1 atm. Thus $Q = 0.0016$ and the viscosity ratio $N = 0.015$. To delineate the transition from convective to absolute instability, a convectively unstable jet in the parameter space above the transition curve was produced. The test rig was released in the drop tower, and the jet image was recorded to ascertain if it remained convectively unstable under microgravity conditions. The corresponding values of Re and We were recorded. Then they reduced the jet velocity, which lowered both We and Re, and repeated the drop test until transition from convective to absolute instability was observed. The images of convective instability and absolute instability are shown respectively in

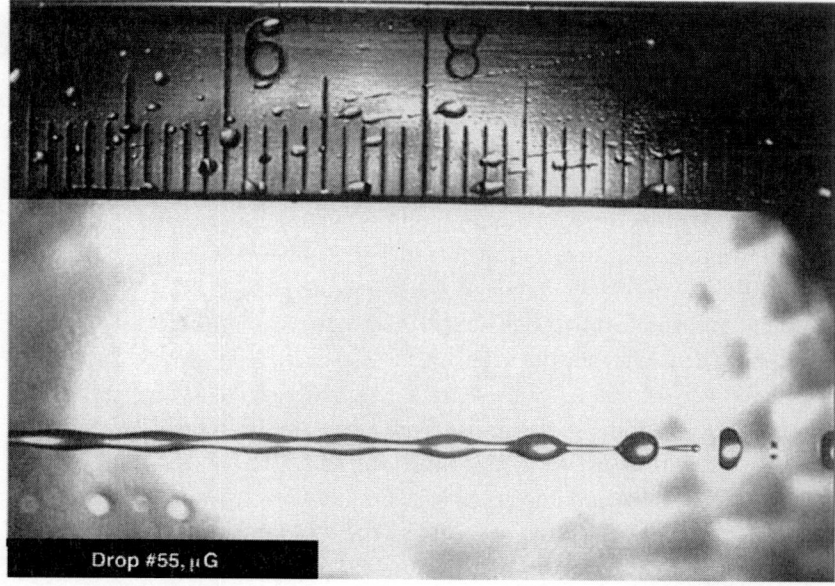

Figure 9.6. A convectively unstable jet. $We = 17.857, Re = 5.40$.

Figures 9.6 and 9.7 (Vihinen et al., 1997). The experimental points corresponding to the convective and absolute instability are respectively shown as the filled circle and square in Figure 9.5. The uncertainties associated with Re and We are indicated respectively with horizontal and vertical bars. These points should be compared with the theoretical curve with $Q = 0.0016$ and $N = 0.015$ corresponding to the experimental condition. All experimental points appear to fall in the region where they belong according to the theory, except the point that lies very close to the transition curve. However, the deviation is within error uncertainty. The amount of experimental data is limited by the availability of the drop tower facilities. Based on these limited experimental points, it appears that absolute instability does occur even at large Reynolds numbers in a viscous liquid jet in a viscous gas in a microgravity environment of $10^{-4}\, g$. The consequence of absolute instability is explosive breakup of the jet. Most of the jet liquid rushed toward the nozzle tip to form a continuously growing spherical liquid ball. No transition from jetting to dripping was observed as a consequence of absolute instability in microgravity. Absolute instability appears to start at a short distance of order $U(a^2/\nu)$ from the nozzle tip. It takes this distance for the parabolic velocity distribution just inside the nozzle to relax to almost flat velocity distribution where the jet is more susceptible to instability than near the nozzle tip. This is consistent with the finding of Leib and Goldstein (1986b) and Debler and Yu (1988).

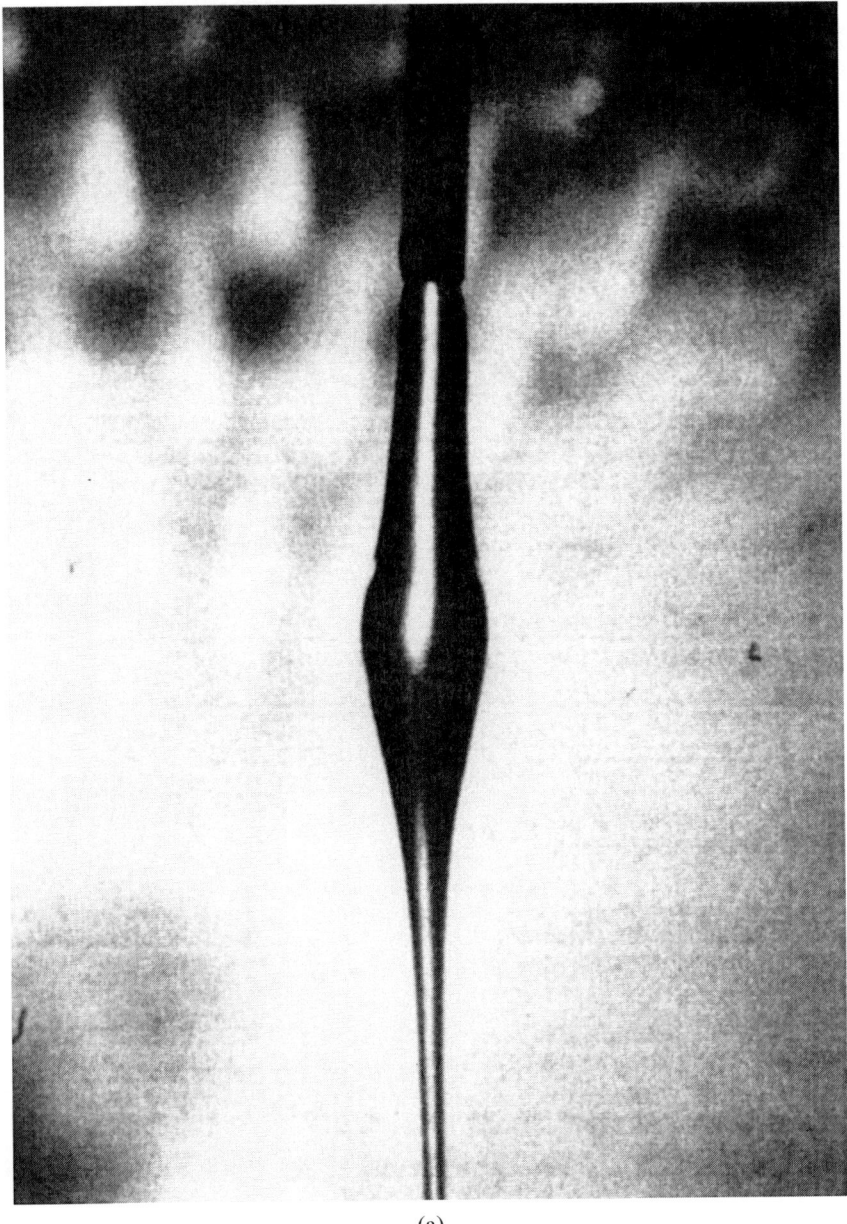

(a)

Figure 9.7. Absolutely unstable jet. $We = 0.082$. (a) 0.2 sec after drop. (b) 0.4 sec after drop.

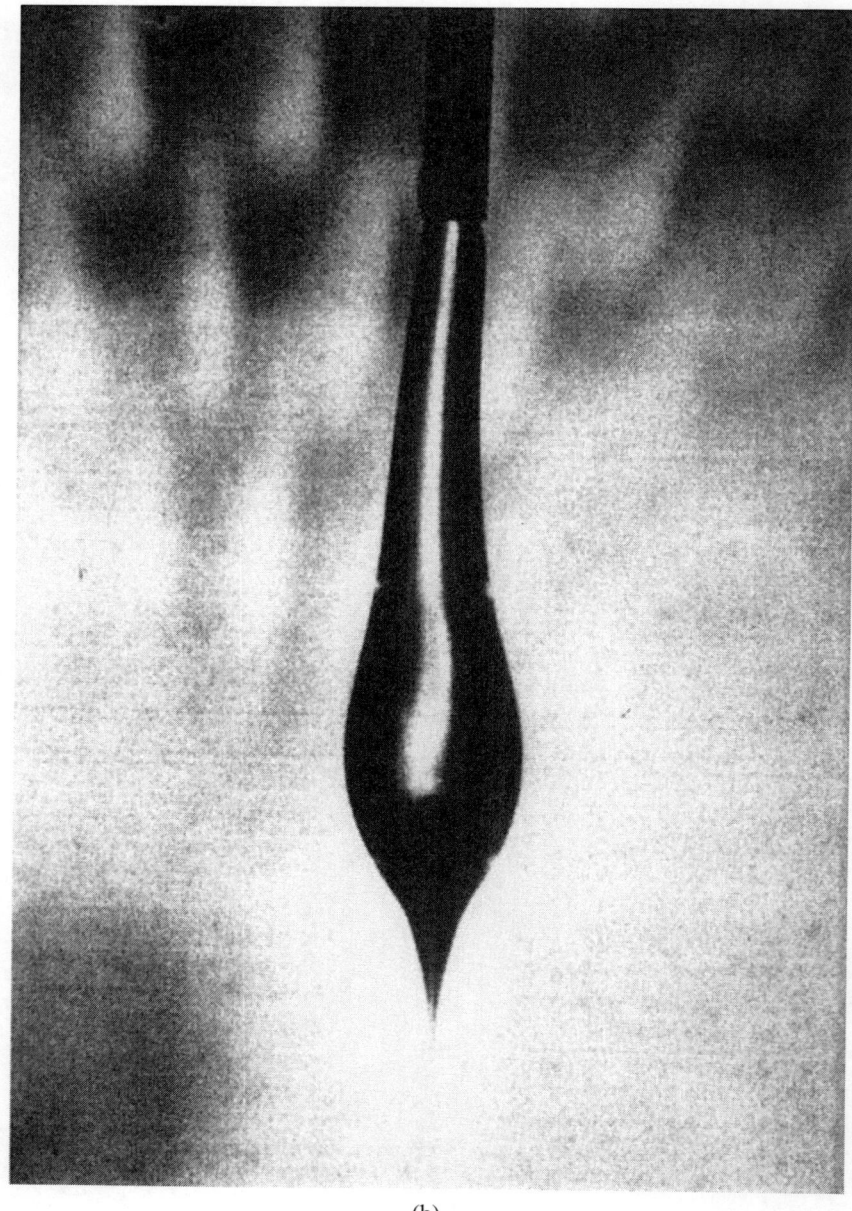

(b)

Figure 9.7. (*continued*)

9.10. Comparisons with Related Work

Clanet and Lasheras (1999) have studied both theoretically and experimentally the transition from dripping to jetting under gravitational influence. They gave the critical Weber number, at which the transition from dripping to jetting occurs, as a function of the Bond number, which is the ratio of gravitational to surface tension force. In the limit of $g \to 0$, their critical Weber number is 4, which is higher than the inviscid limit π of Leib and Goldstein (1986a). The absolute instability of a liquid jet has been conjectured by Lin and Lian (1989) and Monkewitz (1990) to lead to dripping. The transition from jetting to dripping under gravity has been analyzed by Le Dizes (1997) in light of global instability in the sense of Huerre and Monkewitz (1990). Le Dizes found that if the Reynolds number and the Froude number are much larger than 1, a locally uniform jet becomes locally absolutely unstable at the orifice with Weber number 3.125, which is very close to the inviscid limit of π obtained by Leib and Goldstein for an infinitely long uniform jet. The results of O'Donnel et al. (2001) and Vihinen et al. (1997) show, however, that absolute instability can occur at practically any Reynolds number and may not be related to dripping at the nozzle exit.

The relation between local or global absolute instability with dripping jets cannot yet be said to be firmly established. However, the physical mechanism of absolute instability appears relatively clear. Both convective and absolute instability are caused by capillary pinching. When the ratio of the surface energy to the kinetic energy per unit length of the jet (i.e., We^{-1}) is sufficiently large, the disturbance energy can propagate and amplify both in the downstream and upstream directions corresponding to absolute instability; but when this ratio is small, the disturbance can only propagate and amplify in the downstream direction, corresponding to convective instability. This ratio depends more on the viscous force in the jet than the inertial force in the gas reflecting in the Q-effect depicted in Figure 9.5. The larger the viscous effect (i.e., the smaller the Reynolds number), the larger this ratio is.

The classical notion of absolute and convective instability in parallel flows that are homogeneous in the flow direction has been applied locally over sections of other weakly nonparallel flows (see Drazin and Reid, 1981). This approach is reasonable if the wavelengths of the disturbance are small compared to the distance over which the flow inhomogeneity becomes appreciable. Local absolute instability in various flows has been shown to lead to the instability of the entire flow (global instability) if the region of local absolute instability is sufficient large (Monkewitz, 1990; Le Dizes, 1997; Huerre and Monkewitz, 1990; Chomaz, Huerre, and Redekopp, 1988). The physical implication of this approach is still being explored. For example, Yakubenko

(1997) demonstrated with an inviscid inclined jet in a gravitational field that a flow having no absolute instability anywhere locally can sustain a self-excited global instability. This is physically possible due to a feedback effect in a nonhomogeneous flow. The feedback effect may arise from the change in the downstream propagating disturbance wave speed which forces the disturbance upstream to distort the basic flow. This implies that global instability need not be the consequence of local absolute instability. Finally, it should be pointed out that absolute instability is a linear concept related to the onset of instability. The details of the nonlinear evolution between the onset of instability and the final stage of pinching off of the liquid bridge will be taken up in the last two chapters.

9.11. Mechanism of Jet Breakup

We evaluated all items in the energy budget (9.10) by integrating the integrands obtained in the previous section using the Gauss quadrature and Runge–Kutta methods. The two methods were implemented respectively through the IMSL subroutines DQDAGS and DIVPRK. The results obtained by the two methods were compared for possible syntax or program error. All integrals in (9.10) were evaluated independently from one another. The comparison of the sum of the integrals on the right side of (9.10) with the integral on the left side provides an independent check of the overall numerical accuracy. Moreover the number of terms in the eigenfunction expansions were varied in the integrands of (9.10) to ascertain the numerical accuracy. A minimum of three significant digit accuracy is maintained in the results to be presented.

Table 9.2 displays the energy budget in a liquid jet perturbed by the disturbances whose amplification rates are depicted in Figure 9.2. All items of the energy budget, except SHB which is zero in this case of $Fr^{-1} = 0$, as can be seen from (9.1) and (9.10), are listed for various wave numbers for comparisons. The wave numbers cover both stable and unstable disturbances. All items are normalized with the maximum kinetic energy term occurring at $k_{rm} = 0.684$. We see that the positive rates of change of the disturbance kinetic energy are mainly due to the work done by the surface tension on the control liquid volume. Although the viscous normal stress exerted by gas (represented by NVG) as well as normal stress work (represented by PRL and NVL) at the top and bottom of the cylindrical liquid column also contribute to the growth of the unstable disturbance, they are several orders of magnitude smaller than the surface tension term SUT. The major factor that resists disturbance growth is viscous dissipation. The pressure and the shear stress

Table 9.2. *Energy Budget for the Rayleigh Mode:* $Re = 1000$, $Fr^{-1} = 0.0$, $We = 400$, $Q = 0.0013$, $N = 0.018$, $l = 10$

k_r	KE	REY ($\times 10^{-4}$)	SUT	PRL ($\times 10^{-3}$)	PRG ($\times 10^{-3}$)	NVG ($\times 10^{-3}$)	DIS ($\times 10^{-1}$)	SHL ($\times 10^{-5}$)	NVL ($\times 10^{-4}$)	SHG ($\times 10^{-2}$)
0.140	0.127	−0.120	0.129	0.0159	0.0335	0.00939	−0.0128	−0.00791	0.0245	−0.115
0.200	0.248	−0.234	0.254	0.0734	−0.206	0.0252	−0.0362	−0.0225	0.0674	−0.178
0.300	0.503	−0.474	0.520	0.181	−0.323	0.0795	−0.113	−0.0676	0.197	−0.287
0.400	0.771	−0.721	0.804	0.420	−0.506	0.169	−0.241	−0.134	0.380	−0.384
0.500	0.980	−0.910	1.03	0.736	−0.949	0.280	−0.407	−0.210	0.558	−0.452
0.600	1.06	−0.984	1.14	0.650	−1.28	0.389	−0.592	−0.292	0.654	−0.473
0.684	1.00	−0.931	1.09	1.19	−1.43	0.452	−0.680	−0.366	0.622	−0.445
0.700	0.973	−0.908	1.06	2.15	−1.45	0.458	−0.716	−0.380	0.603	−0.434
0.800	0.697	−0.681	0.781	0.882	−1.31	0.433	−0.681	−0.451	0.398	−0.327
0.900	0.296	−0.347	0.353	0.496	−0.718	0.264	−0.457	−0.388	0.131	−0.159
1.100	−0.057	3.00	0.0672	1.47	−2.47	1.47	−1.20	−1.09	0.00318	−0.476
1.200	−0.160	7.71	0.182	5.79	−6.71	3.36	−3.34	−3.70	0.0105	−1.08

exerted by the gas at the liquid-gas interface are also significant factors against instability. Although the liquid tangential viscous stress represented by *SHL* and the bulk Reynolds stress represented by *REY* also contributed to drain the kinetic energy from the disturbance, they are many orders of magnitude smaller than *DIS*. However, the sum of all negative terms is not sufficiently large in magnitude to counter the destabilizing effect of the surface tension. Thus capillary pinching remains the mechanism of the instability of a viscous liquid jet in a viscous gas by the Rayleigh mode, which was demonstrated by Chandrasekhar (1961) who considered an inviscid liquid jet in a vacuum. An inviscid Rayleigh jet is neutral with respect to disturbances of wave number larger than the cut-off wave number $k_{rc} = 1$. Thanks to viscous dissipation these disturbances are actually damped, according to Table 9.2. The stabilizing and destabilizing factors retain their signs in the range of k_r given in Table 9.2, except for the Reynolds stress term. Although some energy is transferred from the mean flow to the disturbances of wavelength shorter than $2\pi R_1$, the growth of the disturbances are suppressed by viscous dissipation. Note that the change of *SUT* with k_r is not monotonic, and its maximum does not occur at k_{rm}. In fact all the other terms also do not change monotonically, and their maxima do not occur at the same k_r. This indicates the significance of interplay among all items in determining the maximum growth rate.

Table 9.3 gives the energy budget in a liquid jet corresponding to Figure 9.3, which displays the amplification of disturbances giving rise to Taylor mode instability. While the surface tension work terms remain positive for wave numbers smaller than one in both Rayleigh and Taylor modes, they become negative for shorter unstable disturbances in the Taylor mode. Hence surface tension is not responsible for initiating the formation of droplets whose radii are smaller than the jet radius. The dominant positive terms are *PRG* and *SHG*. Thus the work done by the fluctuating gas pressure and the tangential shear stress at the interface is mainly responsible for the Taylor mode instability. Other positive terms *PRL*, *SHL*, and *NVN* are at least one order of magnitude smaller than these two terms. Recall that *PRG* and *SHG* are negative in Table 9.2. This is the fundamental difference in the origin of instability for the Rayleigh and the Taylor modes. Another fundamental difference is that while the normal component of the gas viscous stress plays a minor role in destabilizing the Rayleigh jet, it plays a major role together with viscous dissipation and reverse energy transfer to the mean flow through Reynolds stress in stabilizing the disturbance of wave numbers greater than the cut-off wave number $k_{rc} = 5.4$. This is clearly displayed in Table 9.2. Table 9.3 presents the case in which gas pressure fluctuation plays a slightly larger role than the interfacial shear. Table 9.4 presents another case of Taylor mode instability in

Table 9.3. Energy Budget for Taylor Mode: $Re = 1000, Fr^{-1} = 0.0001, We = 4761.9, Q = 0.013, N = 0.019, l = 10$

k_r	KE	REY	SUT	PRL	PRG	NVG	DIS	SHL ($\times 10^{-4}$)	NVL ($\times 10^{-4}$)	SHG ($\times 10^{-4}$)	SHB ($\times 10^{-1}$)
0.1	0.00298	5.12×10^{-5}	2.34×10^{-4}	6.95×10^{-5}	0.00184	2.77×10^{-6}	-1.18×10^{-4}	-1.78×10^{-4}	0.000429	4.4×10^{-4}	0.00551
1.0	0.536	-0.0247	3.92×10^{-4}	0.0193	0.484	-0.00194	-0.0833	-3.87	0.829	0.164	-0.178
2.0	0.993	-0.0544	-0.219	0.0282	1.23	-0.0196	-0.359	-16.7	1.89	0.441	-0.424
2.3	1.00	-0.0573	-0.358	0.0262	1.42	-0.0309	-0.467	-19.2	1.87	0.531	-0.499
3.0	0.826	-0.0458	-0.762	0.0198	1.76	-0.0756	-0.729	-17.4	1.38	0.744	-0.677
4.0	0.507	0.0154	-1.12	0.0161	2.04	-0.223	-1.14	0.391	0.490	1.02	-0.924
5.0	0.262	0.0292	-0.585	0.0101	2.43	-0.553	-2.18	3.66	0.0544	1.26	-1.18
6.0	-0.809	-0.118	1.43	-0.0312	3.07	-1.14	-5.30	-1.60	0.217	1.51	-1.50
8.0	-10.8	-1.03	14.1	-0.406	4.86	-3.30	-26.5	256.0	12.0	2.00	-2.28

Table 9.4. *Energy Budget for Taylor Mode:* $Re = 500$, $Fr^{-1} = 0.0005$, $We = 10^6$, $Q = 0.0013$, $N = 0.018$, $l = 10$

k_r	KE	REY	SUT	PRL $(\times 10^{-2})$	PRG	NVG	DIS	SHL $(\times 10^{-3})$	NVL $(\times 10^{-4})$	SHG	SHB
1.00	0.273	−0.534	-6.06×10^{-7}	1.23	0.251	0.00306	−0.115	−0.612	0.892	0.212	−0.0354
2.00	0.676	−0.162	-5.95×10^{-4}	1.94	0.778	−0.00437	−0.548	−3.86	3.10	0.727	−0.117
4.05	1.00	−0.325	-6.82×10^{-8}	0.0668	1.84	−0.157	−2.28	−12.5	4.92	2.24	−0.297
6.00	0.848	−0.378	−0.0202	−1.41	2.64	−0.481	−4.43	−16.2	3.79	4.02	−0.456
10.0	0.402	−0.321	−0.0607	−1.61	3.26	−1.23	−9.03	−12.1	1.22	8.55	−0.738
15.0	0.191	−0.215	−0.113	−0.823	2.98	−1.59	−16.2	−5.79	0.295	16.5	−1.07
20.0	0.108	−0.134	−0.149	−0.361	2.36	−1.51	−26.4	−2.49	0.0841	27.3	−1.41
25.0	0.0566	−0.0759	−0.149	−0.129	1.75	−1.29	−39.9	−0.826	0.0212	41.4	−1.75
30.0	0.0196	−0.0329	−0.0861	−0.0238	1.06	−0.901	−56.9	−0.111	0.00235	58.9	−2.09
35.0	−0.00717	−0.00180	0.049	0.00198	0.915	−0.987	−77.2	−0.0193	0.000308	79.7	−2.44

which interfacial shear plays a more significant role than pressure fluctuation. Note that the magnitude of *SHG* increases more rapidly than that of *PRG* as k_r is increased and *SHG* eventually dominates over the gas pressure term for this case. Thus the unstable disturbances near the cut-off wave number $k_{rc} = 33.5$ are mainly generated by the shear stress fluctuation with significant help from the gas inertia force manifested in the gas pressure fluctuation.

The energy budget at the saddle point in Figure 9.3 is $KE = 0.3321, REY = -0.2639 \times 10^{-4}, SUT = 0.1957, PRL = 0.1702, PRG = -0.7063 \times 10^{-3}, NVG = 0.1890 \times 10^{-3}, DIS = -0.2914 \times 10^{-1}, SHL = -0.5326 \times 10^{-4}, NVL = 0.1226 \times 10^{-2},$ and $SHG = -0.1027 \times 10^{-2}$. Note that the rate of pressure work done by the liquid at the upper and lower ends of the control volume *PRL* is increased significantly over that in Table 9.2. Otherwise all of the work terms retain their same qualitative roles. Hence the mechanism of the jet breakup by absolute instability is essentially the same as that of the Rayleigh mode by capillary pinching except the axial jet pressure fluctuation plays an equally important role.

9.12. Summary

The onset of instability in a viscous liquid jet in the presence of a surrounding viscous gas can manifest itself as convective or absolute instability depending on the flow parameters. The two different modes of convective instability, the Rayleigh and Taylor modes, are caused by fundamentally different physical mechanisms. The main cause of Rayleigh mode instability is capillary pinching, which is resisted by inertia in the form of pressure fluctuation and the viscous shear stress exerted by the gas at the interface. On the contrary, the gas pressure and shear fluctuations are the main means of supplying energy to the disturbances in Taylor mode instability. The surface tension tends to resist the formation of short waves. The unstable disturbances in an absolutely unstable jet propagate in both the upstream and downstream directions; this propagation is accompanied by a large liquid pressure fluctuation in the axial direction. The pressure fluctuation is of the same order of magnitude as the surface tension term. Capillary pinching remains a dominant source of both absolute instability and the Rayleigh mode of convective instability. The nature of absolute and convective instability revealed in the studies that neglect gas velocity remains qualitatively the same when the gas viscosity is neglected. However, the gas viscosity plays a very significant role in delineating the quantitative nature of jet breakup mechanisms. It even brings about a qualitative change in the effect of gas density on the onset of absolute instability. When gas viscosity is neglected as in Chapter 8, the domain of absolute instability is enlarged as gas density increases (Fig. 8.2). However, the reverse is true when gas viscosity is taken into account, as demonstrated in Figure 9.5.

References

Boyd, J. P. 1989. *Chebyshev and Fourier Spectral Method*. Springer-Verlag.

Canuto, C., Hussaini, N. Y., Quarteroni, A., and Zang, T. A., 1988. *Spectral Methods in Fluid Mechanics*. Springer-Verlag. pp. 55, 295.

Chandrasekhar, S. 1961. *Hydrodynamic and Hydormagnetic Stability*, Oxford University Press, p. 537.

Chomaz, J. M., Huerre, P., and Redekopp, L. G. 1988. Bifurcation to local and global modes in spatially developing flows. *Phys. Rev. Lett.* **60**, 25.

Clanet, C., and Lasheras, J. C. 1999. Transition from dripping to jetting, *J. Fluid Mech.* **383**, 307.

Davey, A., and Drazin, P. G. 1969. *J. Fluid Mech.* **36**, 209.

Debler, W., and Yu, D. 1988. The breakup of laminar liquid jets. *Proc. R. Soc. Lond. A* **415**, 106.

Donnelly, R. J., and Glaberson, W. 1996. *Proc. R. Soc. Lond.* **290**, 547.

Drazin, P. G., and Reid, W. H. 1981. *Hydrodynamic Stability*. Cambridge University Press.

Eggers, J. 1997. Nonlinear dynamics and breakup of free-surface flows. *Rev. Modern Phys.* **69**, 865.

Goedde, E. F., and Yuen, M. C. 1970. *J. Fluid Mech.* **40**, 495.

Gottlieb, D., and Orszag, S. 1977. *Numerical Analysis of Spectral Methods*: *Theory and Applications*. SIAM, Philadelphia; Arrowsmith, Bristol.

Huerre, P., and Monkewitz, P. A. 1990. Local and global instability in spatially developing flows. *Ann. Rev. Fluid Mech.* **22**, 473–537.

Joseph, D. D., and Renardy, Y. Y. 1992. *Fundamental of Two-Fluid Dynamics*. Springer-Verlag.

Keller, J. B., Rubinow, S. I., and Tu, Y. O. 1972. *Phys. Fluids* **16**, 2052.

Lanczos, C. 1956. *Applied Analysis*. Prentice-Hall.

Le Dizes, S. 1997. Global models in falling capillary jets. *Eur. J. Mech B/Fluids* **16**, 761.

Leib, S. J., and Goldstein, M. E. 1986a. The generation of capillary instability on liquid jet. *J. Fluid Mech.* **168**, 479.

Leib, S. J., and Goldstein, M. E. 1986b. Convective and instability of a viscous liquid jet. *Phys. Fluids* **29**, 952.

Lin, S. P., and Kang, D. J. 1987. *Phys. Fluids* **30**, 2000.

Lin, S. P., and Lian, Z. W. 1989. Absolute instability of a liquid jet in a gas. *Phys. Fluids A.* **1**, 490.

Lin, S. P., and Chen, J. N. 1998. Roles played by the interfacial shears in the instability mechanism of a viscous liquid jet surrounded by a viscous gas in a pipe. *J. Fluid Mech.* **376**, 37.

Lin, S. P., and Creighton, B. J. 1990. *J. Aerosol. Sci. Technol.* **12**, 630.

Lin, S. P., and Ibrahim, E. A. 1990. *J. Fluid Mech.* **218**, 641.

Lin, S. P., and Lian, Z. W. 1990. *AIAA J.*, **28**, 120.

Lin, S. P., and Lian, Z. W. 1993. *Phys. Fluids* **5**, 771.

Lin, S. P., and Lian, Z. W. 1994. Absolute and convective instability of a viscous liquid jet surrounded by a viscous gas in a vertical pipe. *Phys. Fluids* **6**, 2545.

Lin, S. P., and Reitz, R. D. 1998. *Ann. Rev. Fluid Mech.* **30**, 85.

Lin, S. P., Phillips, W. R. C., and Valentine, D. T. (Eds.) 1994. *Nonlinear Instability of Nonparallel Flows*. IUTAM Symposium Proceedings, Potsdam, NY, 1993. Springer-Verlag.

Monkewitz, P. A. 1990. *Eur. J. Mech. B/Fluids* **9**, 395.

O'Donnel, B., Chen, J. N., and Lin, S. P. 2001. Transition from convective to absolute instability in a liquid jet. *Phys. Fluids* **13**, 2732–2734.

Preziosi, L., Chen, K., and Joseph, D. D. 1989. *J. Fluid Mech.* **201**, 323.

Sterling, A. M., and Sleicher, C. A. 1975. *J. Fluid Mech.* **68**, 477.

Vihinen, I., Honohan, A., and Lin, S. P. 1997. Absolute and convective instability of a viscous liquid jet in microgravity. *Phys. Fluids* **9**, 3117–3119.

Yakubenko, P. A. 1997. Global capillary instability of an inclined jet. *J. Fluid Mech.* **346**, 181.

Zhou, Z. W., and Lin, S. P. 1992. Effects of compressibility on the atomization of liquid jets. *J. Propul. Power* **8**, 736.

10

Annular Liquid Jets

Annular jets are encountered in many industrial processes. Their stability has been studied in the contexts of ink-jet printing (Hertz and Hermanrud, 1983; Sanz and Meseguer, 1985), encapsulation (Lee and Wang, 1989; Kendall, 1986), gas absorption (Baird and Davidson, 1962), and atomization (Crapper, Dombrowski, and Pyott, 1975; Lee and Chen, 1991; Shen and Li, 1996; Villermaux, 1998). Shen and Li analyzed the spatial-temporal instability of an annular liquid jet surrounded by an inviscid gas. Hu and Joseph (1989) investigated the temporal instability of a three-layered liquid core-annular flow. The instability of annular layeres has been used to model the formation of liquid bridges in microairways in lungs (Newhouse and Pozrikidis, 1992). The related problems of liquid bridge instability are reviewed by Alexander (1998). Annular jet instability is also of considerable theoretical interest because it includes many other flow instabilities as special cases (Meyer and Weihs, 1987). Moreover, it serves to establish knowledge of the fluid physics of flows with two distinctive curved fluid-fluid interfaces subjected to different shear forces, capillary forces, and inertial forces under variable gravitational conditions.

10.1. An Annular Jet

Consider the flow of a fluid in an annulus enclosing another fluid, which is surrounded by yet another fluid inside a circular pipe of radius R_w as shown in Figure 10.1. The axis of the pipe aligns with the direction of the acceleration due to gravity g. All three fluids are incompressible. The governing equations of motion of the fluids are

$$\partial_t \mathbf{V}_j + \mathbf{V}_j \cdot \nabla \mathbf{V}_j = -\frac{1}{\rho_j} \nabla \cdot \boldsymbol{\sigma}_j + \mathbf{g}, \tag{10.1}$$

$$\nabla \cdot \mathbf{V}_j = 0 \qquad (j = 1, 2, 3), \tag{10.2}$$

172

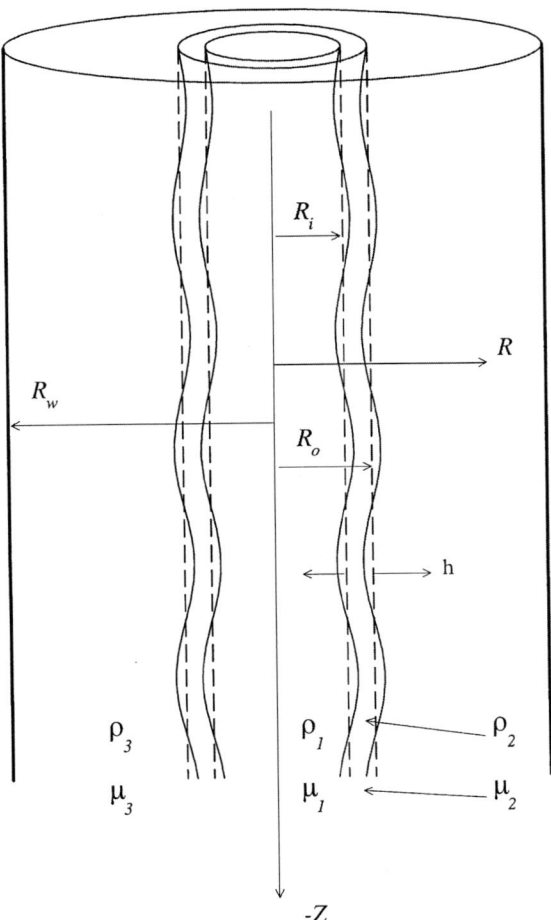

Figure 10.1. Definition sketch.

where $j = 1, 2$, and 3 designates the inner core fluid, the fluid in the annular, and the outer fluid, respectively, t is the time, $\mathbf{V}$ is the velocity, ρ is the density, and σ is the stress tensor. For Newtonian fluids

$$\sigma_j = -P_j \mathbf{I} + \mu_j \left[(\nabla \mathbf{V}_j) + (\nabla \mathbf{V}_j)^T \right],$$

where P is the pressure, $\mathbf{I}$ is the identity matrix, and μ is the dynamic viscosity.

The corresponding boundary conditions are the no-slip conditions at the pipe inner wall, that is,

$$\mathbf{V}_3 = 0 \quad \text{at} \quad R = R_w, \tag{10.3}$$

where R is the radial distance in the cylindrical coordinates (R, θ, Z); the continuity of the velocity at each fluid-fluid interface,

$$\mathbf{V}_1 = \mathbf{V}_2 \quad \text{at} \quad R = R_i, \tag{10.4}$$

$$\mathbf{V}_2 = \mathbf{V}_3 \quad \text{at} \quad R = R_0, \tag{10.5}$$

where R_i and R_0 are, respectively, the inner radius and outer radius of the annulus; the kinematic conditions at the interface i.e.

$$U_{1,3} = (\partial_t + \mathbf{V}_{1,3} \cdot \nabla) R \quad \text{at} \quad R = R_i \quad \text{or} \quad R_0, \tag{10.6}$$

where U is the radial component of the velocity, and the subscript 1,3 designates either fluid 1 or fluid 3; the dynamic boundary condition at the fluid-fluid interfaces,

$$(\boldsymbol{\sigma}_2 - \boldsymbol{\sigma}_1) \cdot \mathbf{n}_i - \mathbf{n}_i \cdot S_i \nabla \cdot \mathbf{n}_i = 0 \quad \text{at} \quad R = R_i, \tag{10.7}$$

$$(\boldsymbol{\sigma}_3 - \boldsymbol{\sigma}_2) \cdot \mathbf{n}_0 - \mathbf{n}_0 \cdot S_0 \nabla \cdot \mathbf{n}_0 = 0 \quad \text{at} \quad R = R_0, \tag{10.8}$$

where S_i and S_0 are, respectively, the surface tensions of the inner and the outer interfaces, and $\mathbf{n}_i$ and $\mathbf{n}_0$ are, respectively, the unit normal vectors of the interfaces defined by

$$F_i = R - R_i = 0 \quad \text{and} \quad F_0 = R - R_0 = 0.$$

The unit normal vectors at the two interfaces are defined to be positive if they point in a positive radial direction. Thus

$$(\mathbf{n}_i, \ \mathbf{n}_0) = (\nabla F_i / |\nabla F_i|, \ \nabla F_0 / |\nabla F_0|). \tag{10.9}$$

We seek an axisymmetric, nonswirling, steady, constant radius, annular basic flow that satisfies exactly the above governing equations and their boundary conditions. For such a flow, the radial and azimuthal velocity components must vanish, and the pressure differences across each fluid-fluid interface must balance exactly the surface tension forces associated with the two different radii at any axial location at all times. Thus, the axial dynamic pressure gradient in each fluid must be constant but of different value because of the different hydrostatic pressure experienced by each fluid. It is easily verified that the governing Equations (10.1) and (10.2) are reduced to

$$\frac{N_j}{Re} \frac{1}{r} \partial_r (r \partial_r W_j) = -K_j, \tag{10.10}$$

$$K_j = \left(-\frac{1}{Q_j} \partial_z \bar{p}_j - Fr^{-1} \right),$$

where r and z are the radial and axial distances normalized with the characteristic length H, and W is the axial velocity normalized with a reference velocity W_m. Both H and W_m are yet to be chosen; $\bar{p} = P/\rho_2 W_m^2$ is the dimensionless pressure; Re, Fr, N_j, and Q_j are respectively the Reynolds number, the Froude number, the kinematic viscosity ratio and the density ratio defined by

$$Re = W_m H/v_2, \qquad Fr = W_m^2/gH, \qquad N_j = v_j/v_2, \qquad Q_j = \rho_j/\rho_2,$$

v being the kinematic viscosity. Integration of (10.10) yields

$$W_j = -\frac{K_j}{4N_j} Re \, r^2 + C_{j1} \ln r + C_{j2}, \tag{10.11}$$

where C_{j1} and C_{j2} are integration constants determined by the boundary conditions to be

$$C_{11} = 0,$$

$$C_{21} = (1 - Q_1) \, Re \, Fr^{-1} r_i^2/2,$$

$$C_{31} = (Q_3 - 1) \, Re \, Fr^{-1} r_0^2/(2Q_3 N_3) + C_{21}/(Q_3 N_3),$$

$$C_{22} = (K_2 - K_3/N_3) \, Re \, r_0^2/4 + (C_{31} - C_{21}) \ln r_0 + C_{32},$$

$$C_{12} = (K_2 - K_1/N_1) \, Re \, r_i^2/4 + C_{21} \ln r_i + C_{22},$$

$$C_{32} = K_3 \, Re \, r_w^2/(4N_3) - C_{31} \ln r_w,$$

$$K_1 = \left[K_2 + (1 - Q_1) \, Fr^{-1}\right]/Q_1, \tag{10.12}$$

$$K_3 = \left[K_2 + (1 - Q_3) \, Fr^{-1}\right]/Q_3, \tag{10.13}$$

where K_2 depends on the choice of the length scale H and the velocity reference W_m. For a better numerical accuracy, H and W_m should be chosen respectively as the annular shell thickness h and the velocity at $r = (r_i + r_0)/2$. For this choice of normalization,

$$K_2 = \{-4 + Re \, Fr^{-1} A_1$$
$$- 4[C_{21} \ln (r_i + r_0)/2 + A_2]\}/Re\left[-(r_i + r_0)^2/4 + A_3\right]. \tag{10.14}$$

$$A_1 = \left(Q_3^{-1} - 1\right)\left(r_0^2 + r_w^2\right)/N_3,$$

$$A_2 = (C_{31} - C_{21}) \ln r_0 - C_{31} \ln r_w,$$

$$A_3 = \left(1 - Q_3^{-1} N_3^{-1}\right) r_0^2 + r_w^2/Q_3 N_3.$$

On the other hand, it is more expedient to normalize the length and velocity, respectively, with the core fluid radius and the core fluid velocity along the pipe axis, for a direct comparison with the known results of the axisymmetric

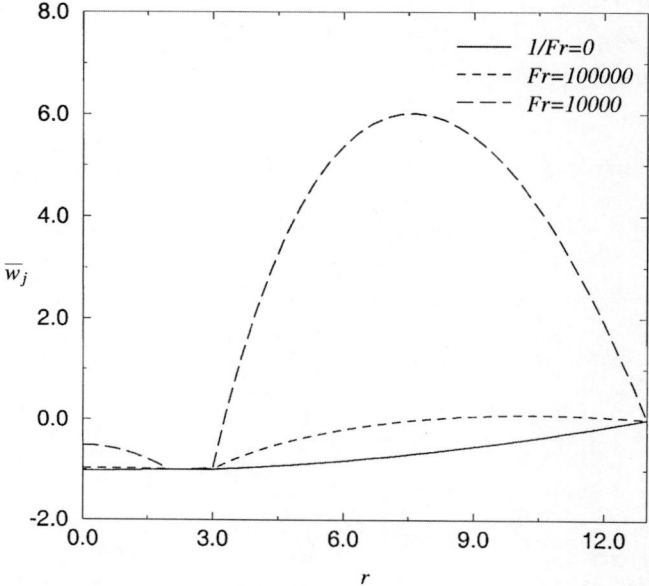

Figure 10.2. The effects of Froude number on the basic velocity profile. $Re = 1000$, $Q_1 = Q_3 = 0.0013$, $\bar{N}_1 = 0.018$, $\bar{N}_3 = 0.18$, $r_i = 2$, $r_w = 13$.

Poiseuille flow and the simple liquid jet. For this latter normalization (scaling I),

$$K_2 = \{ -4 + Re\ Fr^{-1} \left[(Q_1^{-1} - 1)\ r_i^2 / N_1 \right] A_1$$
$$ - 4[C_{21}\ ln\ r_i + A_2] \} / Re \left[(1 - Q_1^{-1} N_1^{-1})\ r_i^2 + A_3 \right].$$

A representative basic flow velocity distribution is given in Figure 10.2.

10.2. Stability Analysis of the Annular Jet

To investigate the onset of instability in this basic flow we introduce disturbances to the basic flow and analyse the spatial-temporal behavior of the disturbance. Related experiments (Kendall, 1986) show that the consequence of instability is the formation of axisymmetric compound drops, unless the Reynolds number is so high that nonaxisymmetric sprays may result (Lee and Chen, 1991). We focus here on the formation of axisymmetric compound liquid capsules and assume the disturbance to be axisymmetric. Hence the governing equation of the disturbance is still given by (9.14), except that the subscript α in (9.14) is now replaced by $j = 1$, 2, or 3. The wave number k and the wave frequency ω retain their physical representations. However, the real part of k is now related to the wavelength λ by $k_r = 2\pi h/\lambda$ because the annular shell thickness h is now used for normalization. The linearized boundary conditions corresponding to (10.3)-(10.8) written out in cylindrical components in the same order are

$$\phi_3(r_w) = 0, \tag{10.15}$$

$$D\phi_3(r_w) = 0, \tag{10.16}$$

$$\phi_{1,3}(r_{i,0}) = \phi_2(r_{i,0}), \tag{10.17}$$

$$\frac{1}{r_{i,0}}[D\phi_{1,3}(r_{i,0}) - D\phi_2(r_{i,0})] = [DW_{1,3}(r_{i,0}) - DW_2(r_{i,0})]\,\xi_{i,0}, \tag{10.18}$$

$$\frac{ik}{r_{i,0}}\phi_{1,3}(r_{i,0}) = [\omega + ikW_{1,3}(r_{i,0})]\,\xi_{i,0}, \tag{10.19}$$

$$Q_{1,3}\,N_{1,3}\,E^{-2}\phi_{1,3}(r_{i,0}) - E^{-2}\phi_2(r_{i,0})$$
$$= \xi_{i,0}\left[Q_{1,3}\,N_{1,3}\,D^2W_{1,3}(r_i, r_0) - D^2W_2(r_{i,0})\right], \tag{10.20}$$

$$Q_{1,3}[\omega + ikW_{1,3}(r_{i,0})]\,E\phi_{1,3}(r_{i,0}) - [\omega + ikW_2(r_{i,0})]\,D\phi_2(r_{i,0})$$

$$- ik\,[DW_{1,3}(r_{i,0})\,\phi_{1,3}(r_{i,0}) - DW_2(r_{i,0})\phi_2(r_{i,0})]$$

$$+ \frac{1}{Re}\left\{\left[L\phi_2(\bar{r}_{i,0}) - 2k^2\left(D\phi_2(r_{i,0}) - \frac{1}{r_{i,0}}\phi_2(r_{i,0})\right)\right]\right.$$

$$\left. - Q_{1,3}\,N_{1,3}\left[L\phi_{1,3}(r_{i,0}) - 2k^2\left(D\phi_{1,3}(r_{i,0}) - \frac{1}{r_{i,0}}\phi_{i,0}(r_{i,0})\right)\right]\right\}$$

$$+ ikWe_{i,0}^{-1}\left(\frac{1}{r_{i,0}} - k^2 r_{i,0}\right)\xi_{i,0} = 0, \tag{10.21}$$

where $r = R/h$, and its subscripts retain the same designations as its dimensional counterparts. $r_{i,0}$ appearing in the argument of a function signifies that the function is to be evaluated at $r = r_i$ or $r = r_0$ depending on whether (10.20) is to be evaluated at the inner or outer interface, and

$$D^2 = \frac{d^2}{dr^2}, \qquad E^{-2} = \frac{d^2}{dr^2} - \frac{1}{r}\frac{d}{dr} + k^2,$$

$$L = \frac{d^3}{dr^3} - \frac{1}{r}\frac{d^2}{dr^2} - \left(\frac{1}{r^2} - k^2\right)\frac{d}{dr},$$

$$We_{i,0} = \rho_2 W_m^2 h/S_{i,0}.$$

$We_{i,0}$ is the Weber number associated with the inner or the outer interfaces. Equations (10.15) and (10.16) are the no-slip conditions at the pipe wall.

Equations (10.17) and (10.18) are respectively the continuity of the radial and tangential components of velocity at the interfaces. Equation (10.19) is the interfacial kinematic condition, (10.20) and (10.21) are, respectively, the tangential and normal component of the dynamic boundary condition. Moreover, we shall demand that the solution is bounded along the cylindrical axis, that is,

$$\phi_3(0) = D\phi_3(0) = 0. \tag{10.22}$$

10.3. A Pseudo-Spectral Method of Solution

We now extend the method used for the two-fluid domains in Chapter 9 to the three-fluid domains encountered here. First, the flow domain of each fluid is mapped into a strip of thickness 2 by use of the linear transformations

$$y_j = a_j r + b_j, \tag{10.23}$$

where

$$a_1 = 2/r_i, \qquad b_1 = -1,$$

$$a_2 = \frac{2}{r_i - r_0}, \qquad b_2 = -\frac{r_i + r_0}{r_i - r_0},$$

$$a_3 = \frac{2}{r_w - r_0}, \qquad b_3 = -\frac{r_w + r_0}{r_w - r_0},$$

Hence

$$r \in [0, r_i] \rightarrow y_1 \in [-1, 1],$$
$$r \in [r_i, r_0] \rightarrow y_2 \in [1, -1],$$
$$r \in [r_0, r_w] \rightarrow y_3 \in [-1, 1].$$

Then we write ϕ_j as the finite sum of a convergent series as

$$\phi_j(y_j) = \sum_{n=0}^{M_j} h_{jn}(y_j)\phi_j(y_{jn}), \tag{10.24}$$

where y_{jn} is the Gauss–Lobatto collocation point given by

$$y_{jn} = \cos\left(\frac{\pi n}{M_j}\right),$$

and h_{jn} is the Lagrange cardinal function

$$h_{jn}(y_j) = \frac{(-1)^{n+1}(1-y_j^2)}{C_n M_j^2 (y_j - y_{jn})} \frac{d}{dy} T_{M_j}(y_j), \qquad (10.25)$$

$$C_0 = C_{M_j} = 2, \qquad C_n = 1 \quad \text{for} \quad 1 \leq n \leq M_j - 1,$$

where $T_{M_j}(y_j)$ is the M_j-th order Chebyshev polynomial

$$T_{M_j}(y) = cos\,(M_j\,cos^{-1}y) \qquad -1 \leq y \leq 1.$$

By substituting (10.24) into (10.14) and evaluating the resulting equation at $M_j - 4$ collocation points, we obtain $M_1 + M_2 + M_3 - 12$ equations. We also have to satisfy the 17 equations from (10.15) to (10.22). Thus we have a system of $M_1 + M_2 + M_3 + 5 = N$ equations,

$$\mathbf{Ax} = \omega\,\mathbf{Bx}, \qquad (10.26)$$

in the same number of unknowns,

$$\mathbf{x} = [(\phi_{10} \cdots \phi_{1M_1}), (\phi_{20} \cdots \phi_{2M_2}), (\phi_{30} \cdots \phi_{3M_3}), \xi_i, \xi_0].$$

The expressions of the $N \times N$ matrices A and B are given in Appendix A.

There are 10 flow parameters (Re, Fr, We_i, We_0, Q_1, Q_3, $\bar{N}_1$, $\bar{N}_3$, r_i, r_w), where $\bar{N}_1 = Q_1 N_1$ and $\bar{N}_3 = Q_3 N_3$. Note that $r_0 = r_i + 1$. The annular liquid shell possesses two interfaces. Thus it has two degrees of freedom, and the linear differential system possesses two independent interfacial modes of solution. Unlike a two-dimensional plane liquid sheet, the two modes are not exactly in phase or 180° out of phase because the two interfaces are not exactly symmetrical with respect to the midsection of the annular sheet. However, they may be almost symmetrical if the curvature difference is small. Hence we call the mode that displaces the two interfaces almost in phase the para-sinuous mode and the mode that displaces the two interfaces nearly 180° out of phase, the para-varicose mode, or simply the sinuous and varicose modes. Sketches of sinuous and varicose modes are given in Figure 10.3. For encapsulation applications, the varicose mode should be avoided, since the shell tends to pinchoff at the location where the core is the thickest. This will lead to part of the core material uncovered.

For purely temporal disturbances k is real, and if it is given together with the complete set of flow parameters, then the complex wave frequency ω can be solved from (10.26) as the eigenvalue. For spatially growing disturbances k is also complex. The numerical algorithm for the eigenvalue solution of (10.26) is given in Appendix B.

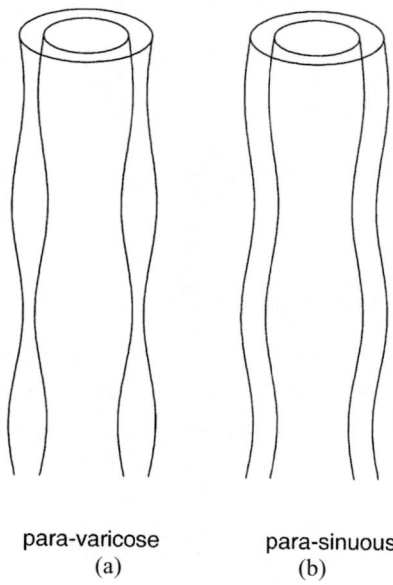

para-varicose   para-sinuous

(a)   (b)

Figure 10.3. Schematic of symmetric and antisymmetric disturbances for an annular liquid sheet.

10.4. Parametric Responses in an Encapsulation Process

The application of the results to the encapsulation of a fluid by another fluid by use of an annular jet will be demonstrated next. Figure 10.4 gives the temporal growth rate ω_r and the spatial growth rate k_i as functions of k_r for the set of flow parameters shown in the caption. The values of Q_1, Q_3, $\bar{N}_1$, and $\bar{N}_3$ correspond to the case of a water annular jet enclosing air and surrounded by air at room condition. The two curves are qualitatively the same except that the wave number corresponding to the maximum growth rates are smaller for the spatial disturbances for the sinuous mode. The two curves for the varicose mode are almost identical except near the cut-off wave number beyond which the amplification rates become negative. Figure 10.5 gives a similar set of two amplification curves for the same parameters as those given in Figure 10.4, except that the Weber numbers are much larger. For relatively large Weber numbers, the temporal and spatial amplification curves fall almost on top of each other. A similar situation was encountered by Keller, Rubinow, and Tu (1972) in their analysis of a simple jet. The spatial amplification curves $\omega_r = 0$ in Figures 10.4 and 10.5 can be approached from the lower half k-plane where $\omega_r > 0$ and $k_i < 0$ without encountering a pinch point singularity in the complex ω-plane. Therefore the causality condition that initially no disturbances exist as $t \to -\infty$ and $z \to -\infty$ in

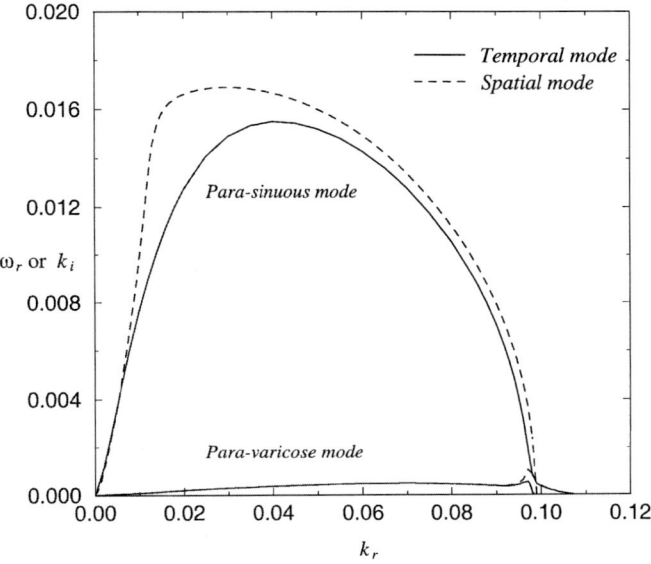

Figure 10.4. Temporal and spatial growth rates. $Re = 500$, $1/Fr = 0.0$, $We_i = We_0 = 50$, $\bar{N}_1 = \bar{N}_3 = 0.018$, $Q_1 = Q_3 = 0.0013$, $r_i = 10$, $r_w = 21$.

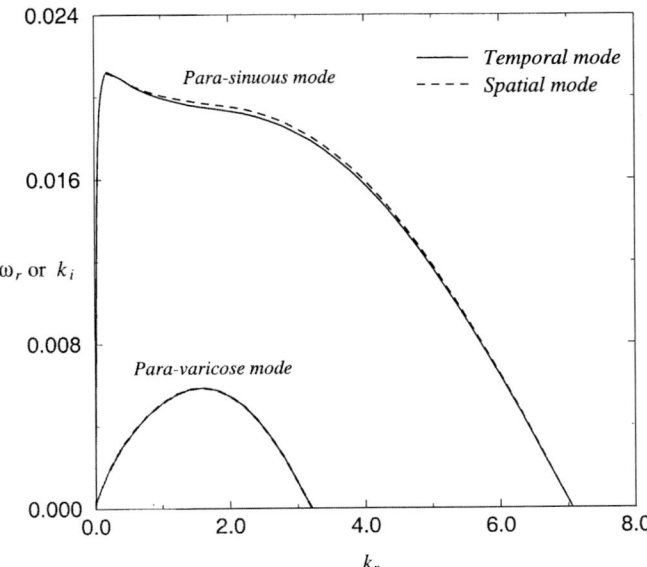

Figure 10.5. Temporal and spatial growth rates. $Re = 2000$, $Fr = 50000$, $We_i = We_0 = 200000$, $\bar{N}_1 = \bar{N}_3 = 0.018$, $Q_1 = Q_3 = 0.0013$, $r_i = 10$, $r_w = 21$.

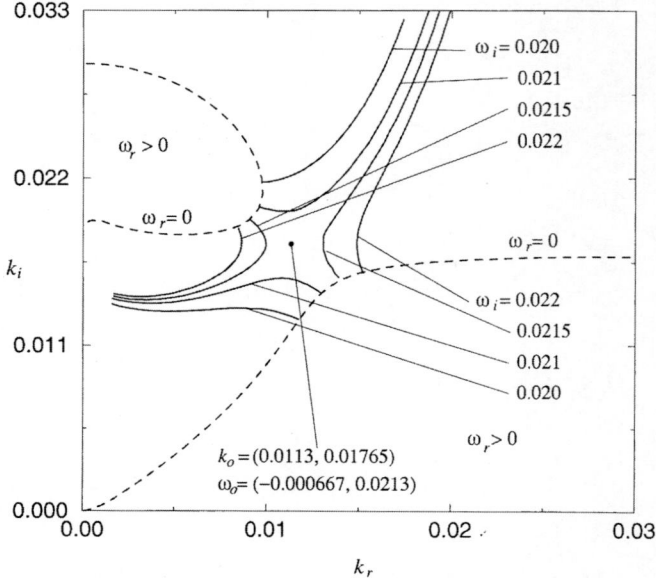

Figure 10.6. Convective instability, sinuous mode. $Re = 500$, $1/Fr = 0.0$, $We_i = We_0 = 50$, $\bar{N}_1 = \bar{N}_3 = 0.018$, $Q_1 = Q_3 = 0.0013$, $r_i = 10$, $r_w = 21$.

the downstream region is satisfied. Moreover, ω_i increases with k_r along the spatial amplification curves. Hence, these curves represent downstream propagating convectively unstable disturbances whose amplitudes decay in time at a given point in space but grow without bound with respect to an observer traveling at the disturbance group velocity.

In some parameter space, the sinuous mode encounters in the k-plane a saddle point singularity. An example of the emergence of a saddle point is illustrated in Figures 10.6 and 10.7. There are two branches of the spatial amplification curves $\omega_r = 0$ in Figure 10.6. The upper and lower branches can be reached from the upper and lower half k-planes, respectively. Along the upper branch $d\omega_i/dk_r < 0$, and along the lower branch $d\omega_i/dk_r > 0$. The lower branch represents the downstream propagating convectively unstable disturbance. The upper branch represents the upstream propagating evanescent waves. The saddle point in Figure 10.6 does not prevent the deformation of the Fourier integral counters for superposing all Fourier components of either upstream or downstream propagating disturbances. As the Weber number is reduced from 50 (Fig. 10.6) with the rest of the parameters fixed, the upper and the lower branches move toward each other, and at a critical Weber number they meet at a saddle point. The Weber number 45 in Figure 10.7 slightly overshoots the critical Weber number. Figure 10.7 shows that when

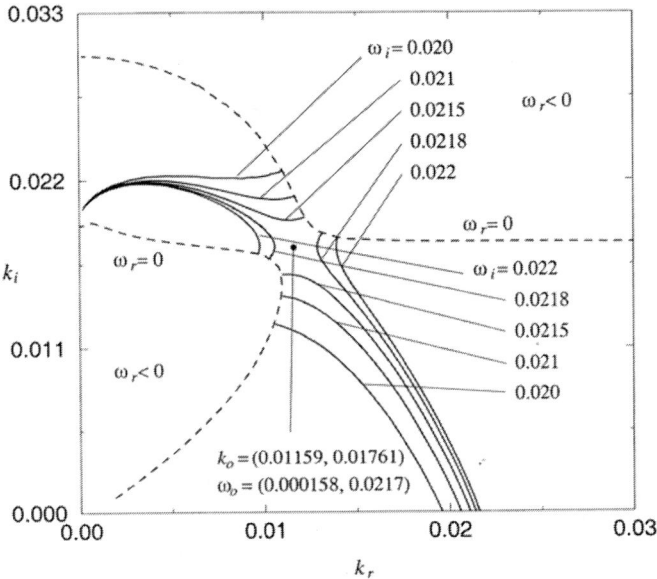

Figure 10.7. Absolute instability, sinuous mode. $Re = 500$, $1/Fr = 0.0$, $We_i = We_0 = 45$, $\bar{N}_1 = \bar{N}_3 = 0.018$, $Q_1 = Q_3 = 0.0013$, $r_i = 10$, $r_w = 21$.

the critical Weber number is slightly exceeded, the upper and lower halves of the upper and the lower branch remain connected as they split off from the saddle point at the critical Weber number to form two new branches. The image of the saddle point in the k-plane is a pinch point singularity in the complex ω-plane with $\omega_r > 0$. Exact locations of the saddle points in the parameter space of 10 dimensions is numerically tedious and expensive to obtain. The appearance of a saddle point signals the occurrence of absolute instability in a flow in which a disturbance grows unbounded with time as it spreads in both the upstream and downstream directions.

Figure 10.8 shows three curves, each with a given value of Q_1 and Q_3, that separate the parameter space into convective and absolute instability. The fixed parameters are shown in the figure caption. Below each curve, the flow is absolutely unstable; above each curve, it is convectively unstable. In the parameter space of absolute instability, encapsulation by breaking an annular jet cannot be achieved because the disturbance would propagate upstream to interrupt the process. However, in the convectively unstable regime, capsules can be formed if appropriate measures are taken, as will be explained shortly. Note that the region of absolute instability, that is, the region in which encapsulation is prohibited, increases with increases in the density changes across the interfaces.

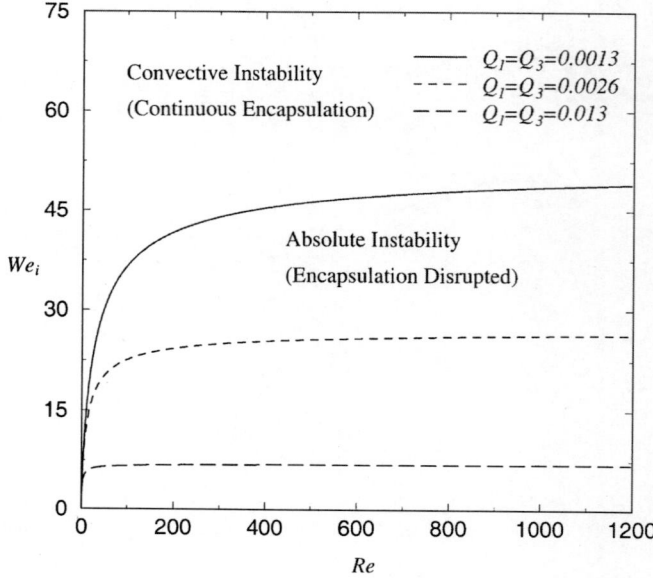

Figure 10.8. The critical Weber number for absolute instability. $1/Fr = 0.0$, $We_i = We_0$, $\bar{N}_1 = \bar{N}_3 = 0.018$, $r_i = 10$, $r_w = 21$.

Figure 10.9 shows the effect of the Weber number variation on the rates of sinuous and varicose mode disturbances. For each Weber number set, the varicose mode has a slightly larger cut-off wave number above which the disturbance decays. Amplification curves of different Weber numbers intersect at point e for both modes. For sinuous disturbances with wave numbers smaller than that corresponding to point e, the amplification rate can be raised by decreasing the Weber number by, for example, increasing the surface tension. Hence the capillary force is destabilizing in this range of wave numbers. The same conclusion applies to the varicose mode except in a very small wave number range. In this range as well as in the range of k_r greater than that corresponding to point e of both modes, the amplification curve for the larger Weber number is above that for the smaller one. Therefore in these ranges the capillary force becomes stabilizing. Yet the jet is unstable in these ranges up to the cut-off wave number. Therefore in these ranges the instability of both modes is caused by forces other than capillary force. These forces will be discussed later. Note that the varicose mode dominates over the sinuous mode only in a very small range of wave numbers between the cut-off wave numbers of the two modes for each Weber number in Figure 10.9. In almost all of the rest of the unstable wave spectrum the amplification rate of the sinuous mode is one to two orders of magnitude larger than that of the

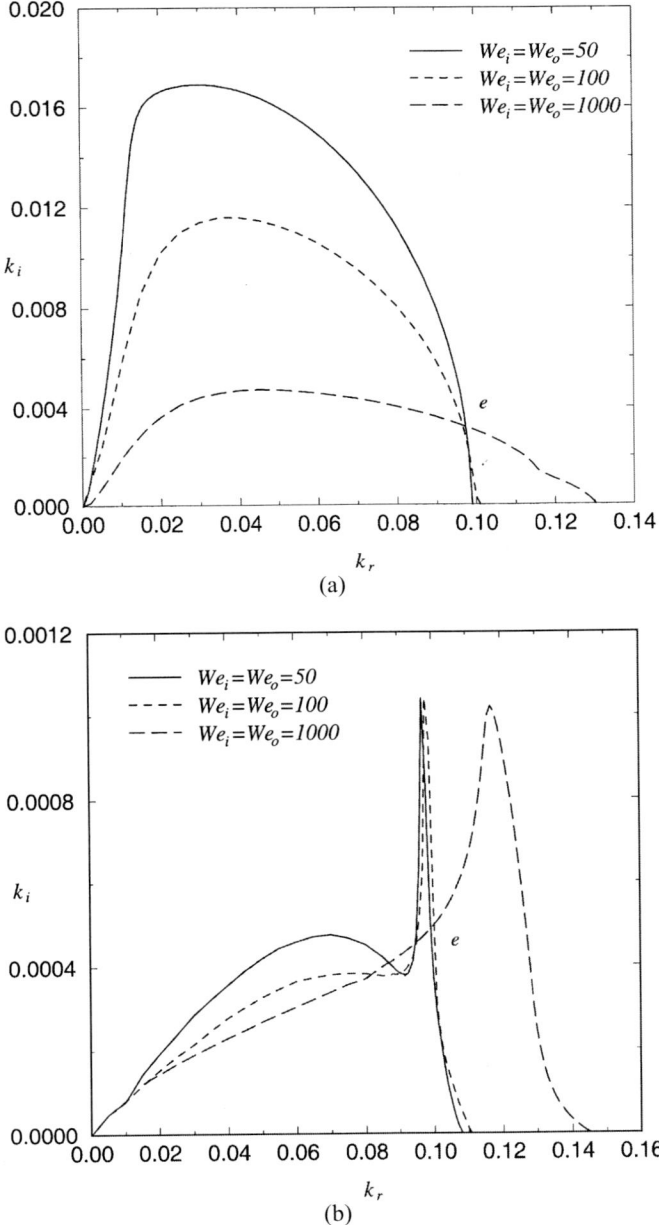

Figure 10.9. Effects of Weber number on the spatial disturbance growth rate of convective instability. $Re = 500$, $1/Fr = 0.0$, $\bar{N}_1 = \bar{N}_3 = 0.018$, $Q_1 = Q_3 = 0.0013$, $r_i = 10$, $r_w = 21$. (a) Para-sinuous mode. (b) Para-varicose mode.

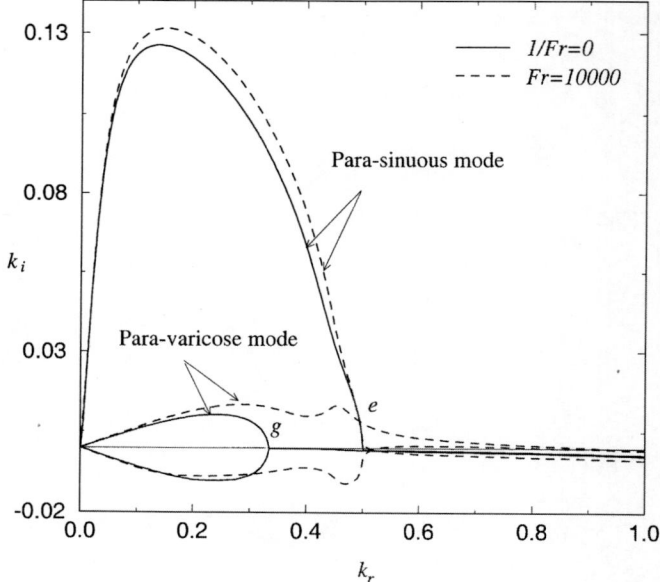

Figure 10.10. Effects of Froude number on the spatial disturbance growth rate for both modes. $Re = 1000$, $We_i = We_0 = 20$, $\bar{N}_1 = \bar{N}_3 = 0.018$, $Q_1 = Q_3 = 0.0013$, $r_i = 2$, $r_w = 13$.

varicose mode. However, a successful encapsulation is not easily achieved for the parameter specified in the figure, even if we impart an external forcing with frequencies corresponding to the wave numbers in the range where the sinuous mode dominates over the varicose mode. This is because the varicose mode, however feeble, will still be excited together with the sinuous mode, thereby contaminating the encapsulation process.

The example given in Figure 10.9 is for the case of zero gravity. The occurrence of a small range of wave spectra, in which the undesirable varicose mode is dominant at finite gravity, is shown in Figure 10.10. The para-sinuous and para-varicose mode amplification curves for $Fr = 10000$ intersect at point e. Beyond this point the varicose mode dominates over the sinuous mode. However, when gravity is zero, the dominance of the varicose mode is completely eliminated. Moreover, between the cut-off wave numbers of the varicose and sinuous modes, even the presence of the varicose mode is completely eliminated. Thus a sinuous disturbance of a chosen wavelength can be isolated in this range of wave numbers where the varicose mode is stable, by imparting an external forcing with a frequency corresponding to the chosen wavelength $\lambda = 2\pi h / k_r$. Using the method of Rayleigh, we can estimate the radius of a capsule by equating the its volume with the volume of the annular jet of

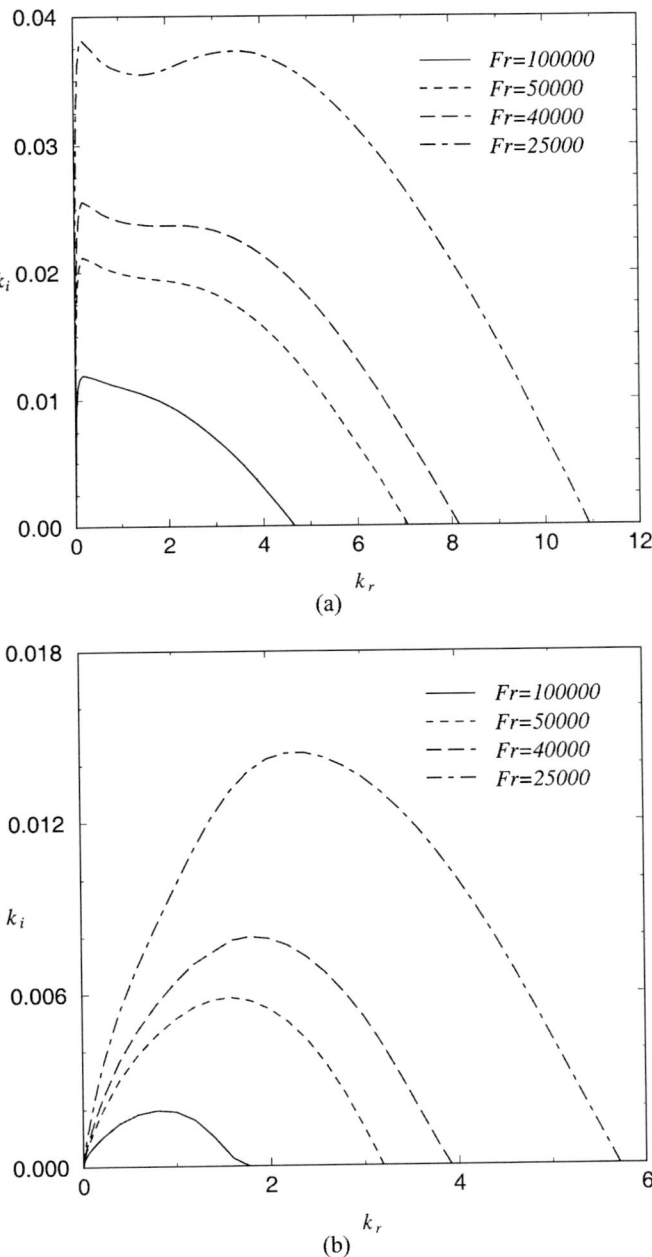

Figure 10.11. Effects of Froude number on the disturbance growth rate. $Re = 2000$, $We_i = We_0 = 200000$, $\bar{N}_1 = \bar{N}_3 = 0.018$, $Q_1 = Q_3 = 0.0013$, $r_i = 2$, $r_w = 13$. (a) Para-sinuous mode. (b) Para-varicose mode.

length λ from which the capsule is formed. This method of estimation gives the capsule radius as $R = [3\pi h(r_i + h)^2/2k_r]^{1/3}$.

It is clear from the results presented up to this point that a successful uniform encapsulation is possible if (i) the process is carried out outside of the absolute instability domain in the parameter space and (ii) a monochromatic external excitation is introduced in the wave number range where the varicose mode instability is absent but the sinuous mode is convectively unstable. We define the parameter space in which these two conditions are satisfied to be the encapsulation domain. In the remaining part of this chapter, we will discuss how the encapsulation domain can be enlarged by varying different flow parameters.

We can enlarge the encapsulation domain by reducing the Froude number (Fig. 10.11) or by increasing Re (Fig. 10.12). Figures 10.13 and 10.14 show respectively the effects of variations of We_i and We_0. Encapsulation between $k_r = 1/3$ and $1/2$ without varicose mode interference is still possible. Encapsulation domain can also be widened by use of a less viscous gas for the outer gas (Fig. 10.15) or it can be enlarged as N_1 is increased with N_3 fixed (Fig. 10.16). Thus, it is easier to encapsulate more viscous fluids.

Figure 10.17 shows how the encapsulation domain can be enlarged and the difference between the sinuous and varicose mode amplification rates increased by use of a larger pipe radius r_w. In Figure 10.18 we see that it is easier to encapsulate core fluid with a smaller pipe diameter relative to the shell thickness because the configuration has a larger encapsulation domain. As r_w is increased in Figure 10.17, the amplification rate of the para-sinuous disturbance increases. Whereas the para-varicose disturbance amplification decreases. An increase in the inner radius has the completely opposite effect (Fig. 10.18). We can demonstrate with the basic flow expression that an increase in r_w is accompanied by a thinning of the shear layer in the outer gas and thickening of the shear layer in the inner gas core. An increase in r_i has the opposite effect on the interfacial shear strain rate. Hence, both Figures 10.17 and 10.18 are the manifestation of the destabilizing effect of interfacial shear noted earlier by Yih (1967), Hooper and Boyd (1983), Hinch (1984), Kelly et al. (1989), Tilley, Davis, and Bankoff (1994), Coward and Renardy (1996), and many others.

Figure 10.19 shows the effect of the gas density variation on the amplification rates. The growth rates of both modes decrease as the gas density is increased. This trend is opposite to that found by Shen and Li (1996) who neglected the effects of the gas viscosity. The trend of the effect cannot be reversed by reducing $\bar{N}_1$ and $\bar{N}_3$. In fact when N_1 and N_3 (instead of $\bar{N}_1$ and $\bar{N}_3$) are fixed so that the effect of density is decoupled from that of viscosity,

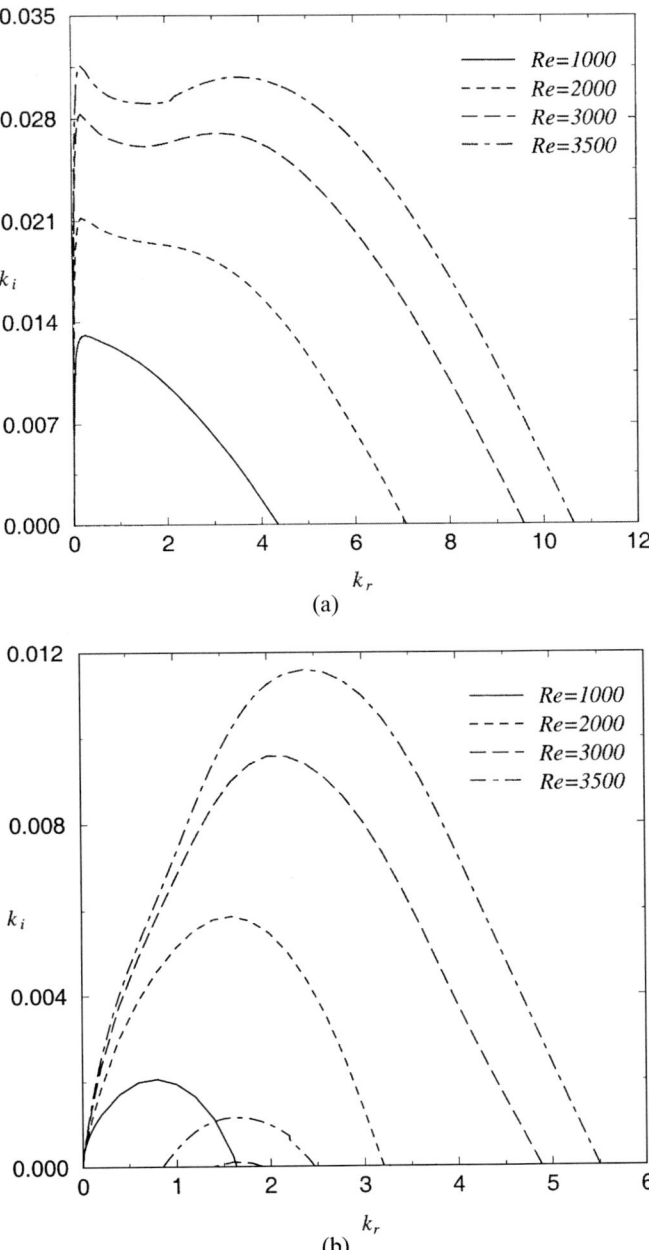

Figure 10.12. Effects of Reynolds number on the disturbance growth rate. $Fr = 50000$, $We_i = We_0 = 200000$, $\bar{N}_1 = \bar{N}_3 = 0.018$, $Q_1 = Q_3 = 0.0013$, $r_i = 2$, $r_w = 13$. (a) Para-sinuous mode. (b) Para-varicose mode.

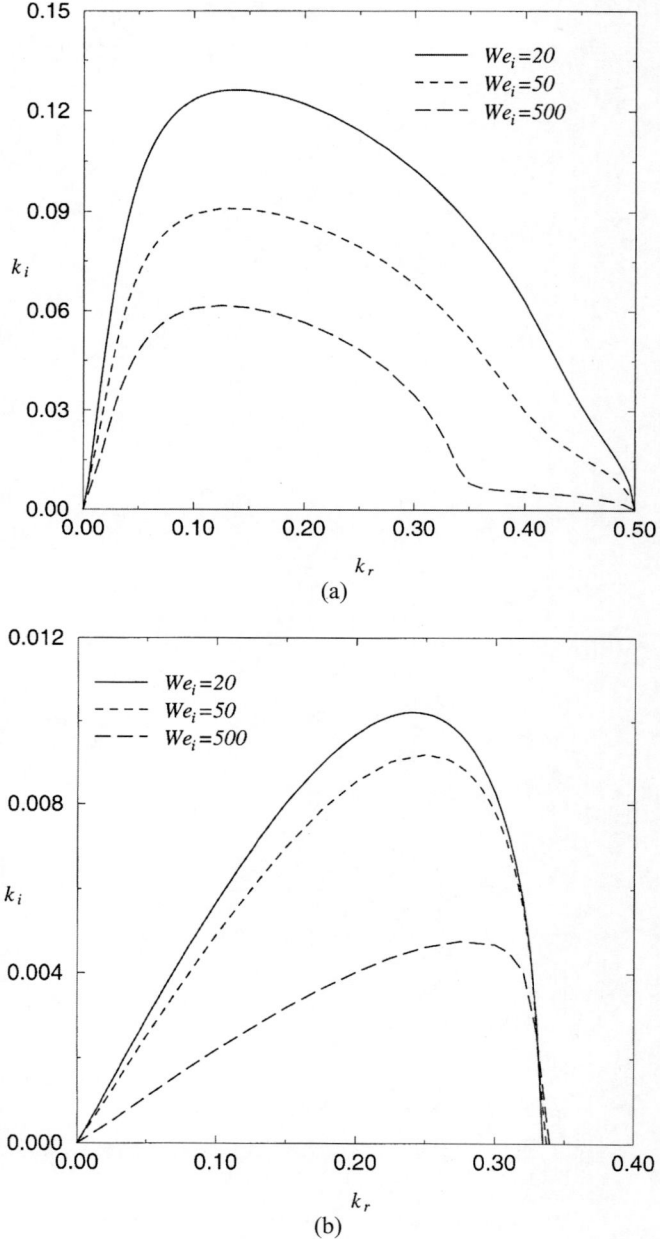

Figure 10.13. Effects of We_i on the disturbance growth rate. $Re = 1000$, $1/Fr = 0.0$, $\bar{N}_1 = \bar{N}_3 = 0.018$, $Q_1 = Q_3 = 0.0013$, $r_i = 2$, $r_w = 13$. $We_0 = 20$. (a) Para-sinuous mode. (b) Para-varicose mode.

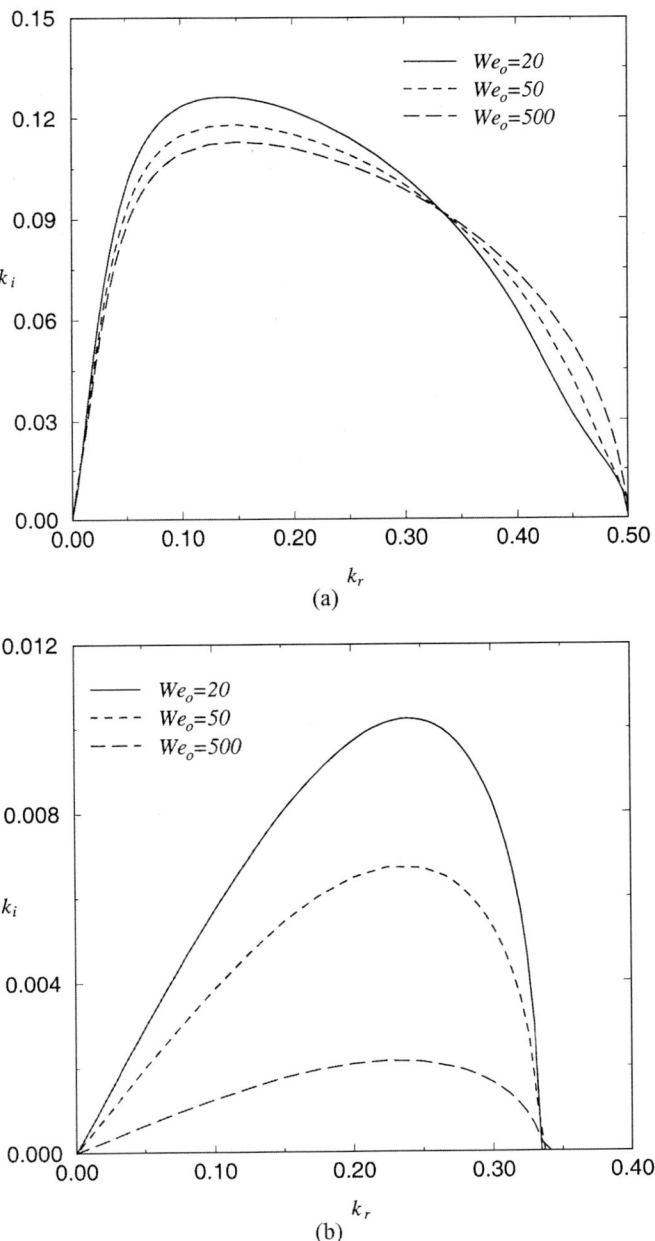

Figure 10.14. Effects of We_0 on the disturbance growth rate. $Re = 1000$, $1/Fr = 0.0$, $\bar{N}_1 = \bar{N}_3 = 0.018$, $Q_1 = Q_3 = 0.0013$, $r_i = 2$, $r_w = 13$. $We_0 = 20$. (a) Para-sinuous mode. (b) Para-varicose mode.

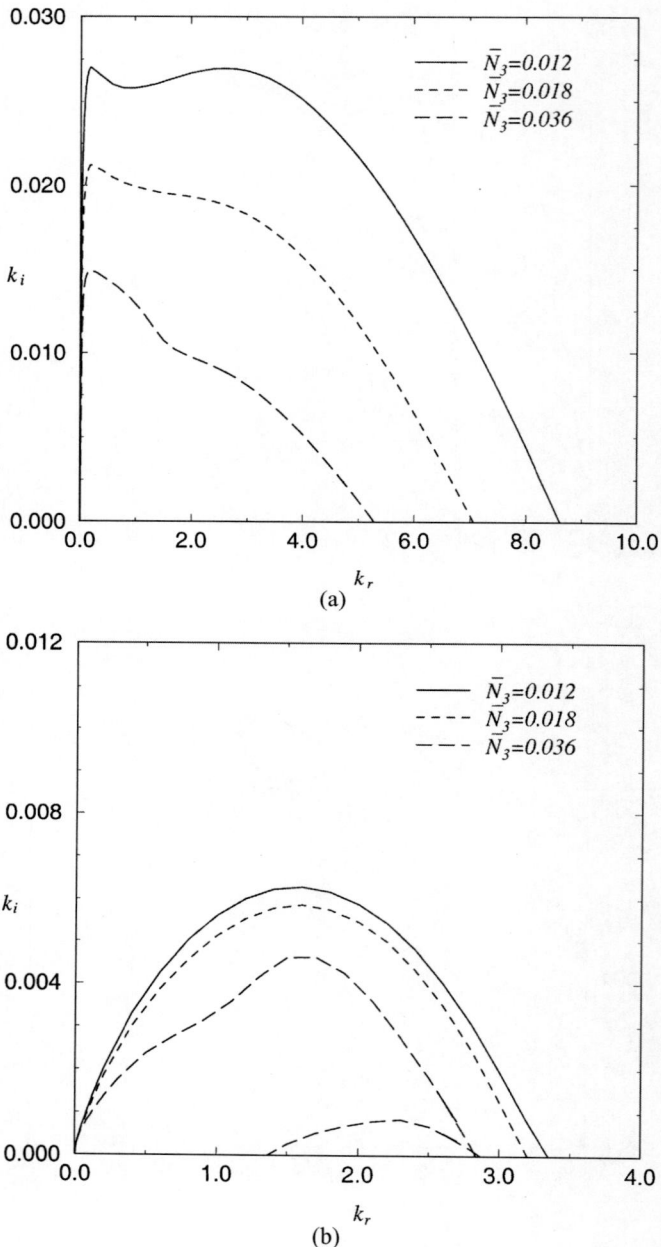

Figure 10.15. Effects of viscosity ratio $\bar{N}_3$. $Re = 2000$, $Fr = 50000$, $We_i = We_0 = 200000$, $Q_1 = Q_3 = 0.0013$, $r_i = 5$, $r_w = 16$, $\bar{N}_1 = 0.018$. (a) Para-sinuous mode. (b) Para-varicose mode.

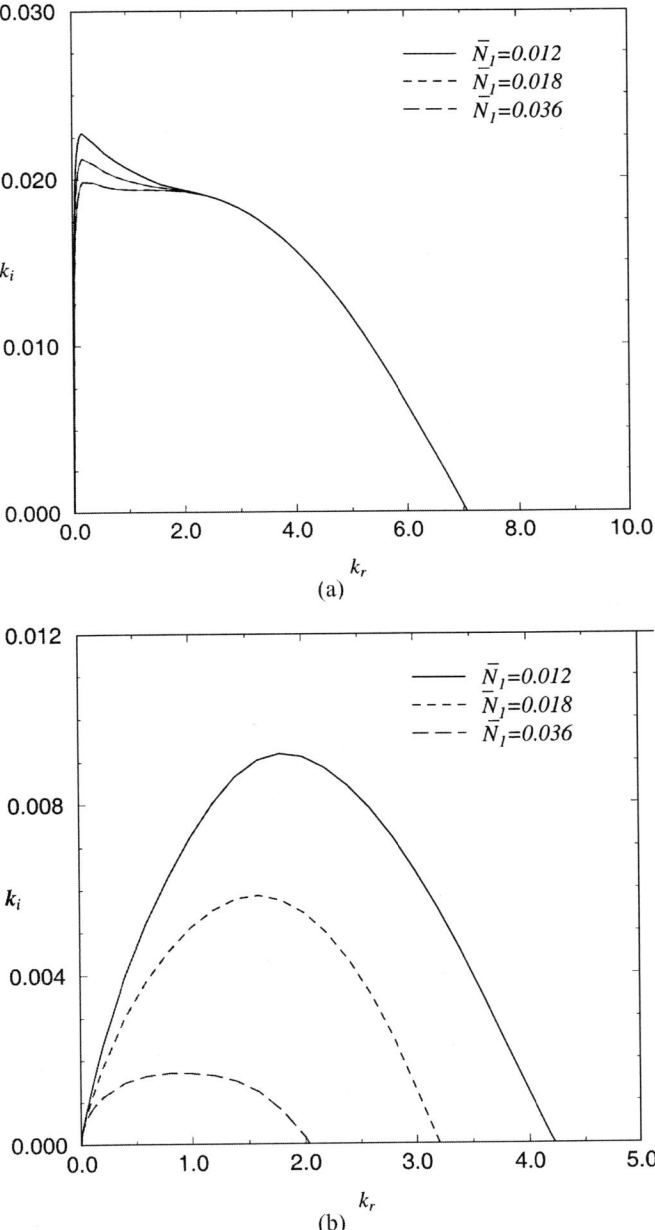

Figure 10.16. Effects of viscosity ratio $\bar{N}_1$. $Re = 2000$, $Fr = 50000$, $We_i = We_0 = 200000$, $Q_1 = Q_3 = 0.0013$, $r_i = 5$, $r_w = 16$, $\bar{N}_1 = 0.018$. (a) Para-sinuous mode. (b) Para-varicose mode.

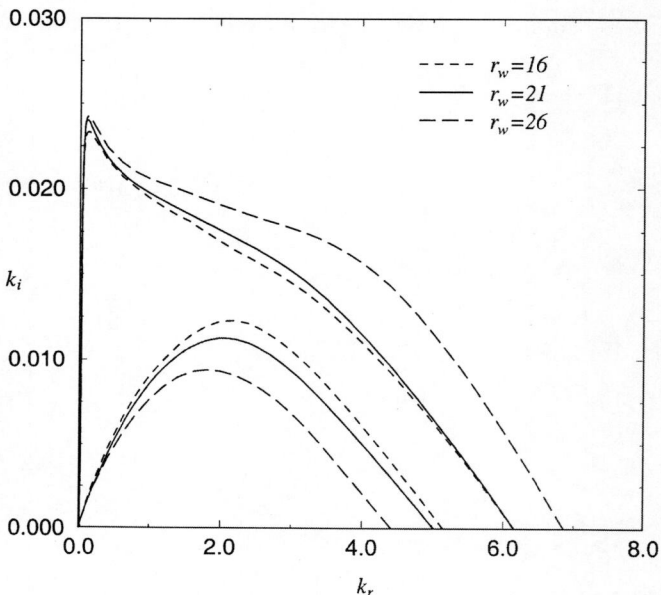

Figure 10.17. Effects of pipe radius on the disturbance growth rate for both modes. $Re = 2000$, $Fr = 50000$, $We_i = We_0 = 200000$, $\bar{N}_1 = \bar{N}_3 = 0.018$, $Q_1 = Q_3 = 0.0013$, $r_i = 10$. The upper three curves are the para-sinuous mode; the lower three curves are the para-varicose mode.

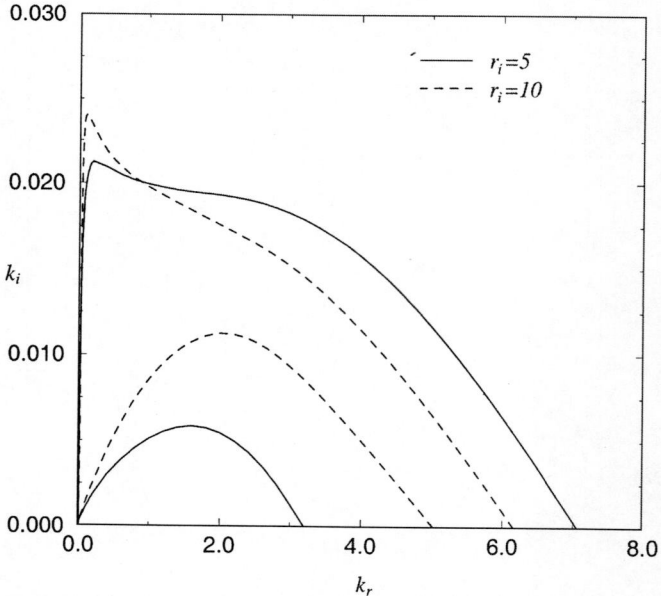

Figure 10.18. Effects of curvature on the disturbance growth rate for both modes. $Re = 2000$, $Fr = 50000$, $We_i = We_0 = 200000$, $\bar{N}_1 = \bar{N}_3 = 0.018$, $Q_1 = Q_3 = 0.0013$, $r_i = 16$. The upper two curves are the sinuous mode; the lower two curves are the varicose mode.

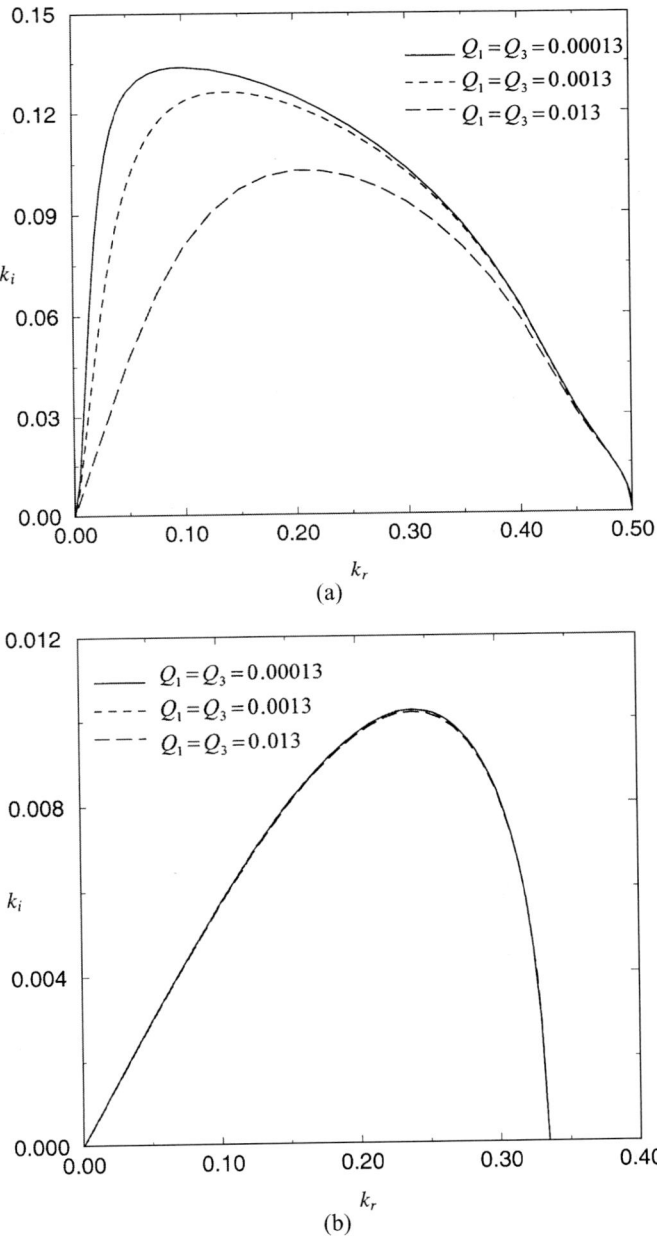

Figure 10.19. Effects of density ratio on the disturbance growth rate. $Re = 1000$, $1/Fr = 0.0$, $We_i = We_0 = 20.0$, $\bar{N}_1 = \bar{N}_3 = 0.018$, $r_i = 2$, $r_w = 13$. (a) Para-sinuous mode. (b) Para-varicose mode.

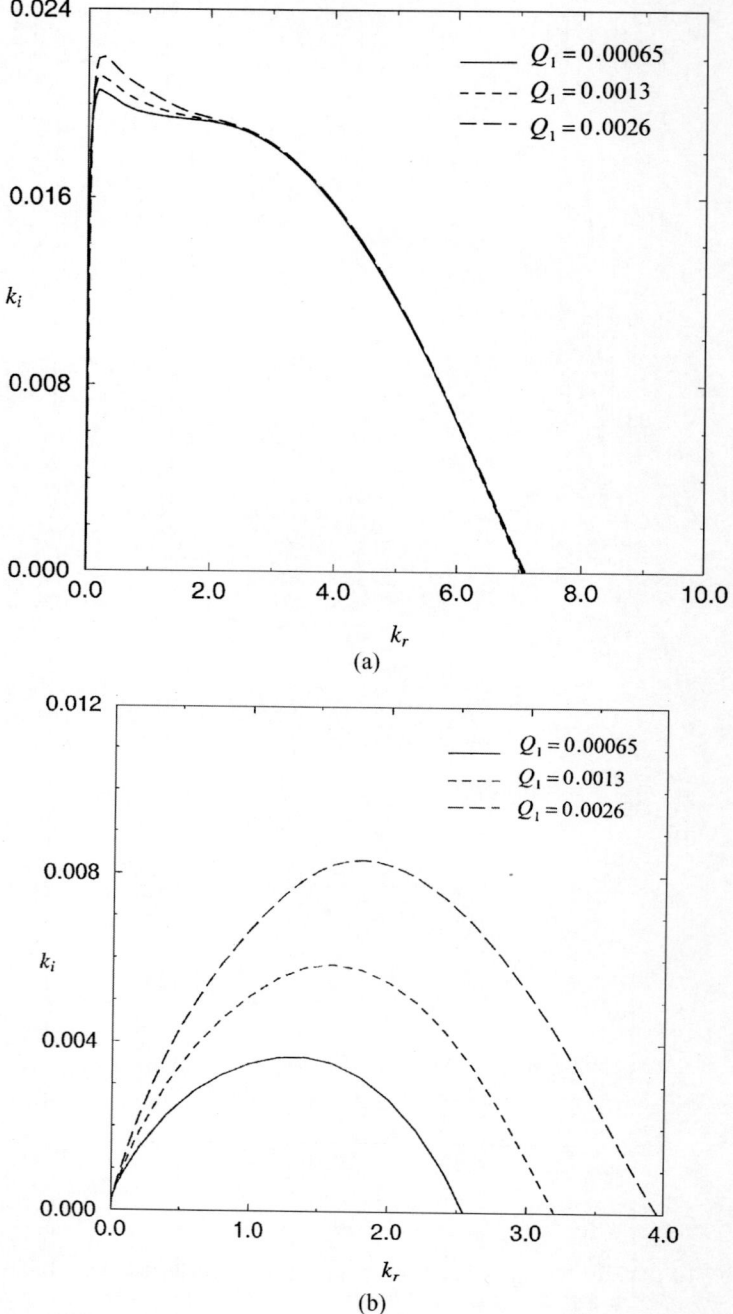

Figure 10.20. Effects of density ratio Q_1. $Re = 2000$, $Fr = 50000$, $We_i = We_0 = 200000$, $\bar{N}_1 = \bar{N}_3 = 0.018$, $r_i = 5$, $r_w = 16$, $Q_3 = 0.0013$. (a) Para-sinuous mode. (b) Para-varicose mode.

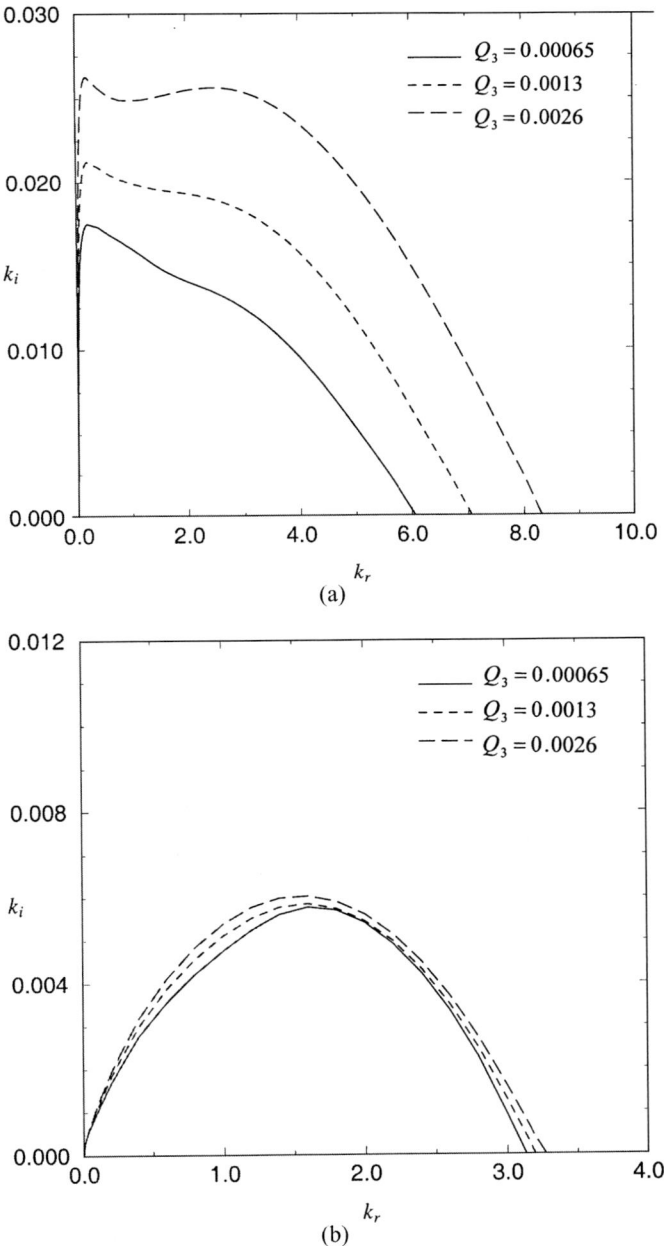

Figure 10.21. Effects of density ratio Q_3. $Re = 2000$, $Fr = 50000$, $We_i = We_0 = 200000$, $\bar{N}_1 = \bar{N}_3 = 0.018$, $r_i = 5$, $r_w = 16$, $Q_1 = 0.0013$. (a) Para-sinuous mode. (b) Para-varicose mode.

the same trend remains. The growth rate of the sinuous mode remains one order of magnitude larger than that of the varicose mode even when the gas density is increased by 100 times. Meanwhile, the amplification rate of the varicose mode is hardly changed. When the density of the outer gas is fixed, the amplification rates of both modes increase with the core gas density, although more dramatically for the varicose mode (Fig. 10.20). As the core gas density increases, the encapsulation domain becomes narrower (Fig. 10.21). Thus, a wider range of drop sizes is possible for lighter core fluids.

10.5. Summary

This chapter gave a detailed investigation of the instability of an annular liquid shell encapsulating a gas and surrounded by another gas with respect to spatial-temporal disturbances. The annulus is inherently unstable. It is susceptible to unstable para-sinuous and para-varicose modes of convective instability at the onset of instability. The amplification rate of the para-sinuous mode is generally much larger. In a certain parameter space the sinuous mode disturbance can cause the annulus to become absolutely unstable. The transition Weber number below which the annulus is absolutely unstable and above which the flow is convectively unstable is obtained as a function of the Reynolds number, with the rest of the flow parameters held constant. We saw that a successful encapsulation is possible if the process is carried out outside the parameter space of absolute instability, which causes the disturbance to propagate upstream to interrupt the continuous process. An excessive increase in interfacial tension, with the aim of obtaining larger capsules, might lead to an interruption of the process by entering into the parameter space of absolute instability. The convectively unstable sinuous mode must be promoted and the varicose mode suppressed for a possible encapsulation of a core material inside a uniform shell. The encapsulation process is impossible in certain unfavorable parameter spaces in a narrow band of wave numbers in which varicose mode instability dominates the sinuous mode. However, this range of wave numbers can be eliminated by making the proper choices of flow parameters: by reducing the level of gravity, for example. There are 10 relevant flow parameters that can be varied to create a condition under which the varicose mode is stable, but the sinuous mode is convectively unstable in a range of wave numbers. In this range, a monochromatic external forcing can be applied to amplify a selected sinuous mode disturbance, thus facilitating the production of capsules of uniform size and shell thickness.

The capillary force destabilizes sinuous as well as varicose mode disturbances of wavelengths longer than certain transition wavelengths, but it stabilizes relatively short wavelength disturbances. The transition wave

number depends on the flow parameters. Yet the short wavelength disturbances may remain convectively unstable, and therefore their instability is
caused by factors other than capillary force. These factors include interfacial
pressure and shear fluctuations, Reynolds stresses, viscous energy dissipation, and the work done by the pressure and viscous stress fluctuation along
the flow direction. The quantitative relative importance of these factors can be
determined only after a comprehensive energy budget is obtained for a wide
range of flow parameters along the lines of Lin and Chen (1998) who elucidated the physical mechanism for the case of a simple jet based on an energy
consideration. Experimental verification of the present theoretical prediction
is not yet available. The encapsulation of a core fluid with two different liquids can be investigated with the present theoretical results, but that requires
additional numerical data.

Extensions of the present analysis to the problems of micro- and nano-
encapsulations will find many applications in medical and biological sciences
(Benita, 1996). Our analysis can also be extended to investigate the problems
related to other applications mentioned at the beginning of this chapter. For
example, in the airblast atomization of a liquid rocket propellant, the core fluid
is a dense liquid oxygen and the annulus is a fast-moving hydrogen gas (Beir
and Chigier, 1972; Lasheras and Hopfinger, 2000). A similar situation exists in
a coaxial prefilming air-assist atomizer (Glathe, Wozniak, and Richter, 2001).
Certain ink jets consist of two liquid compound annular jets injected into air
(Hertz and Hermanrud, 1983). While the compound ink jet breakup is due
to capillary pinching as in the encapsulation process, the airblast atomization
process is due to interfacial shear and pressure fluctuations. It is worthwhile to
record that a particular para-sinuous wave motion in a viscous liquid annular
sheet flowing in an inviscid gas was studied by Meyer and Weihs (1987).
They obtained a critical shell thickness above which the annular jet behaved
as a simple jet. In the limiting case of an infinitely thick shell, they recovered
the air jet in water of Rayleigh, who gave the most amplified wavelength to
be $\lambda = 2\pi R_0/0.484$.

Exercises

10.1. Using the computer algorithm described in Appendixes A and B, re-
produce Figure B.1 for a simple jet.

10.2. Using the computer algorithm described in Appendixes A and B, re-
produce Figure B.1 for the circular Poiseuille flow.

10.3. For $Q_1 = 0.0013$, $Re = 2,000$, $Fr = 50,000$, $We_i = We_0 = 200,000$,
$\bar{N}_1 = \bar{N}_3 = 0.018$, $r_i = 5$, $r_w = 16$, show that the encapsulation domain
increases with increasing Q_3, and hence the gas inertial force outside

the annular shell can be raised to increase the possible size range (Chen and Lin, 2002).

References

Alexander, J. I. D. 1998. In *Free Surface Flows*. Ed. by C. Hendrik, and H. J. Rath. p. 209. Springer, New York.

Baird, M. H. I., and Davidson, J. F. 1962. *Chem. Eng. Sci.* **17**, 467–472.

Beir, J. M., and Chigier, N. A. 1972. *Combustion Aerodynamics*. Applied Science Publishers London.

Benita, S. 1996. *Microencapsulation*, Marcel Dekker.

Bers, A. 1983. *Handbook of Plasma Physics*. Vol. 1, pp. 452–516. North Holland.

Briggs, R. J. 1964. *Electron Stream Interaction with Plasma*. MIT Press.

Chen, J. N., and Lin, S. P. 2002. *J. Fluid Mech.* **450**, 235–258.

Coward, A. V., and Renardy, Y. Y. 1996. *J. Fluid Mech.* **334**, 87–109.

Crapper, G. D., Dombrowski, N., and Pyott, G. A. D. 1975. *J. Fluid Mech.* **68**, 497–502.

Davey, A., and Drazin, P. G. 1969. *J. Fluid Mech.* **36**, 209–218.

Drazin, P. G., and Reid, W. H. 1985. *Hydrodynamic Stability*. Cambridge University Press.

Glathe, A., Wozniak, G., and Richter, T. 2001. *Atomiz. Sprays* **11**, 21.

Hertz, C. H., and Hermanrud, B. 1983. *J. Fluid Mech.* **131**, 271–287.

Hinch, E. J. 1984. *J. Fluid Mech.* **144**, 463–468.

Hooper, A. P., and Boyd, W. G. C. 1983. *J. Fluid Mech.* **128**, 507–528.

Hu, H. H., and Joseph, D. D. 1989. *J. Fluid Mech.* **205**, 359–396.

Kelly, R. E., Goussis, D. A., Lin, S. P., and Hsu, F. K. 1989. *Phys. Fluids A* **1**, 819–828.

Keller, J. B., Rubinow, S. I., and Tu, Y. O. 1972. *Phys. Fluids* **16**, 2052–2055.

Kendall, J. M. 1986. *Phys. Fluids* **29**, 2086–2094.

Lanczos, C. 1956. *Applied Analysis*. Prentice Hall.

Lasheras, J. C., and Hopfinger, E. J. 2000. *Ann. Rev. Fluid Mech.* **32**, 275.

Lee, J. G., and Chen, L. D. 1991. *AIAA J.* **29**, 1589–1595.

Lee, C. P., and Wang, T. G. 1988. *J. Fluid Mech.* **188**, 411–435.

Lefebvre, A. H. 1989. *Atomization and Sprays*. Hemisphere.

Lin, S. P., and Chen, J. N. 1998. *J. Fluid Mech.* **376**, 37–51.

Meyer, J., and Weihs, D. 1987. *J. Fluid Mech.* **179**, 531–545.

Newhouse, L. A., and Pozrikidis, C. 1992. *J. Fluid Mech.* **242**, 193.

O'Donnell, B., Chen, J. N., and Lin, S. P. 2001. *Phys. Fluids* **13**, 2732–2734.

Sanz, A., and Meseguer, J. 1985. *J. Fluid Mech.* **159**, 55–68.

Shen, J., and Li, X. 1996. *Acta Mechanica* **114**, 167–183.

Tilley, B. S., Davis, S. H., and Bankoff, S. G. 1994. *Phys. Fluids* **6**, 3906–3922.

Villermaux, E. 1998. *J. Propul. Power* **14**, 807.

Yih, C. S. 1967. *J. Fluid Mech.* **67**, 337–352.

11

Nonlinear Capillary Instability of Liquid Jets and Sheets

In the previous chapters we saw that if $We < Q^{-1}$, the disturbance of wave numbers smaller than a cut-off wave number is unstable at the onset of instability. The cut-off wave number increases with Q. The capillary force is then shown with linear theories to be responsible for the onset of instability in the presence or absence of fluid viscosities. Subsequent to the onset, the amplitude of disturbances grows rapidly and the neglected nonlinear terms in the linear theory are no longer negligible. Thus the nonlinear evolution of disturbances that lead to the eventual pinching off of drops from a liquid jet can only be described with nonlinear theories. Similarly the pinching off of small droplets from the interface caused by interfacial pressure and shear fluctuations at the onset of instability, when $We > Q^{-1}$, requires nonlinear theories to describe. Experimental observations of the nonlinear phenomena are presented first.

11.1. Experiments

11.1a. Capillary Pinching

Linear theory predicts that unstable disturbances of different wavelengths grow at different rates and different natural frequencies corresponding to the different wavelengths. Figure 11.1 shows the nonlinear evolution of the disturbances when external sinusoidal forcings are introduced at three different natural frequencies. The forcing frequency for Figure 11.1(a) corresponds to $k = 0.683$, which is close to the Rayleigh's most amplified disturbance. Figures 11.1(b) and (c) correspond to the cases of lower forcing frequencies corresponding to $k = 0.25$ and $k = 0.075$, respectively. The disturbances of wavelengths shorter than that of the fastest growing disturbance appear to grow more slowly as they are convected downstream from the nozzle exit, as predicted by linear theory. However, the detail of the eventual breakup further downstream and the emergence of the bulges between the wave crest are not

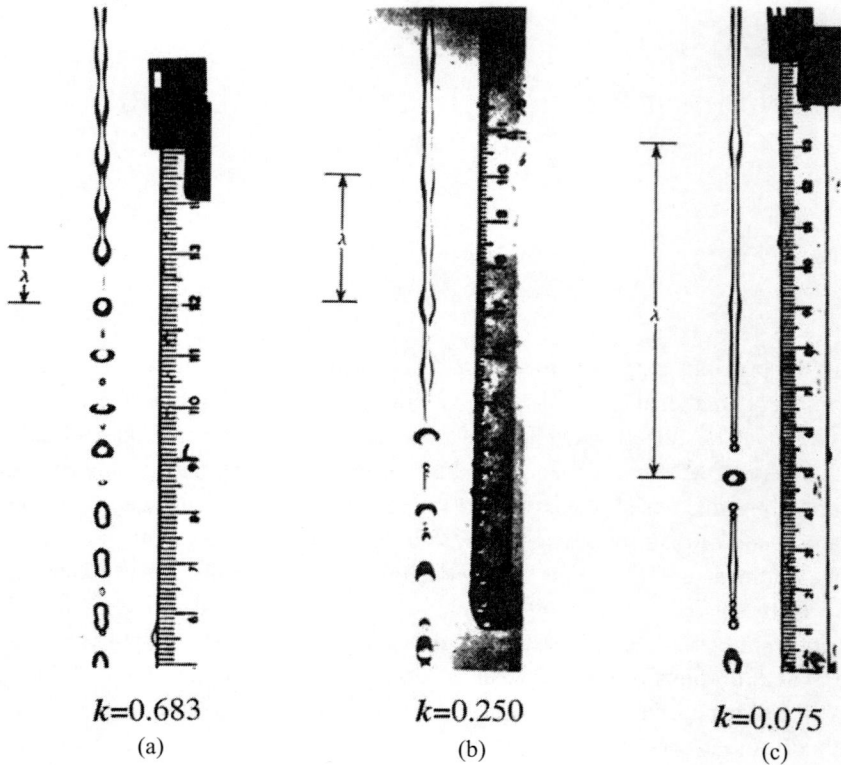

$$k=0.683 \qquad k=0.250 \qquad k=0.075$$

(a)   (b)   (c)

Figure 11.1. Photographs of unstable disturbances at three wave numbers. Secondary swellings are clearly visible between the wave peaks of the applied perturbation, whose wavelength is λ. (From Rutland and Jameson, 1970.)

predicted by linear theory. The neglected nonlinear convective acceleration term in the linear theory can induce higher harmonics through self-interaction of lower harmonics even if the introduced external forcing is of a pure harmonic. The induced higher harmonics appear to lead to the formation of satellites between the main drops.

The axial growth rates of the amplitudes of the fundamental and the first three harmonics were obtained by use of Fourier decomposition of the measured signal by Taub (1976). Comparisons of his results with the theoretical calculations based on Yuen's (1968) analysis are given in Figure 11.2. Notice that the amplitudes of higher harmonics are smaller at the same distance from the nozzle exit, probably reflecting that the energy in the higher harmonics are drained from the lower harmonics through weakly nonlinear self-interactions of harmonics. Furthermore, the amplitudes of all four harmonics approach

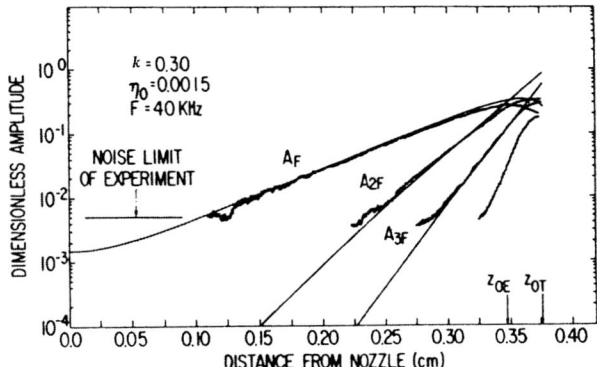

Figure 11.2. Measured amplitudes of first through fourth harmonics, and comparison with Yuen's (1968) analysis. (From Taub, 1976; K and η_0 correspond to k and ε, respectively.)

each other near the breakoff point Z_{OE}. Unfortunately no phase relation between the harmonics was measured. Nevertheless the results seem to suggest that the amplitudes and phases of the harmonics may be controlled by external forcing to allow all harmonics to peak at the same point to avoid the formation of satellites, which is facilitated by peaking of the different harmonics at different axial positions. (Chaudhary and Maxworthy, 1980a,b; Chaudhary and Redekopp, 1980).

The satellites formed from the ligament broken off from the main drops have been observed to either merge with the downstream main drops or upstream drops or travel at the same velocity as the main drops depending on the disturbance amplitude (Pimbley and Lee, 1977). Figure 11.3 divides the region into forward- and rear-merging zones in the $(\lambda/2a)$ versus t_b space, t_b being the time to breakoff after the fluid leaves the nozzle. At a given $\lambda/2a$ or equivalently at a given frequency, disturbances of smaller amplitudes will take a longer time or a longer distance downstream of the nozzle exit to pinchoff a satellite. Thus the initial amplitude along the 10λ-merge line is larger than that along the 4λ-merge line in Figure 11.3. When the disturbance amplitude is so large that the breakoff time is below the lower solid line in Figure 11.3, no satellite formation is observed. The breakup time is probably too short to allow the higher harmonics to develop fully. Above the upper solid line in the figure, the satellites have been observed to merge rearward. In this region a ligament breaks off first on the downstream side. The capillary force excited on the ligament at this end continues to decelerate the satellite after the upstream side of the ligament is broken off. This allows the upstream main drops to catch up and merge. In the region between the two solid lines, a ligament

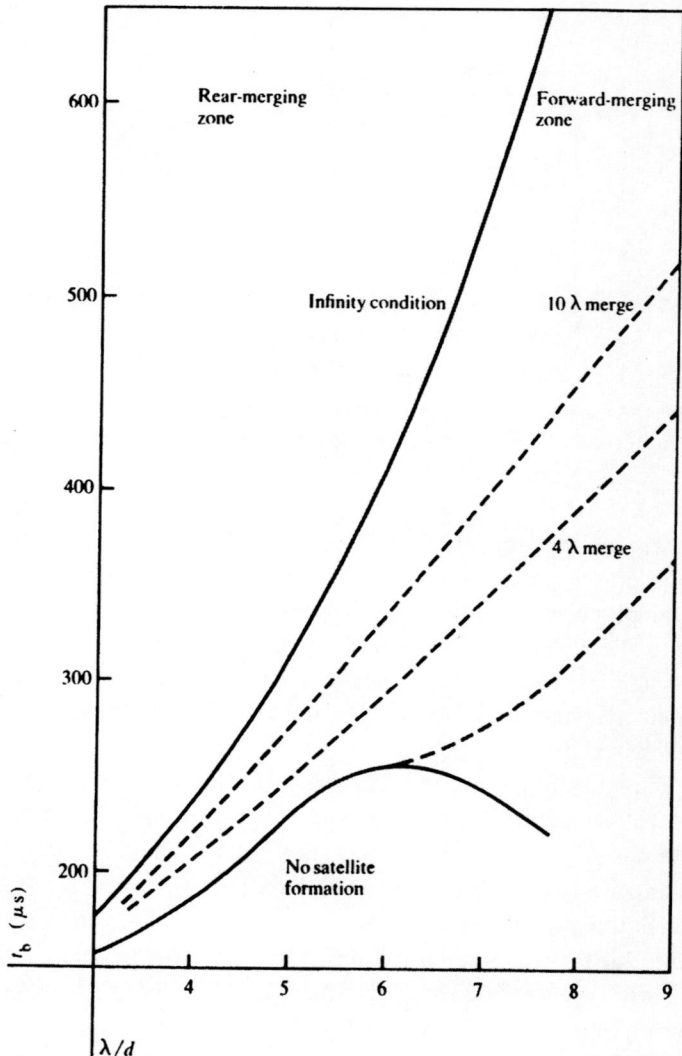

Figure 11.3. Map of the satellite condition. (From Pimbley and Lee, 1977).

breaks off first at the upstream end, and the capillary force accelerates the
satellite forward to merge with the downstream main drop. Along the upper
solid line a satellite appears to be able to maintain its identity without merging
with main drops. A nonlinear theory that describes these phenomena must
consider the initial phase and amplitude of the disturbance in a jet of known
We and *Re* even in the absence of surrounding gas.

11.1b. Atomization

The transition from drop formation to atomization from a liquid jet was described in Chapter 6. The highly nonlinear physical mechanism of atomization is very difficult to describe rigorously. The engineering applications of atomization have up to now mainly relied on phenomenological correlations with the help of linear stability analysis and limited numerical simulations. The nonlinear mechanisms of the shedding of small drops from the edge of a liquid sheet depicted in Figures 3.9 and 5.6 and the atomization of a liquid jet described in Figures 6.3 and 12.9 remain to be fully understood. A rigorous analysis of these phenomena must include the effect of interfacial shear. Referring to a swirl atomizer (Fig. 5.6), Taylor (1959) stated, "So far no mechanism has been put forward to describe the separation of the fluid contained in these edges from the continuous part of the sheet." This statement probably has not yet become an overstatement. The status can be appreciated from the review articles by Przekwas (1996), Chigier and Dumouchel (1996) and Ruff and Faeth (1966). We shall focus on the nonlinear mechanism of jet breakup.

11.2. Nonlinear Perturbation Theories

The time and length scales of the entire nonlinear process from the beginning to the end do not remain the same. The intricate dynamic response of the interface introduces additional numerical difficulties. An accurate numerical simulation of the complete nonlinear phenomenon requires extremely high temporal and spatial resolutions. Although significant progress has been made thanks to computer power improvement, the task remains a challenging one. Right after the onset of instability, the disturbance amplitude becomes finite but remains small during the initial stage of nonlinear evolution. Then the solution can be expanded in powers of small amplitude to approximately describe the initial evolution. In certain stages of the evolution, a small parameter can be frequently identified, and a perturbation solution obtained. The perturbation solutions can be sometimes extrapolated beyond their intended regions of validity. Perturbation theories will be described first, before numerical simulation.

11.2a. Potential Theories

If nonlinear behavior exhibits itself in a time shorter than the viscous diffusion time, viscosity can be neglected. Moreover if the flow is assumed to be irrotational the governing equation is

$$\nabla^2 \phi = 0,$$

where ϕ is the velocity potential related to $\mathbf{v}$ by $\mathbf{v} = \nabla \phi$. The boundary conditions are given by (1.5) and (1.7). For the solution, the initial free surface displacement and velocity must be given along the entire length of the jet. The axisymmetric solution can be expanded in power series of the small wave amplitude ε,

$$\phi = \sum_n \varepsilon^n \phi_n (r, z, \tau), \qquad h = \sum_n \varepsilon^n h_n (r, z, \tau),$$

where h is the perturbed jet radius.

Substituting the amplitude series solution into the governing system of equations and collecting the terms of the same power of ε produces a system of equations for each order of ε. The solution can then be obtained order by order. The approximate solution may be advanced sequentially until an approximate order when the solution yields the description of the nonlinear behavior under investigation. However, the perturbation solution, in general, encounters singularities that make the solution not uniformly valid. To obtain a valid solution, the variables must be renormalized properly at appropriate steps. The same method of solution has also been applied to a semi-infinite jet for which the kinematic conditions at the nozzle exit must be specified (Bogy, 1979).

The perturbation solutions for the inviscid jets have been partially successful in describing the satellite formation observed in experiments. A list of representative works in this area is given in the Chapter 11 references. The perturbation series cannot be expected to be convergent even if $\varepsilon \to 0$ asymptotically. Singular perturbation indicates that the solution cannot be constructed by a simple superposition of all Fourier components. For example, to obtain a consistent solution near the linear cut-off wave number $k_c = 1$, we need to not only introduce a new time scaling as $T = \varepsilon^2 \tau$ but also diffuse the wave number $k = 1 + \varepsilon^2 k_2 + \cdots$, so that the slow temporal and spatial modulation is allowed for consistency. Considerations of temporal and spatial modulation that cannot be taken care of by a simple superposition of the harmonics of the introduced fundamental mode allowed Nayfe (1970) to estimate the cut-off wave number for a finite, small-amplitude disturbance to be $k_c = 1 + \varepsilon^2(3/4)$. The effect is quite small, however; even for $\varepsilon = 0.1$ the increase is only less than 1%. Recall that the cut-off wave number can be raised rather significantly by the presence of gas based on the linear theory alone (Fig. 7.4).

11.2b. One-Dimensional Approximation

Near the onset of instability the axisymmetric flow in a jet is locally parallel in a sense that the flow varies slowly in the axial direction compared with the radial direction. Thus the solution may be constructed with basis functions that are endowed with this property. Such a basis function with the radial and axial component w_r and w_z is given by (Eggers, 1997)

$$\mathbf{W} = \begin{bmatrix} w_r^{(2i,\bar{z})} (R, Z) \\ w_z^{(2i,\bar{z})} (R, Z) \end{bmatrix} = \begin{bmatrix} -\dfrac{R^{2i+1}}{2i+2} \delta' (Z - \bar{Z}) \\ R^{2i} \delta (Z - \bar{Z}) \end{bmatrix}, \tag{11.1}$$

where i is a discrete index, $\bar{Z}$ denotes the axial location of the segment of a slender jet, the prime denotes differentiation with Z, and Z and R are respectively the axial and radial distance in the cylindrical coordinate system. The velocity field can then be approximated by

$$\mathbf{V} (R, Z, t) = \sum_{i=0}^{\infty} \int_{-\infty}^{\infty} V^{(2i)} (\bar{Z}, \tau) \, \mathbf{W}^{(2i\bar{Z})} (R, Z) \, d\bar{Z}. \tag{11.2}$$

Since $\mathbf{W}$ is divergence free, $\mathbf{V}$ satisfies the continuity equation for incompressible fluids.

The Galerkin projection of the solution to the Navier–Stokes equation may be accomplished on the basis of (11.1). Substituting (11.2) into the Navier–Stokes equations, multiplying the resulting equations with $\mathbf{W}$, integrating over a local volume within the free surface $R = H(Z, t)$, that is, $0 \le R \le H$, and using the free surface boundary condition (1.5) without the surface gradient term, one obtains

$$\int_V [\partial_t \mathbf{V} + (\mathbf{V} \cdot \nabla) \mathbf{V}] \cdot \mathbf{W} dV = \int_V \mathbf{g} \cdot \mathbf{W} dV$$

$$- S \int_{\partial V} K (\mathbf{n} \cdot \mathbf{W}) \, ds - 2\mu \int_V \mathbf{D} (\mathbf{V}) \mathbf{D} (\mathbf{W}) \, dV, \tag{11.3}$$

where K is the free surface curvature (see Exercise 1.4), and

$$\mathbf{D} (\mathbf{V}) = \frac{1}{2} [(\nabla \mathbf{V}) + (\nabla \mathbf{V})^{\mathrm{T}}].$$

Hence the projection onto the first $n + 1$ components of the basis function with $i = 0, \ldots, n$, results in a set of $n + 1$ equations that no longer depend on R. For example, truncating the projection at $i = 0$, we have

$$\mathbf{W}(R, Z) = \mathbf{W}^{(0, \bar{Z})} (R, Z) = \begin{bmatrix} -\dfrac{R}{2} \delta' (Z - \bar{Z}) \\ \delta (Z - \bar{Z}) \end{bmatrix}, \tag{11.4}$$

and thus

$$\mathbf{V}(R, Z, t) = \begin{bmatrix} -\dfrac{R}{2} V_0'(Z, t) \\ V_0(Z, t) \end{bmatrix}. \tag{11.5}$$

With these expressions of **W** and **V**, the integrals in (11.3) can be evaluated to yield

$$H^2 V_{0t} - \frac{1}{8} \left(H^4 V_{0t}' \right)' + H^2 (V_0 V_0') + \frac{1}{16} \left(H^4 V_0'^2 \right)' - \frac{1}{8} \left(H^4 V_0 V_0'' \right)'$$

$$= -S H^2 K' + \mu \left[3 \left(H^2 V_0' \right)' - \frac{1}{8} \left(H^4 V_0'' \right)'' \right] + H^2 g. \tag{11.6}$$

Note that we have chosen the positive Z-direction to be in the direction of gravity. Substituting (11.5) into the kinematic boundary condition gives

$$H_t + V_0 H' = -\frac{1}{2} V_0 H. \tag{11.7}$$

Without the gravity term, Equations (11.6) and (11.7) are the dimensional Cosarat equations given by Bogy (1979) from considerations of conservation of mass and linear and angular director momenta with the constitutive relations in a continuum. Bogy used the average velocity U_0 and the jet radius a to nondimensionalize the equations.

Similar but more general methods have been developed by Entov and Yarin (1984), Dewynne, Howell, and Wilmott (1994) for jets or sheets whose axes need not be along straight lines. The book of Yarin (1993) may be consulted for their applications.

Bogy (1979) carried out the perturbation solution of (11.6) and (11.7) without the gravity term. In particular he obtained the perturbation solution to the third order, with the boundary conditions

$$h(0, \tau) = 1, \, v_0(0, t) = 1 + \varepsilon \cos \omega \tau,$$

and the radiation condition that the growth rate does not approach infinity far downstream. Good agreements between theory and experiments occur at $\lambda/(2a) = 6 \, (\omega = 0.525)$ in Figure 11.2. However, the solution is singular at $\omega = 1/2$ and 1. The agreement is not as good for small values of $(\lambda/2a)$ and $(\lambda/2a) > 2\pi$. Numerical solutions of (11.6) and (11.7) have been performed with a finite difference method by Bogy, Shine, and Talke (1980) who included the viscous effect, and with a finite element method by Schulkes (1993) who compared the results of earlier workers including Lee (1974) for inviscid cases.

11.2c. Ligament Pinchoff

Just before and after the breakoff of the ligaments from the main drops, the characteristic length in the axial direction appears to be considerably larger than that in the radial direction. Therefore rescaling is required to better describe the pinchoff phenomenon. Let us scale the axial distance, radial distance, time, and velocity with l, $\varepsilon\,l$, T_0, and U_0, respectively,

$$Z = lz, \qquad H = \varepsilon lh, \qquad t = T_0\tau, \qquad V_0 = U_0v, \qquad (11.8)$$

where ε is the slenderness ratio, that is, the ratio of the radius of the ligament to its length. Substituting (11.8) into (11.6) and (11.7) and omitting the ε^2 terms, ε being much smaller than one, produces

$$\left(\frac{l}{U_0 T_0}\right) h^2 \frac{\partial v}{\partial \tau} + h^2 v v' = +\frac{gl}{U_0^2} h^2 - \left(\frac{S}{\rho U_0^2 l}\right) h^2 \kappa' + 3\left(\frac{\mu}{\rho U_0^l}\right)(h^2 v')', \tag{11.9}$$

where primes now denote differentiation with z, and κ is given by

$$\kappa = \frac{1}{\varepsilon h\,(1 + \varepsilon^2 h'^2)^{1/2}} - \varepsilon^2 \frac{h''}{(1 + \varepsilon^2 h'^2)^{3/2}}. \tag{11.10}$$

If the surface force and viscous force terms on the right side of (11.8) are to balance with each other, we must have

$$U_0 = \frac{1}{\varepsilon}\left(\frac{S}{\mu}\right), \qquad l = \varepsilon\left(\frac{\rho v^2}{S}\right). \tag{11.11}$$

If the acceleration term is to balance with the rest of the terms, then the corresponding time scale is

$$T_0 = \varepsilon^2\left(\frac{\rho^2 v^3}{S}\right). \tag{11.12}$$

Note that the characteristic variables contain only the physical properties of the fluid and the slenderness ratio ε. Thus the initial jet radius and velocity are no longer remembered by the ligament. With these scalings, (11.8) reads

$$\partial_\tau v + vv' = \varepsilon^3\,(gv^4\rho^3/S^3) - \kappa' + 3\left(v_0'\,h^2\right)'/h^2. \tag{11.13}$$

Thus the gravity term must be kept only when both g and v are extremely large. The kinematic equation (11.7) written in the same variable is

$$h_\tau + vh' = -\frac{1}{2}v'h. \tag{11.14}$$

Equations (11.13) and (11.14) are precisely the same first-order set of one-dimensional equations obtained by Eggers (1997) who expanded the solution in power of the radial distance.

Strictly speaking the above one-dimensional equations obtained from the lubrication theory approach (sometimes called long wave expansion or slender body theory depending on the application) are applicable only when the flow varies more rapidly in a direction perpendicular to the flow direction. Therefore the region near the pinchoff point between the ligament and the main drop cannot be expected to be described adequately by (11.13) and (11.14), where the interfacial slopes on both sides of the pinchoff point are not small. Using a finite difference method, Eggers and Dupont (1994) solved (11.13) and (11.14), with ε^2 terms in the curvature fully included, to simulate the experiment of Chaudhary and Maxworthy (1980a,b). The comparison between the theory and the experiment is good almost up to the formation of the pinch point. However, the detachment of the ligament from a main drop as observed in the experiment was not reproduced. Note that the slope on the drop-neck junction is not small on the drop side.

Just before pinchoff, a neck between a ligament and a main drop eventually becomes so thin that the continuum theory is no longer valid. The breakup of a nanometer scale jet will be discussed in the last section of this chapter.

11.3. Numerical Simulation

The numerical simulation of an inviscid liquid jet in an inviscid gas and of a viscous Newtonian jet in a vacuum are contrasted in this section. The numerical simulations based on the Laplace equation and the Stokes equation for related problems are given in the next chapter. In the present discussions, we will emphasize the physical concept extracted from the simulation results. Technical details of numerical simulation will be referred to in the known works in the references at the end of this chapter.

11.3a. An Inviscid Jet in an Inviscid Gas

The volume integration of the Laplace equation of the velocity potential ϕ over the entire volume of the irrotational flow within a given flow boundary Γ can be converted to a surface integral by means of the divergence theorem to yield (Ligget and Liu, 1983)

$$\beta\phi\left(\mathbf{r}\right) + \int_{\Gamma} (\phi\partial_n G - G\partial_n\phi)\, d\Gamma = 0, \qquad (11.15)$$

where $\mathbf{r}$ is the position vector of a point on the flow boundary, G is the free space Green's function of the Laplace equation, ∂_n denotes the differentiation in the outward unit normal direction at the flow boundary, and β is the solid angle of the singularity at the nodal points of the boundary elements (BE) covering the surfaces of the flow field. Adjacent boundary elements share common sides of an area. The end points of each side are the nodal points of the net covering the entire flow boundary. For an axisymmetric flow, the Green function and its normal derivative in the cylindrical coordinate (r, θ, z) are given by

$$G = 4rK(-p)/\sqrt{\beta},$$

$$\partial_n G = \frac{-1}{2\sqrt{\beta}} \left[n_r K(p) + \frac{E(p)}{(r - r_i)^2 + (z - z_i)} \right.$$
$$\left. \times \left\{ \left[(r - r_i)^2 - (z - z_i)^2 \right] n_r + 2rn_z (z - z_i) \right\} \right],$$

where

$$\beta = (r + r_i)^2 + (z - z_i)^2,$$

$$p = \frac{(r - r_i)^2 + (z - z_i)^2}{(r + r_i)^2 + (z - z_i)^2},$$

and $K(p)$ and $E(p)$ are respectively the complete elliptic integrals of the first and second kind. G and $\partial_n G$ represent, respectively, the influence of the source of strength 2π and the doublet at a point (r_i, z_i) on the boundary. Equation (11.15) is applicable to each flow domain of liquid and gas.

The spatial variation of ϕ in each element can be approximated locally by an appropriate function (a linear function, for example). The evaluation of (11.15) for all elements results in a system of algebraic equations of the form

$$D_{ij}\phi_j = S_{ij}q_j \qquad (i, j = 1, 2, \ldots, N), \qquad (11.16)$$

where N is the total number of nodes, and $q = \partial_n \phi$. To start the solution, the interface of the originally stationary liquid is displaced from $r = 1$ to $r = 1 + h\cos(kz)$. The uniform gas flow originally parallel to the jet is assumed to be unperturbed far from the jet axis in the radial direction. The flow is also assumed to be periodic in the axial direction. Then the initial gas pressure P_g and gas velocity potential are given respectively by (Sears, 1960).

$$P_g = -Qhk\cos(kz)\frac{K_0(kr)}{K_1(k)}, \qquad (11.17)$$

$$\phi_g = z - h\sinh(kz)\frac{K_0(kr)}{K_1(k)},$$

Where K_0 and K_1 are respectively the modified Bessel function of the zeroth and first orders. The subsequent time evolution must satisfy the dynamic boundary condition

$$\partial_\tau \phi + \frac{1}{2}(\nabla\phi)^2 + P_g + \frac{\kappa}{We} = 0, \tag{11.18}$$

$$Q\partial_\tau \phi_g + \frac{1}{2}(\nabla\phi_g)^2 + P_g = 0,$$

where ϕ without the subscript g now stands for the liquid phase. κ is the curvature of the interface assigned to the liquid. Thus the interface moves with the liquid particles on the liquid surface. If $D/D\tau$ denotes the time rate of change following the surface fluid particle, (11.18) can be written as

$$\frac{D\phi}{D\tau} = \frac{1}{2}(\nabla\phi)^2 - P_g - \frac{\kappa}{We}, \tag{11.19}$$

$$P_g = -Q\left[\frac{1}{2}(\nabla\phi)^2 + \frac{D\phi_g}{D\tau} - \nabla\phi\nabla\phi_g\right].$$

The motion of the surface particle must satisfy the kinematic condition

$$\frac{Dr}{D\tau} = \partial_r\phi, \qquad \frac{Dz}{D\tau} = \partial_z\phi. \tag{11.20}$$

The details of the numerical procedure for obtaining numerical results from (11.16)-(11.20) can be found in Spangler, Hilbing, and Heister (1995).

Figure 11.4 shows an example of the time evolution of a disturbance of a fixed wavelength, a given initial radial displacement amplitude of $\eta = 0.004$, and a gas perturbation amplitude $\varepsilon = 0.00129\eta$ in the flow parameter range where the gas inertia force starts to become important according to linear theory (first wind-induced regime). The formation of the precursor of the satellite drop is clearly demonstrated. The nonlinear evolution in a liquid jet with the Weber number considerably larger than Q is given in Figure 11.5. In this so-called second wind-induced regime, the surface tension is stabilizing for a disturbance of $k = 4.23$ according to the linear theory. (See Chapter 7.) The foundation of an axisymmetric disk protruding from the liquid into the gas and the formation of a small ring at the tip of the disk are seen in the figure. The breakup of this ring into small droplets may be presumed to form atomized droplets. The viscosity of fluids is neglected in the simulation. Had the viscosity been included, the shear stress exerted by air at the interface would have bent the disk toward the direction of relative velocity of the air to the liquid. Moreover the viscous force would work in concert with the pressure fluctuation at the interface to promote atomization as suggested by linear theory.

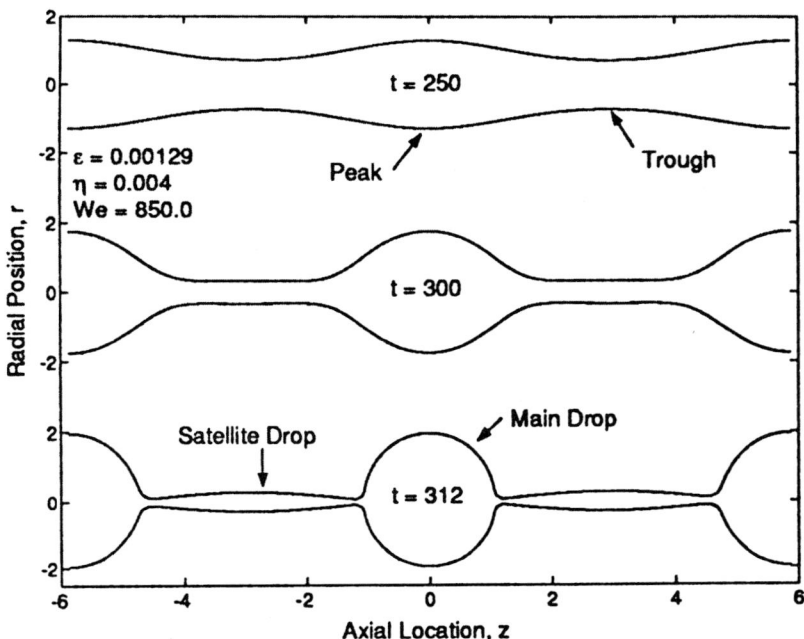

Figure 11.4. Nonlinear jet evolution in the first wind-induced regime. $Q = 0.00129$, $k = 1.07$, $\eta = 0.004$, $We = 850$.

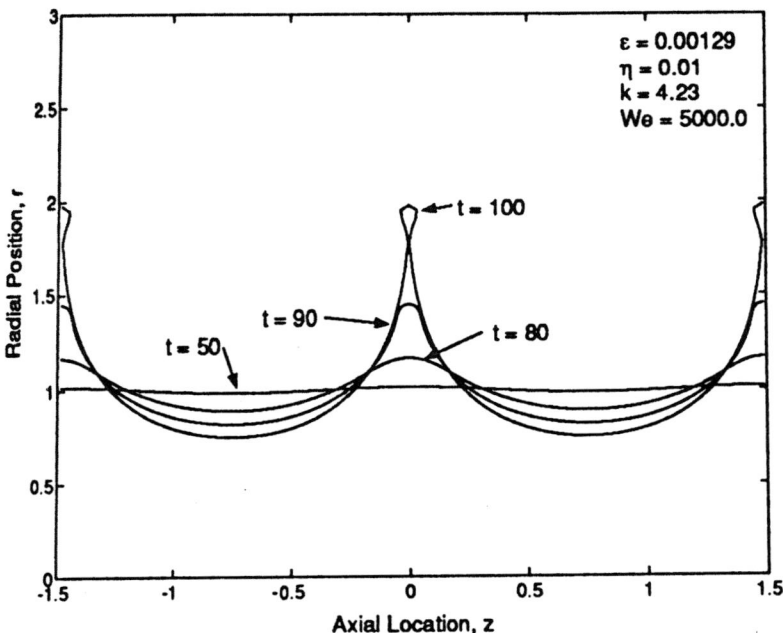

Figure 11.5. Nonlinear jet evolution in the second wind-induced regime. $Q = 0.00129$, $\eta = 0.01$, $k = 4.23$, $We = 5000$.

11.3b. Navier–Stokes Simulation

The numerical simulation of practically important flow processes including liquid jet breakup often involves a rapidly developing area. References are given at the end of this chapter to indicate the diverse methods and accomplishments. An example given here provides a reference point for future progress. The most commonly used numerical methods for the Navier–Stokes equations are probably the finite difference, finite volume, and finite element methods. Typical results obtained with a finite element method are given here to demonstrate what has been achieved and what challenges lie ahead.

Consider the breakup of an incompressible Newtonian liquid jet in the absence of surrounding gas. To make the numerical computation tractable, we assume the disturbance is axisymmetric and periodic in the axial direction. Thus the axial extent of the computational domain can be reduced to within $0 \le z \le \lambda/2$, where λ is the dimensionless wavelength $\lambda = 2\pi a/k$, the distance being normalized with the jet radius a. If we assume that the fluid volume within the computational domain remains constant and the flow has an aft-fore symmetry at the end points of one wavelength, then additional boundary conditions

$$w = 0, \qquad \partial_z u = 0 \quad \text{at } z = 0, \lambda/2$$

must be imposed, where w and u are, respectively, the axial and radial velocity components normalized by $(S/\rho a)^{1/2}$. This normalization (Ashgriz and Mashayek, 1995) implies that the characteristic time of the disturbance evolution is $(\rho a^3/S)^{1/2}$ and the original unperturbed jet is stationary. Hence the Reynolds number appearing in previous chapters is now replaced by the Ohnesorge number $Oh = (aS/\rho v^2)^{1/2}$. In the finite element method of numerical simulation of the axisymmetric flow, only the flow domain in $0 \le z \le \lambda/2$ and $0 \le r < h$ needs to be considered on a plane passing through the z-axis at any azimuthal location. The flow domain is then covered with a finite number of sufficiently small elements that share common sides with their neighboring elements inside the free surface. The velocity field in each element is approximated by the sum of appropriately chosen shape functions $N_i(z, r, \tau)$

$$\mathbf{u} = \sum_{i=1}^{M} \mathbf{u}_i(\tau) N_i(z, r, \tau), \tag{11.21}$$

where M is the number of corners or sides of each element. For example, for a four-cornered element $M = 4$, if the liquid jet is cut into small slices along the jet axis and finite elements are imbedded into each slice, then the rotation of each element about the axis forms a finite element of volume. The nodal amplitude functions $\mathbf{u}_i(\tau)$ are determined to satisfy, on the average in each element, the governing equations using the shape functions as the weighting functions. Multiplying the Navier–Stokes equations with N_j, integrating over each finite element volume, and applying the divergence theorem to the term involving the stress tensor τ yields

$$Oh \int_{V(t)} N_j \frac{D\mathbf{u}}{Dt} dV + \int_{V(t)} (\nabla N_j)^{\mathrm{T}} \cdot \tau dV = \int_{\partial V} N_j \tau \cdot \mathbf{n} dS, \quad (11.22)$$

where $V(t)$ and ∂V are the volume and the enclosing surface of each volume element. Along the free surface, we can apply the dynamic boundary condition to reduce the surface integral on the right side of (11.22) to

$$\int_{\partial V(t)} Oh N_j \kappa \mathbf{n} dS.$$

The pressure term in (11.22) requires special attention for incompressible fluids (Aidun and Lin, 1986, 1987). Ashgriz and Mashayek use the assumption $p = -\gamma \nabla \cdot \mathbf{u}$ where $\gamma = O(10^9)$. They also divide the jet into slices with fixed end positions along the axis. The nodes of each element in each slice are allowed to move radially according to $r_i(\tau + \delta\tau) = c r_i(\tau)$, where $c = h_i(z, \tau + \delta\tau)/h_i(z, \tau)$. Thus $D\mathbf{u}/D\tau$ in (11.22) is given by

$$\frac{D\mathbf{u}}{Dt} = \partial_\tau \mathbf{u} + W \partial_z \mathbf{u} + \left(u - \frac{c-1}{\delta t} r\right) \partial_r \mathbf{u}. \quad (11.23)$$

The position of the free surface at each time step is determined by requiring the volume flux across each slice to be equal to the negative of the volume swept by the free surface in the same time interval. The outcome of this finite element approximation is a system of algebraic equations equal in number to the number of the unknown nodal amplitude functions. The system can be solved at each time starting from a given initial free surface geometry.

Figure 11.6 gives some representative results obtained with this simulation method. The nonlinear evolution of a disturbance with a fixed wave number $k = 0.2$ and an initially sinusoidal free surface displacement of amplitude $\eta = 0.05$ is depicted in Figure 11.6(a) for the case of $Oh = 200$ and in Figure 11.6(b) for the case of $Oh = 0.1$. The nonlinear evolution for both cases results in the production of thin ligaments as the satellite precursor. The ligament is thinner and takes longer for the more viscous jet to evolve to the stage of ligament detachment. Note that while the neck at the pinchoff point has a cone-cap shape for the case of the less viscous jet, the neck shape is a cone-thread type for the more viscous jet. Simulations for shorter wavelengths

(a)   (b)

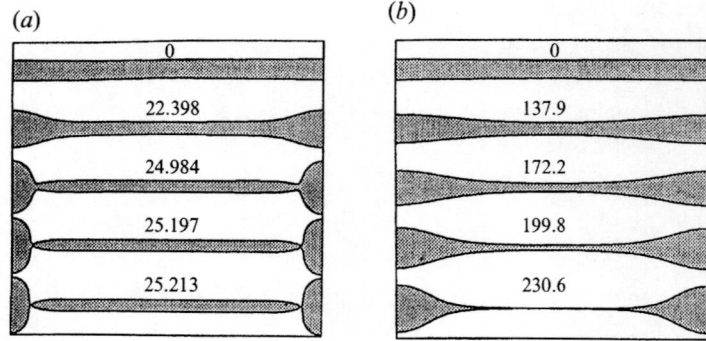

Figure 11.6. Time evolution of the instability of a capillary liquid jet. $\varepsilon_0 = 0.05$, the numbers on the figures indicate the corresponding times. (a) $k = 0.2$, $Oh = 200$. (b) $k = 0.2$, $Oh = 0.1$.

depict the same behavior. More will be said about this in the context of a nanojet in the next section.

In contrast to the inviscid jet in the previous section, the jet velocity does not appear in this Navier–Stokes simulation. Nonlinear evolution is played out by the competition between the surface tension and the jet viscosity. The additional results of Ashgriz and Mashayek show that the surface tension continues to overpower the viscous resistance to break the jet in vacuum. The cut-off wave number of instability is slightly larger than one, in contrast with the prediction by the perturbation theory of Nayfeh (1970). On the other hand the jet velocity relative to the ambient gas appears explicitly in the previous section. It was shown that if the surface tension force is greater than the inertial force, the interfacial force, from the onset of instability to eventual formation of the ligament, is responsible for the jet breakup. Note that the neck at the pinch point for the inviscid case is of cone-cap shape (compare Fig. 11.4 with Fig. 11.6(b)). However, if the inertial force is larger than the interfacial force, liquid sheets protruding into the gas are formed at the interface as shown in Figure 11.5. This cannot be predicted with the model in Section 11.3b. The formation of the protruding liquid sheet is initiated by the pressure fluctuation in the gas with or without viscosity according to the linear theory in Chapter 8. In both examples of simulation, a periodic boundary condition along the jet axis is imposed. Thus only the temporal nonlinear evolution of a disturbance of a given wavelength can be examined. The spatial nonlinear evolution of amplitude and wavelength of disturbances in a viscous jet in a viscous gas remains a challenging task. If we start the simulation of the jet breakup from the introduction of the liquid flow into the nozzle to the formation of the jet and to its subsequent breakup beyond the intact length, then we have no need to assume the periodic boundary condition. The spatial evolution of the amplitude and wavelength consisting of Fourier components of continuously varying wavelength will emerge from the simulation. An example of this approach in the context of a nanojet will be given next.

11.4. Nanojets

When the neck thickness decreases to nanometer scale in the last stage of jet breakup, or when the jet is issued from a nozzle of nanometer scale, the details of the dynamics depend crucially on the molecular interaction. Continuum theory has crossed the border of validity at this stage. Therefore molecular dynamic simulation (MD) is needed for elucidation of the dynamic of the final breakoff process.

When the neck thickness or the jet diameter itself is of order of the thermal capillary length l_T defined by $l_T = (k_B T/S)^{1/2}$, k_B being the Boltzman constant, the stress associated with the molecular thermal fluctuation is expected to be the same order as the surface tension. To make a connection between MD and continuum theories, Moseler and Landman (2000) added a thermal fluctuation stress $-\partial_z f/\pi h^2 \rho$ to the dimensional form of (11.9), where f is a Gaussian white noise variable given by

$$\langle f(z, t) f(z', t') \rangle = 6 k_B T \mu \pi h^2 \delta (z - z') \delta (t - t'),$$

according to the fluctuation dissipation theorem (Fox and Uhlenbeck, 1970; Landau and Lifshitz, 1984). Thus, the one-dimensional equation (11.13) with the additional stochastic stress, when nondimensionalized in the same way (11.13) was, reads

$$\partial_t v + v \partial_z v = -\partial_z \kappa + 3\partial_z \left(h^2 \partial_z v \right) / h^2 - M \sqrt{6/\partial_z} (hN)/h^2, \quad (11.24)$$

where $M = l_T/l$, and (11.14) remains applicable. Equation (11.24) is referred to as the stochastic lubrication equation (SLE) and (11.13) as the lubrication equation (LE). The results of numerical simulation based on SLE and LE by Moseler and Landman for a propane jet issued from a 6-nm-diameter gold nozzle are given in Figure 11.7. They also give results obtained with a large-scale MD simulation. They used a propane molecule model of Jorgenson, Madura, and Swenson (1984), supplemented by angle-bending potentials (Ploeg and Berensen, 1982). For the gold nozzle atoms, they used a many-bodied embedded-atom interaction (Oh and Johnson, 1988). The interaction potentials between the propane and the metal atoms were adopted from Xia et al. (1992). Their MD simulation results are also given in Figure 11.7

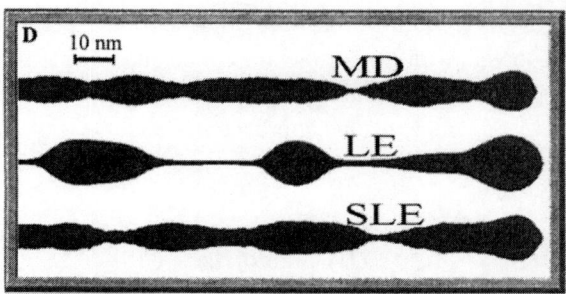

Figure 11.7. Results pertaining to breakup events of the propane jet for a 6-nm-diameter nonwetting nozzle were obtained by the following: an atomistic MD simulation, a simulation using the deterministic LE formulation, and a simulation using the SLE. Double-cone neck shapes are exhibited in the MD and SLE simulations, in contrast to the long thread obtained with the LE simulation.

for comparison. For propane at $210\ K$, $l_T = 0.56$ nm and $l = 2.3$ nm. A thin thread of diameter approximately 1 nm and length more than 10 times as long are produced by LE, which is internally consistent within (11.13) and (11.14) even if the continuum LE is stretched beyond its limit. However, the MD simulation shows that the long thin neck between the two cones cannot be maintained because of thermal fluctuation, which has a length scale l_T close to the neck thickness as well as to the viscous-capillary length l. The neck structure produced by SLE appears to be closer to the MD results than to the LE results. This is probably because the molecular fluctuation is incorporated statistically in the SLE model.

For glycerol at $300\ K$, $l_T = 0.26$ nm and $l = 3.9$ cm $\gg l_T$. Thus the cone-thread-cone structure observed in the previous sections may evolve for several decades before the neck thickness reaches the order of l_T. Then the double-cone neck may result in the nanoscale. For water at $300\ K$, $l_T = 0.24$ nm and $l = 13.7$ nm. In comparison with the glycerol jet, l_T and l are quite close for water. Thus, in the last stage of the evolution of the water jet breakup in its own vapor, after the main ligament is detached from the main drop, there may exist a thin nanoscale thread between the ligament and the main drops. The breakup of the nanometer water thread will probably quickly evolve to a structure resembling that indicated by MD in Figure 11.7.

The MD simulation also discloses an interesting nonlinear behavior of convective instability exhibited in Figure 11.8. Formation of droplets and molecular clusters is observed at an initial transient stage ($t \leq 1$ ns), reaching a steady flow at 200 m/s at $t \sim 1$ ns. Molecular evaporation and formation of necking instabilities are observed in breakup events and drop formation. In

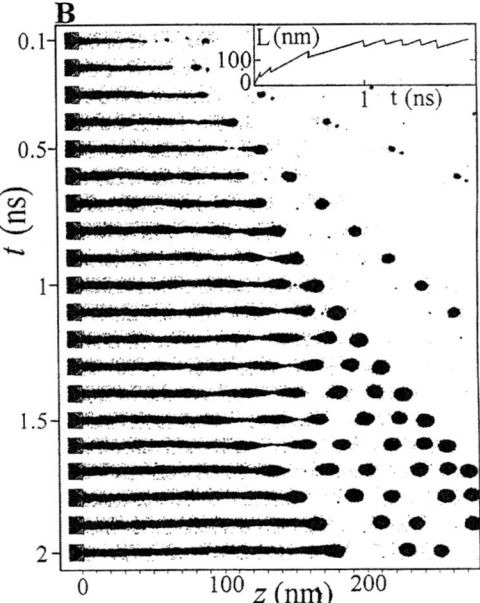

Figure 11.8. Convective instability of a nanojet.

the inset, the evolution of the intact length L is shown. Each sawtooth-shaped discontinuity corresponds to a pinchoff of a drop.

The nonlinear convective instability depicted in Figure 11.8 cannot be simulated with the method described in previous sections.

11.5. Nonlinear Instability of Liquid Sheets

Nonlinear instability of planer liquid sheets in the presence of ambient gas was briefly mentioned near the end of Section 3.1 where attempts were made to compare theories with experiments with limited success.

In the absence of ambient gas, Mehring and Sirigano (1999) investigated the linear and nonlinear wave motion in an inviscid sheet. The sheet thickness considered was so thin compared to the wavelength that the solution could be expanded in powers of distance from the sheet centerline to a fluid particle in the sheet. This allowed them to reduce the problem to one spatial dimension as discussed in Section 11.2b. In fact (11.6) can also be derived by expanding the solution in powers of radial distance (Eggers, 1997). Among many nonlinear wave phenomena, Mehring and Sirigano found from numerical solution of the one-dimensional equation as well as from the two-dimensional numerical simulation by use of a discrete-vortex method that nonlinear sinuous waves in a semi-infinite sheet with $We > 1$ can propagate only in the downstream direction. Thus the criteria of transition from convective to absolute instability established in Section 2.3 (see Exercises 2.6 and 2.7) with linear theory are upheld by the nonlinear theory. For the varicose wave, they found frequency-dependent critical Weber numbers below which the finite amplitude wave decays as it propagates downstream. On the other hand varicose linear waves decay for all Weber numbers as they are convected downstream (see Exercise 2.8) in the absence of ambient gas. They also demonstrated how finite amplitude waves can be maintained by external harmonic excitation even when the wave is linearly stable. This finding supports the linear concept of wave generation discussed in Chapter 5. However, they showed that two finite amplitude varicose waves can be modulated to make the sheet unstable. The modulational instability was shown earlier by Matsuuchi (1974, 1976) with weakly nonlinear analyses.

Mehring and Sirigano also investigated wave motion in an inviscid annular sheet in inviscid gas. They demonstrated how an introduced sinuous finite amplitude wave can induce varicose waves as time proceeds. It remains to be verified whether this is true in the wave number range in which the sinuous wave amplifies but the varicose wave decays according to the linear viscous theory discussed in Chapter 10.

The results obtained for inviscid sheets in a vacuum cannot be easily compared with experiments (Meier, Klöpper, and Grabitz, 1992).

References

Aidun, C. K., and Lin, S. P. 1986. *Proc. 8th International Heat Transfer Conference*, p. 373. Elsevier.

Aidun, C. K., and Lin, S. P. 1987. *Int. J. Heat Mass Transfer* **30**, 408.

Ashgriz, N., and Mashayek, F. 1995. *J. Fluid Mech.* **291**, 163.

Bogy, D. B. 1979. *Ann. Rev. Fluid Mech.* **11**, 207.

Bogy, D. B., Shine, S. J., and Talke, F. E. 1980. *J. Comput. Phys.* **38**, 294.

Chaudhary, K. C., and Maxworthy, T. 1980a. *J. Fluid Mech.* **96**, 275.

Chaudhary, K. C., and Maxworthy, T. 1980b. *J. Fluid Mech.* **96**, 287.

Chaudhary, K. C., and Redekopp, L. G. 1980. *J. Fluid Mech.* **96**, 257.

Chigier, N., and Dumouchel, 1996. In Recent Advances in Spray Combustions, Spray Atomization and Drop Burning Phenomena, Vol. 1 (Ed. K. K. Kuo). pp. 214–259.

Dewynne, J. N., Howell, P. D., and Wilmott, P. 1994. *Q. J. Mech. Appl. Math.* **47**, 541.

Eggers. J., 1993. *Phys. Rev. Lett.* **71**, 3458.

Eggers, J., and Dupont, T. F. 1994. *J. Fluid Mech.* **262**, 205.

Eggers, J. 1997. *Rev. Mod. Phys.* **69**, 865.

Entov, V. M., and Yarin, A. L. 1984. *J. Fluid Mech.* **140**, 91.

Fox, R. F., and Uhlenbeck, G. 1970. *Phys. Fluids* **13**, 1893.

Gipson, G. S. 1987. *Boundary Element Fundamentals, Topics in Engineering*, Vol. 2 (Eds. C. A. Brebbia and J. J. Conner). Computational Mechanics, Boston.

Jorgenson, W. L, Madura, J. D., Swenson, C. J. 1984. *J. Am. Chem. Soc.* **106**, 6638.

Landau, L. D., and Lifshitz, E. M. 1984. *Fluid Mechanics* Pergamon, Oxford.

Lee, H. C. 1974. Drop formation in a liquid jet. *IBM J. Res. Dev.* **18**, 364.

Liggett, J. A., and Liu, P. L. F. 1983. *The Boundary Integral Equation Method for Porous Media Flow.* Allen and Unwin.

Matsuuchi, K. 1974. *J. Phys. Soc. Japan* **37**, 1680.

Matsuuchi, K. 1976. *J. Phys. Soc. Japan* **41**, 1410.

Mehring, C., and Sirigano, W. A. 1999. *J. Fluid Mech.* **388**, 69.

Mehring, C., and Sirigano, W. A. 2000. *Phys. Fluids* **12**, 1417.

Meier, G. E. A., Klöpper, A., and Grabitz, G. 1992. *Exper. Fluids* **12**, 173.

Moseler, M., and Landman, U. 2000. *Science* **289**, 1165.

Nayfe, A. H. 1970. *Phys. Fluids* **14**, 841.

Oh, D. J., and Johnson, R. A. 1988. *J. Mater. Res.* **3**, 471.

Orme, M., and Muntz, E. P. 1990. The manipulation of capillary stream breakup using amplitude-modulated disturbances: A pictorial and quantitative representation. *Phys. Fluids* A **2**, 1124.

Pimbley, W. T., and Lee, H. C. 1977. *IBM J. Res. Dev.* **21**, 21.

Ploeg, P., and Berensen, H. J. C. 1982. *J. Chem. Phys.* **76**, 3271.

Prztkwas, A. J. 1996. In Recent Advances in Spray Combustions. Spray Atomization and Drop Burning Phenomena, Vol. 1. (Ed. K. K. Kuo). 211–239.

Ruff G. A., and Faeth, G. M. 1996. In Recent Advances in Spray Combustion. Spray Atomization and Drop Burning Phenomena, Vol. 1 (Ed. K. K. Kuo), pp. 263–296.

Rutland, D. F., and Jameson, G. J. 1970. *J. Fluid Mech.* **46**, 267.

Schulkes, R. M. S. M. 1993, *Phys. Fluids.* **5**, 2121–2130.

Sears, W. R. 1960. *Small Perturbation Theory.* Princeton University Press.

Spangler, C. A., Hilbing, J. H., and Heister, S. D. 1995. *Phys. Fluids* **7**, 964.

Taub, H. H., 1976. *Phys. Fluids* **19**, 1124.

Taylor, G. I. 1959. *Proc. R. Soc. Lond.* A **253**, 289.

Xia, T. K., Ouyang, J., Ribarsky, M. W., and Landman, U. 1992. *Phys. Rev. Lett.* **69**, 1967.

Yarin, A. L., 1993. *Free Liquid Jets and Films: Hydrodynamics and Rheology.* John Wiley & Sons.

Yuen, M. C., 1968. *J. Fluid Mech.* **33**, 151.

Additional References on Simulation of Jet Breakup

Bousfield, D., Stockel, I. H., and Nanivadekar, C. K. 1990. The breakup of viscous jets with large velocity modulations. *J. Fluid Mech.* **218**, 601–617.

Cline, H. E., and Anthony, T. R. 1978. The effects of harmonics on the capillary instability of liquid jets. *J. Appl. Phys.* **49**, 3203–3208.

Crane, L., Birch, S., and McCormack, P. D. 1964. The effect of mechanical vibration on the breakup of a cylindrical water jet in air. *Brit. J. Appl. Phys.* **15**, 743–750.

Fenn, III, R. W., and Middleman, S. 1969. Newtonian jet instability: The role of air resistance. *AIChE J.* **15**, 379.

Helenbrook, B. T. 2001. A two fluid spectral method. *Comp. Meth. Appl. Mech. Eng.* **191**, 273.

Hughes, T. J. R., Liu, W. K., and Brooks, A. 1979. Finite element analysis of incompressible viscous flows by penalty function formulation. *J. Comput. Phys.* **30**, 1–60.

Kakutani, T., Inoue, Y., and Kan, T. 1974. Nonlinear capillary waves on the surface of liquid column. *J. Phys. Soc. Japan* **37**, 529–538.
Lafrance, P. 1970. Nonlinear breakup of a laminar liquid jet. *Phys. Fluids* **18**, 428.
Lafrance, P., 1975. *Phys. Fluids*. **18**, 428.
Mansour, N. N., and Lundgren, T. S. 1990. Satellite formation in capillary jet breakup. *Phys. Fluids*. A **2**, 1141.
Rutland, D. F., and Jameson, G. J. 1970. *Chem. Eng. Sci.* **25**, 1689.

Work Related to Potential Flow Surface Integral Method

Baker, G. R., Meiron, D. I., and Orszag, S. A. 1980. *Phys. Fluids*. **23**, 1485.
Longuet-Higgins, M. S., and Cokelet, E. D. 1976. *Proc. R. Soc. Lond.* A **350**, 1.
Mansour, N. N., and Lundgren, T. S. 1990. *Phys. Fluids* A **2**, 1141.
Oguz, H. N., and Prosperetti, A. 1990. *J. Fluid Mech.* **219**, 143.

Review Articles on Breakup of Liquid Jets and Sheets

Chigier, N. 1996. Atomization of liquid sheets. In *Recent Advances in Spray Combustion* (Ed. K. K. Kuo). Vol. 1, p. 241. AIAA.
Eggers, J. 1997. Nonlinear dynamics and breakup of free surface flows. *Rev. Modern Physi.* **69**, 865.
Lin, S. P., and Reitz, R. D. 1998. Drop and spray formation from a liquid jet. *Ann. Rev. Fluid Mech.* **30**, 85.
Przekwas, A. J. Theorectical modeling of liquid jet and sheet breakup processes. In *Recent Advances in Spray Combustion* (Ed. K. K. Kuo) Vol. 1, p. 211. AIAA.

12

Epilogue

In the previous chapters, we investigated the fairly well studied phenomena of breakup of liquid sheets and liquid jets. The basic flows were assumed to be steady in the continuum theories. Also, they were either of infinite or of semi-infinite extent in the flow direction. Physically such infinite and semi-infinite steady jets or sheets cannot exist, as predicted by stability analysis. The analytical predictions enjoyed fairly good agreement with many known experiments. However, breakup of a liquid body into smaller parts often takes place under an unsteady situation from the beginning. The examples include the formation of satellites and subsatellites from the ligaments after detaching themselves from the main drops, the formation of drops from a dripping faucet, shaped-charge jets, the formation of micro-drops by external forcing, intermittent fuel sprays, and the phenomenon of jet branching induced by external excitation. These are the subjects to be touched upon in this last chapter.

12.1. Satellite Formation

When a stretched liquid ligament is relaxed, the capillary force associated with the large surface curvature at both ends of the ligament tends to compress and fragment the ligament into small drops. We saw the formation of the ligament during the last stage of nonlinear evolution of instability. The stretching of a liquid ligament submerged in another fluid can be achieved by pure straining or shearing or a combination of both. Figure 12.1 (Stone et al., 1986) shows how a spherical drop is stretched in two purely straining external flows with two different viscosities. N in this figure is the ratio of the dynamic viscosity of the external fluid to that of the liquid drop. The lower half of this figure shows how the drops and satellite are formed when the external flows are stopped. The mechanism of liquid ligament fragmentation appears to be capillary pinching in accordance with the capillary wave generated by the recoil force from the both ends.

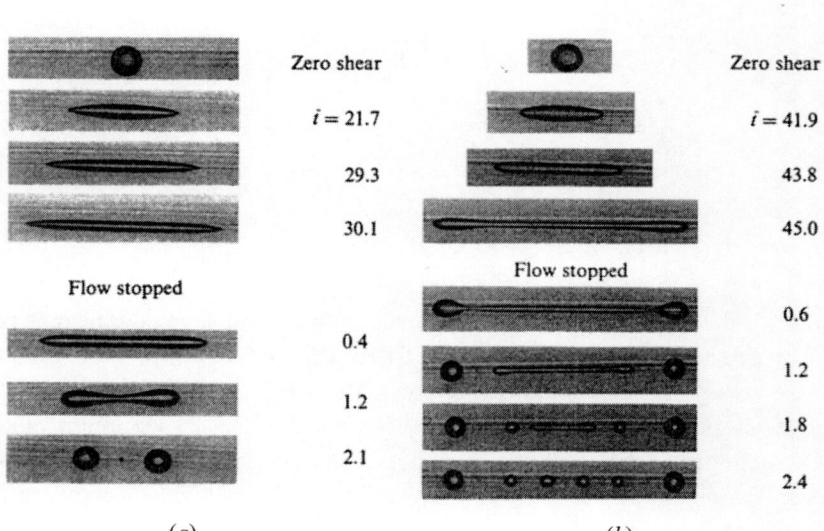

Figure 12.1. The effect of viscosity ratio on the elongation and breakup phenomena.
(a) $N^{-1} = 0.01$. (b) $N^{-1} = 0.046$.

Tjahjadi, Stone, and Ottino (1992) investigated both experimentally and numerically the breakup of a liquid ligament stretched in another fluid between two differentially rotating cylinders. The numerical simulation is based on the Stokes equation

$$\nabla^2 \mathbf{v}_i = \nabla p_i, \qquad \nabla \cdot \mathbf{v} = 0 \qquad (i = 1, 2), \qquad (12.1)$$

where $i = 1$ and 2 stands respectively for the ligament fluid and the surrounding fluid. In (12.1) the length is normalized with the filament mean radius a, time with $a\mu_2/S$, internal pressure with S/a, and external pressure with S/aN. The fluid filament is assumed to deform in an otherwise quiescent external fluid. Thus

$$\mathbf{v}_2(\mathbf{r}) \to 0 \quad \text{as } \mathbf{r} \to \infty.$$

The interfacial kinematic conditions at $\mathbf{r} = \mathbf{r}_s$ are

$$\mathbf{v}_1 = \mathbf{v}_2,$$

$$\frac{D\mathbf{r}_s}{D\tau} = \mathbf{n}(\mathbf{v}_1 \cdot \mathbf{n}),$$

where $\mathbf{r}_s$ is the position vector at the interface. The dynamic boundary condition is

$$\mathbf{n} \cdot \sigma_2 - N^{-1}\sigma_1 = S\mathbf{n}\,(\nabla_s \cdot \mathbf{n}),$$

where $\mathbf{n}$ is a unit normal vector pointing from the filament fluid to the suspending fluid. Green's function for the Stokes equation exists (Ladyzhenskaya, 1969), and thus the solution for the interfacial velocity $\mathbf{v}(\mathbf{r}_s)$ can be expressed as (Rallison and Acrivos, 1978)

$$\frac{1}{2}\,(1 + N^{-1})\,\mathbf{v}_s\,(\mathbf{r}_s) + (1 - N^{-1})\int_s \mathbf{n} \cdot \mathbf{K} \cdot \mathbf{v}_s ds\,(\mathbf{y})$$

$$= -\int_s \mathbf{n} \cdot J\,(\nabla_s \cdot \mathbf{n})\,ds\,(\mathbf{y}), \qquad (12.2)$$

where $\mathbf{y}$ is the integration variable over the interface s and

$$J = \frac{1}{8\pi}\left[\frac{I}{\mathbf{r} - \mathbf{y}} + \frac{(\mathbf{r} - \mathbf{y})(\mathbf{r} - \mathbf{y})}{|\mathbf{r} - \mathbf{y}|^3}\right],$$

$$K = -\frac{3}{4\pi}\frac{(\mathbf{r} - \mathbf{y})(\mathbf{r} - \mathbf{y})(\mathbf{r} - \mathbf{y})}{|\mathbf{r} - \mathbf{y}|^3}.$$

The filament is assumed to be axisymmetric. The azimuthal integration can be evaluated analytically to reduce these surface integrals to line integrals. The details of numerical procedures of solution by this boundary integral method can be found in Tjahjadi et al. (1992) and Pozrikidis (1992). Figure 12.2 gives representative results of the numerical calculation. Corresponding experimental results are also put side by side in the figure to show the excellent comparisons. Breakup was assumed to take place in the simulation. Note that the viscosity of the external fluid is almost 15 times more viscous than the filament fluid. The reverse is true in jet breakup in a gas as discussed in Chapter 11. The number of satellite droplets and their relative sizes depend strongly on the viscosity ratio and are influenced, to a lesser degree, by the initial disturbance wave number. In general, the less viscous the filament is relative to the external fluid, the less able the filament is to resist capillary pinching, which generates several generations of thinner ligaments and smaller satellites. The terminal state of the pinching process obtained with simulation based on the Stokes equation has been confirmed by one-dimensional asymptotic theory (Papageorgiou, 1995).

Many kinds of deformation of simple and compound drops in an evolving extensional flow with finite Reynolds number have been simulated by Shyy et al. (1999) using the Navier–Stokes equations. The breakup of a viscous thread in another viscous fluid was investigated by Tomotika (1935) through linear stability analysis.

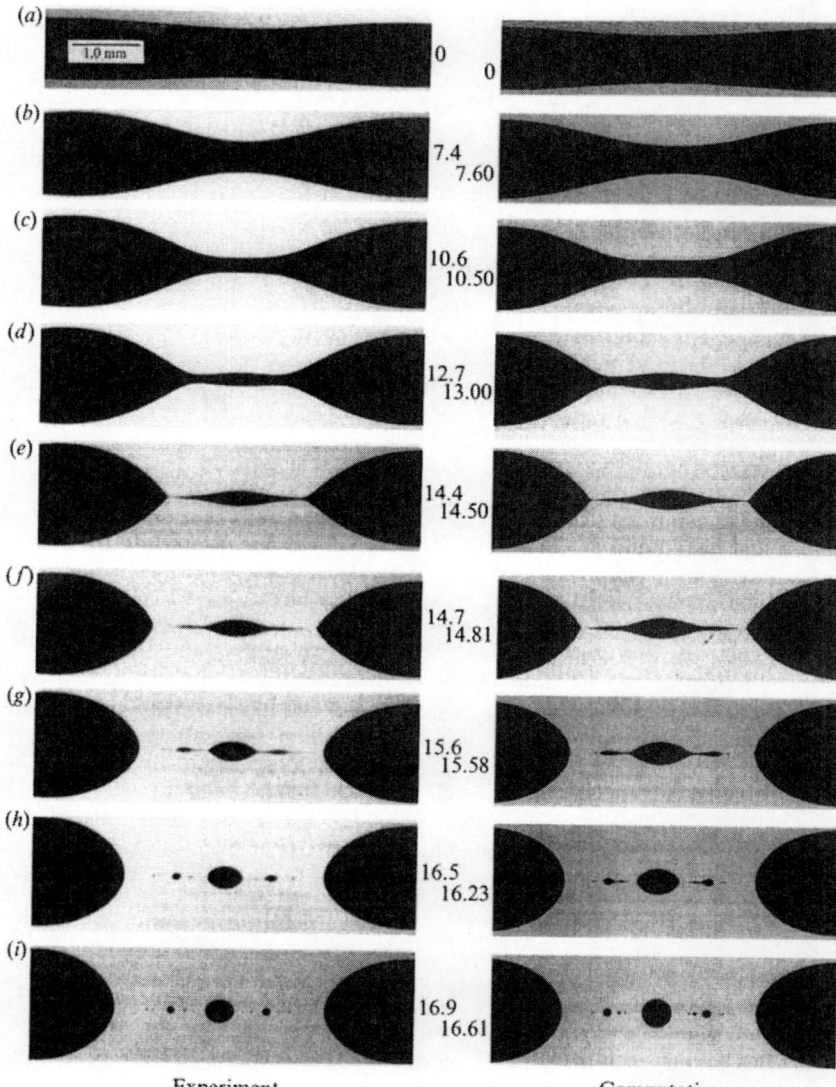

(a)

1.0 mm

0

0

(b)

7.4
7.60

(c)

10.6
10.50

(d)

12.7
13.00

(e)

14.4
14.50

(f)

14.7
14.81

(g)

15.6
15.58

(h)

16.5
16.23

(i)

16.9
16.61

Experiment   Computation

Figure 12.2. Time evolution leading to multiple pinchoffs of a segment of a highly extended fluid filament. The left-hand column shows the experimental results; the right-hand column, the computational simulation. N^{-1} and wave number are 0.067 and 0.50 respectively. The dimensionless times shown correspond to experiment (top) and computation (bottom). (From Tjahjadi et al., 1992).

12.2. Breakoff of a Drop

Consider a pendant drop at the end of a nozzle, under the action of gravity. The pendant remains attached if the volume is sufficiently small. When the volume exceeds a certain value, the pendant surface is unable to balance the body force, and the pendant suddenly accelerates. A neck is quickly formed between the nozzle and the main pendant drop. This gives the surface tension an opportunity to constrict the neck. As the neck diameter grows smaller, capillary pinching becomes stronger. Eventually a drop under the narrowest part of the neck is pinched off. The ligament front separated from the drop at the bifurcation point retract toward the nozzle, while a new bifurcation point forms at the upper part of the ligament.

12.2a. Potential Theory

The formation of the drop is caused by surface tension, although the formation is initiated by gravity. The characteristic time of the drop formation is $T_s = (\rho a^3/S)^{1/2}$, where a is the jet radius. For water at 20°C, T_s is 1.6×10^{-2} sec if $a = 0.26$ cm, which is the nozzle radius in the experiment of Peregrine, Shoker, and Symon (1990). This value of T_s as well as those for other fluids are given in Table 12.1 for comparison. The time scale T_v for diffusing momentum through viscosity across the same characteristic length $a = 0.26$ cm is 6.7 sec, which is two orders of magnitude larger than the capillary time T_s. For mercury the difference is even greater. However, for a more viscous fluid such as glycerol $T_s > T_v$. Therefore for the less viscous fluids the capillary pinching process may be completed before the viscosity can diffuse the momentum to effectively interfere with the capillary pinching. Hence in the first-order approximation, the viscosity may be neglected.

The flow that starts from zero velocity in the liquid above a flat surface at the nozzle remains approximately irrotational. The governing equation is

Table 12.1. *Characteristic Variable of Viscous Threads*

	Mercury	Water	Glycerol
ν [cm²/s]	0.0012	0.01	11.8
$\gamma = S/\rho$ [cm³/s²]	34.7	72.9	50.3
$l_s = (\gamma/g)^{1/2}$ [cm]	0.188	0.273	0.226
$l = (\nu^2/\gamma)$ [cm]	4.2×10^{-8}	1.38×10^{-6}	2.79
$t_v = \nu^3/\gamma^2$ [s]	1.4×10^{-12}	1.91×10^{-10}	0.652
$T_v = R^2/\nu$ [s]	56.33	6.76	0.0057
$T_s = (R^3/\gamma)^{1/2}$ [s]	2.3×10^{-2}	1.6×10^{-2}	1.9×10^{-2}

the Laplace equation as in Section 11.3a. The kinematic boundary condition remains the same as in (11.20), except that time and velocity are, respectively, normalized here with T_s and $(\gamma/a)^{1/2}$. In place of (11.19), the dynamic boundary condition reads

$$\frac{D\phi}{D\tau} = -\frac{1}{2}(\nabla\phi)^2 + P_0 - B_0 z - \kappa, \tag{12.3}$$

where P_0 is a reference pressure, z is the axial distance measured in the opposite direction of gravitational acceleration g, and $B_0 = \rho g a^2/S$ is the Bond number. The dynamics of the problem is characterized by the Bond number and the volumetric flow rate

$$\dot{Q} = -2\pi \int (\partial_z \phi)_{z=0} \, r \, dr = \frac{\pi}{3} W e^{1/2}. \tag{12.4}$$

Initially at $\tau = 0$, the whole system is at rest so that $P_0 = -B_0$, and the free

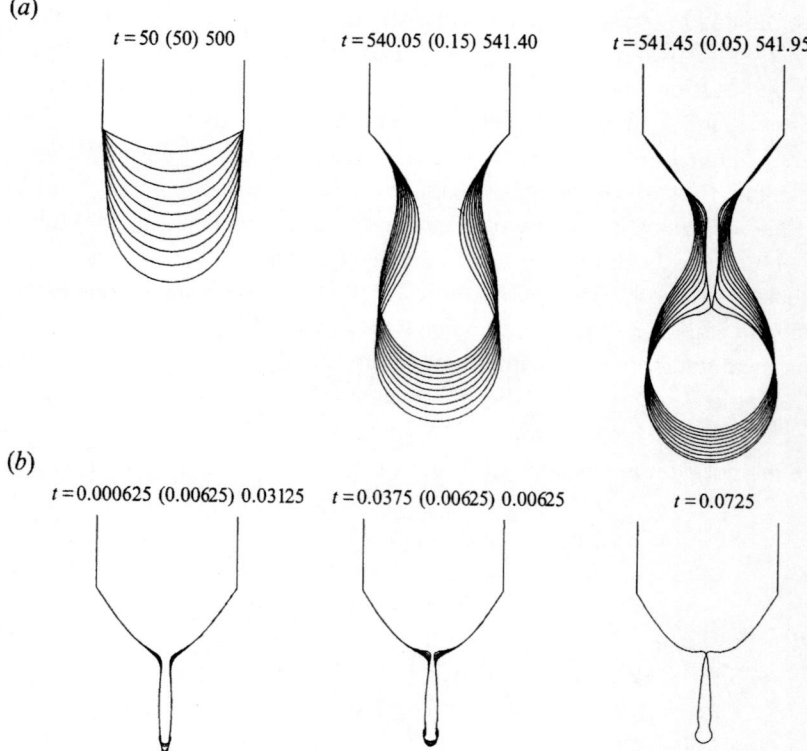

Figure 12.3. Plots of the free-surface shape of the pendant drop during different stages of the evolution. $B_0 = 1$, $We = 9.12 \times 10^{-5}$. (From Schulkes, 1994). (a) Before the bifurcation. (b) After the bifurcation.

surface is flat at the nozzle outlet located at $z = -1$. Along the wall of the nozzle of unit length the radial velocity is zero.

Schulkes (1994) solved the problem by using a boundary element method similar to that described in Chapter 11.3a. Figure 12.3 depicts the slow initial growth of the pendant drop, the instability of the pendant once its volume exceeds a limit, and the formation of a ligament just before the pinchoff of a suspending drop. Note that once the pendant exceeds a certain volume, it takes time of order one of T_s to reach the pinchoff stage.

By means of the boundary element method, the pinchoff region is shown by Day, Hinch, and Lister (1998) to shrink as fast as $t^{2/3}$, where t is the remaining time until pinchoff. They also found that, independent of the initial condition, the similarity solution near the pinchoff point exhibits interfacial slope reversal (Fig. 12.4). Similar slope reversal has been found by Chen and

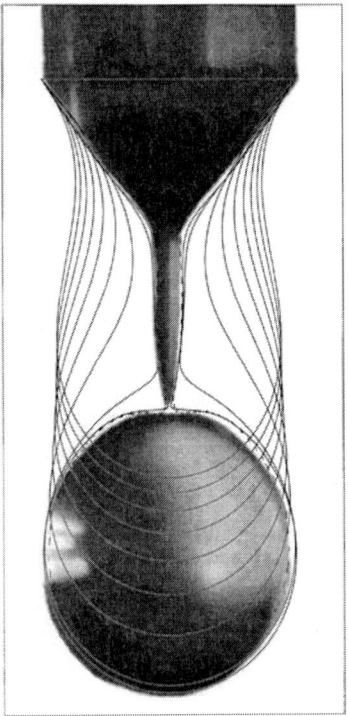

Figure 12.4. Simulation of a drop of water suspended from a circular orifice of radius $R_0 = 0.26$ cm. This gives parameter values $l_s/R_0 = 0.992$ and $l_v/r_0 = 4.89 \times 10^{-6}$. The time distance between profiles is $0.4 (R_0^3/\gamma)^{1/2}$, starting from a point where the drop is already falling. A profile at the snap-off and the corresponding experimental picture from Peregine et al. (1990) are superimposed. The inviscid result of Schulkes (1994) is indicated with a dashed line. (From Eggers and Dupont, 1994).

Steen (1997) for a soap-film bridge. Near the pinchoff point, the neck radius becomes much smaller than a, and the neglected viscous effect definitely becomes important. The relevant characteristic length l and time $t_v = T_0$ discussed in Section 11.2c are given in Table 12.1 for three liquids. These internal scales of length and time are several orders of magnitude smaller than the external scales based on the nozzle radius.

12.2b. Viscous Theory

The same problem has been investigated by Eggers and Dupont (1994). They obtained the finite difference solution of (11.7) and the dimensional (11.9) using a slender body approximation. Their numerical results compare very well with the experimental observation of Peregrine et al. (1990) (Fig. 12.4). The results of Schulkes presented in the previous section are also superimposed in this figure. The dimple on the drop right below the ligament that appears in the Schulkes results does not appear in the experiments or in the results based on viscous theory. Note that the internal scales given in Table 12.1 for water are almost of molecular scale. Near the pinch point, viscous theory most likely will give a more accurate description. However at molecular scales, the neglected intermolecular force must become important, as discussed in Section 11.4. This will probably be reflected as a singularity in the continuum theory. The nature of singularity is discussed by Eggers and Dupont and by Eggers (1995).

Shi, Brenner, and Nagel (1994) obtained the numerical solution of the one-dimensional equation for a very viscous fluid. Figure 12.5 shows the results of numerical simulations for a liquid with a dynamic viscosity of 1 poise falling from a nozzle with a radius $a = 0.225$ cm. Figure 12.5(a) gives a view of the entire pendant system at 10^{-8} sec before snap-off. Near the bottom of the long neck there is a region where the thickness decreases and forms a secondary neck. Contrasting Figure 12.4 with Figure 12.5(a), we see the dependence of the neck structure on the viscosity. Figure 12.5(b) shows an enlargement of the first neck enclosed by brackets when the interface is perturbed by a noise of amplitude on the order of 10^{-4} times the minimum thickness of the neck. Their numerical results show a process of successive necking until the neck is of order 2 Angstrom. (See also Brenner et al., 1977.) The numerical results appear to agree qualitatively with their own experimental observations. The agreement between the theoretical predictions based on the one-dimensional equations and the experiments is remarkable given that the entire pendant system is not truly one-dimensional and that the molecular scale is already reached in the final stage.

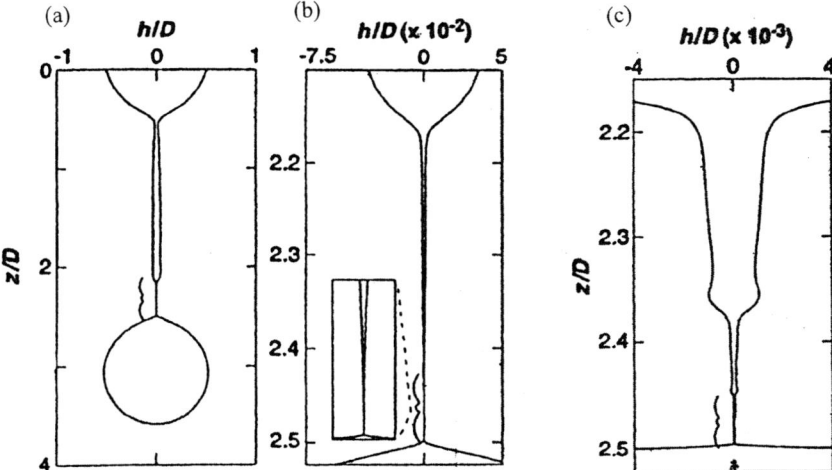

Figure 12.5. Simulation of the profile for a drop of a glycerol-water mixture (85 weight %; viscosity = 1 P) falling from a nozzle of diameter $D = 4.5$ mm. (From Shi et al., 1994). (a) A view of the entire drop, at approximately 10^{-8} sec before snap-off, shows that near the bottom of the long neck there is a region where the thickness decreases and forms a secondary neck. (b) An enlargement of the region in (a) enclosed by brackets. Note that the scales are different along two axes, so that the shape can be easily seen. The inset shows another, greater magnification. (c) Simulation of the same drop as in (a) in the presence of noise. An enlargement of the first neck shows third and fourth necks sprouting from the second neck.

Lister and Stone (1998) considered a drop creeping through an external fluid of the same dynamic viscosity. Numerical simulation with $N^{-1} = 1$ in (12.2) revealed that the neck thickness, R_{min}, decreases linearly with time τ^* from the breakup, and that the Stokes flow solutions are self-similar for various unstable initial conditions if the logarithmic singularity in τ^* is removed. The analytic behavior of the singularity is described by Cohen et al. (1999).

12.2c. From Dripping to Jetting

The previous two sections emphasized the physical breakoff of the drop from the faucet. When the nozzle discharge rate exceeds a certain critical value, uniform drops drip periodically from the faucet until a second flow rate is reached beyond which the drop size begins to vary from one to the next in a quasi-periodic or chaotic way (Clanet and Lasheras, 1999). The former dripping regime is often termed periodic dripping (PD) and the latter regime termed dripping faucet (DF). With the one-dimensional model (11.13) and (11.14), Ambravaneswaran, Phillips, and Basaran (2000) showed that the transition from the first periodic dripping to jetting actually goes through a number of period doubling (halving) bifurcations and hysteresis. (References on classical and more recent works in this area are listed at the end of this chapter.) When the flow rate increases farther from DF, the detachment point of the droplets suddenly moves downstream from the exit of the nozzle, and a continuous jet is formed.

Extending Taylor's (1959) method, Clanet and Lasheras (1999) obtained the critical Weber number for the transition from a "periodic" dripping to jetting. After the detachment of the drop from the ligament, the ligament recoils toward the nozzle and forms a new drop. While the drop moves upward with velocity dz/dt, the liquid continues to flow into the drop at a constant speed V_0 through the nozzle of inner radius R, where z is the axial distance measured from the point where the drop and the ligament first detach from each other at $t = 0$, in the direction opposite to gravity. Hence the mass of the drop grows with time as

$$M = \pi R^2 \rho (z + V_0 t).$$

Equating the time rate of change of momentum in the drop to the force acting on drop, we have

$$\frac{d}{dt}\left(M\frac{dz}{dt}\right) + \pi R^2 \rho V_0 \left(\frac{dz}{dt} + V_0\right) = -Mg + 2\pi R_0 S, \quad (12.5)$$

where R_0 is the nozzle outer radius, assuming the nozzle is wetable. The solution of this equation gives

$$z(t) = -\frac{g}{6}t^2 + (V - V_0)t,$$

where $V = (2SR_0/R^2)$, and $z(t)$ is zero at $t = 0$ and $6(V - V_0)/g$. Between these two times, z reaches a maximum $z_m = 3(V - V_0)^2/2g$ at $t = t_m/2$. Clanet and Lasheras observed experimentally that it takes time t_n to pinchoff a drop, where

$$t_n = 3.16 \left(\rho R_0^3/S\right)^{1/2}.$$

During this time, the point of detachment travels down a distance $l_d = V_0 t_n$. When the maximum distance reachable by the drop is less than this detachment distance, the drop moves progressively downward from cycle to cycle, and a continuous jet is formed. Thus the condition $z_m = l_d$ gives the onset of transition from dripping to jetting,

$$\left(\frac{V_0}{V}\right)^2 - 2\left(\frac{V_0}{V}\right)\left(1 + \frac{gt_n}{3V}\right) + 1 = 0.$$

Because $z(t) > 0$ and thus $V > V_0$, the physically meaningful solution in terms of the Weber and Bond numbers is

$$W_{ec} = 4\frac{Bo_0}{Bo}\left[1 + KBo_0Bo - \left((1 + KBo_0Bo)^2 - 1\right)^{1/2}\right]^2, \qquad (12.6)$$

where $K = 0.372$ and $We = 2\rho V_0^2 R/S$, $Bo = \rho g R^2/S$, $Bo_0 = \rho g R_0^2/S$.

The theoretical prediction agrees very well with their experiments. Comments on the relation between this transition and absolute instability were already made in Chapter 10.

12.3. Microdrop Formation by Suction

In the experiment of Ganan-Calvo (1998) a liquid drop attached to the mouth of a capillary is placed above a hole in a plate perpendicular to the capillary. The drop is stretched by suction from below the hole. The breakup of the stretched thin thread is due to capillary force. If a liquid is supplied through the capillary at a steady rate depending on the suction power, then microdrops can be produced steadily. Typical experimental results are given in Figure 12.6; Figure 12.7 shows a typical size distribution histogram with a standard

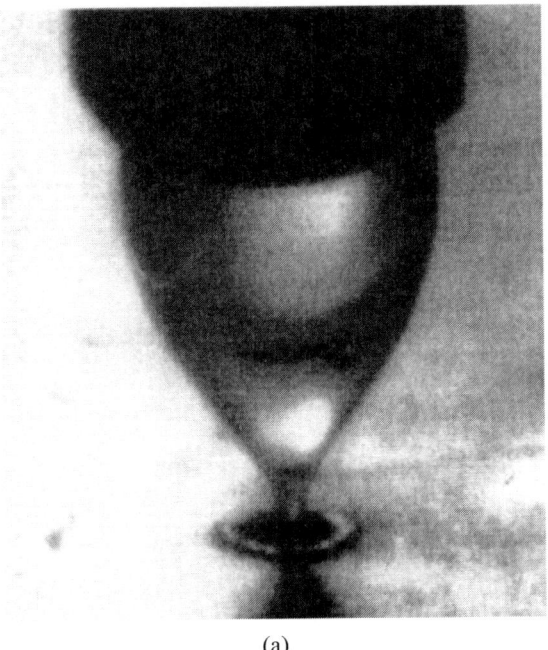

(a)

Figure 12.6. Formation of microsized drops. (From Gannan-Calvo, 1998). (a) Draining of liquid through a hole. (b) A thinning liquid thread drawn down by suction. (c) Formation of microdrops.

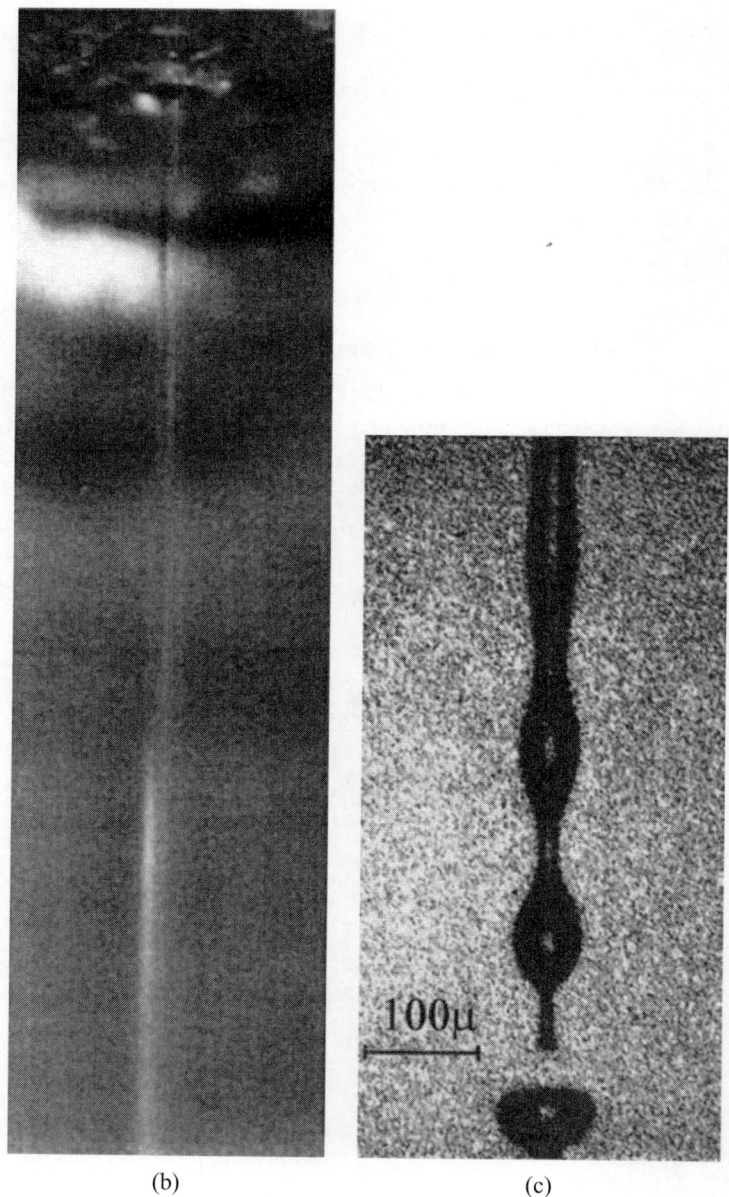

(b)   (c)

Figure 12.6. (*continued*)

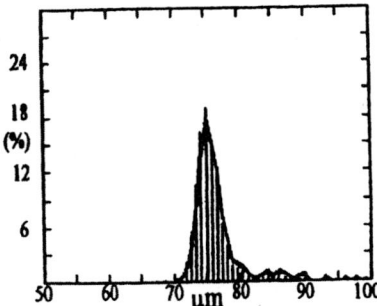

Figure 12.7. Drop size histogram. (From Gannan-Calvo, 1998).

deviation of 6.7%. In this experiment the liquid supply rate is 4.9 μliter/s, and the pressure below the plate is 4.67 kpa lower than that above the plate. When the liquid supply rate is increased to 8.2 μliter/s and the suction is increased to 32.4 kpa, the fairly good uniformity of the drop distribution is lost.

12.4. Microsized Monodispersed Sprays

Woods (1990) discovered that an intact liquid jet can be excited to atomize and form sprays from the liquid-gas interface. In his experiment, a liquid jet is issued from a piezoelectric circular cylinder of 0.05-cm inner diameter. The piezoelectric property of the nozzle is exploited to execute a periodic extension and contraction along the nozzle axis. The axial vibration of the nozzle is regulated by a function generator. When the frequency of the function generator is increased gradually from zero, the jet remains intact until the first critical frequency is reached. An intact jet before this critical frequency is shown in Figure 12.8. In this figure, the jet is not long enough for Rayleigh mode instability to exhibit itself. When the first critical frequency is reached, the jet suddenly atomizes to form a fine spray from the interface (Fig. 12.9). As the frequency increases, the atomization suddenly ceases when the frequency of the excitation passes a narrow band beyond the first critical frequency. The jet will not atomize until a second critical frequency band is reached. Then the jet atomizes to form finer sprays within this second band. Beyond this second band the jet ceases again to atomize until the appearance of the third frequency band, which produces even finer sprays.

The frequency bands within which atomization is excited depend on the flow parameters including the Weber, Reynolds, and Froude numbers, the gas to liquid density ratio Q, the kinematic viscosity ratio M, and the ratio of the forcing amplitude to the jet radius A. The three bands found for

Figure 12.8. A jet remains intact outside of the excitable frequency bands.

an ethyl alcohol jet of the flow parameters given in the figure caption are shown in Figure 12.10. The Strouhal number on the abscissa of the figure is defined as $S_t = U/\omega a$, with U, ω, and a being respectively the average jet velocity, the excitation frequency, and the jet radius. The ordinate D_{10} is the arithmetic mean diameter of the droplet measured at a point 3.0 mm radially and 5.0 mm axially from the nozzle exit, by use of a phase Doppler particle analyzer (Bachalo and Houser, 1984). The three bands of S_t, from left to right, correspond to the frequency bands 76–82, 35–41, and 18–19 kHz, respectively.

The histogram of the size distribution at a point mentioned above for the jet excited at 41 kHz is shown in Figure 12.11. Note that the size distribution in Figure 12.11 is slightly wider than that in Figure 12.7, which was obtained

Figure 12.9. Induced atomization in the excitable frequency bands.

from a different method of spray formation. However, the average diameter of 21 μm here is about one quarter of that in Figure 12.7. Note that the smallest size corresponding to an S_t near 0.2 and producible in the range of 76–82 kHz, is less than one half of 21 μm. In comparison, the present method not only produces smaller microsized droplets but also at a faster rate because the spray is formed drop by drop in the former method. However the monodispersity of the present method is not as good. The other advantage of the present method is that the same nozzle flow can be excited at different frequency ranges to produce fairly monodispersive tailored sprays of different average diameters.

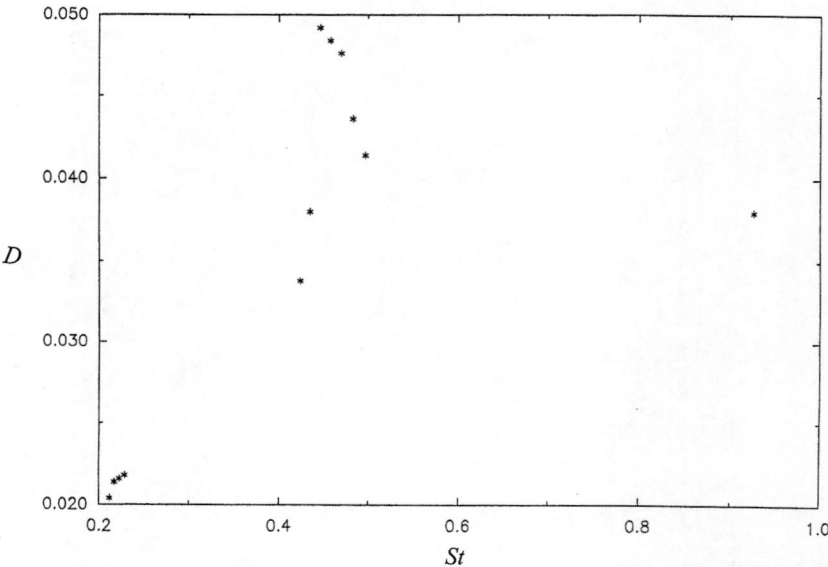

Figure 12.10. Parametric resonance of an ethyl alcohol jet. $Q = 0.001525$, $M = 10.42$, $A = 0.004$, $Re = 750$, $We = 162.63$, $Fr = 7724$.

We should point out that before external forcing, the jet is in the Rayleigh regime of jet breakup. The formation of sprays by external excitation is not in the flow parameter ranges of the Taylor mode discussed earlier. It appears that the present spray formation is caused by parametric resonance (Arnold, 1989; Nayfeh and Mook, 1979). Although the resonant frequency bands in Figure 12.10 have ratios of 1:2:3, it is not the manifestation of the generation

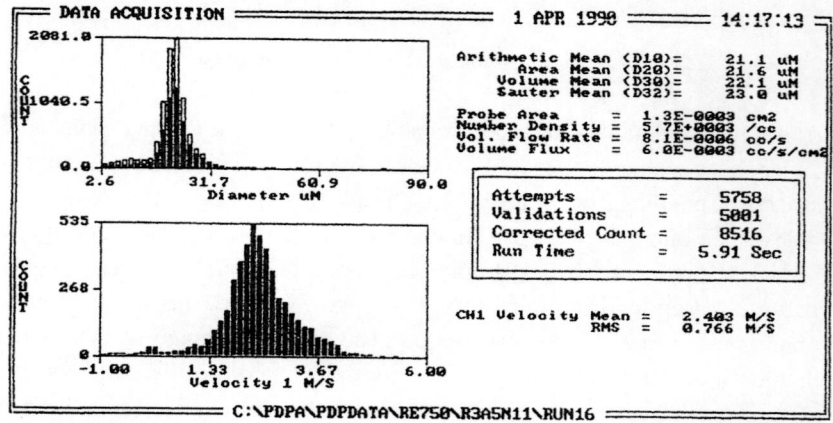

Figure 12.11. The size distribution of droplets formed in the middle band of Figure 11.5.

of higher harmonics at the natural frequency of the original Rayleigh mode. If it were, the jet would not atomize but would form smaller satellite drops within the Rayleigh mode. Atomization of a spherical liquid drop by pulsation is discussed by Lin and Zhou (1996).

12.5. Branching Liquid Jets

Lin and Woods (1991) reported an interesting new phenomenon of branching of liquid jets. They injected a liquid jet vertically downward through a piezoelectric nozzle into the atmosphere. A high-frequency oscillation of the voltage difference between the piezoelectric nozzle inlet and outlet was imparted to induce an oscillation in the length of the nozzle. As the frequency of vibration increased beyond a critical value, the intact jet suddenly forked into two branches. When the frequency of oscillation increased to exceed a second critical value, the two-pronged jet suddenly trifurcated into a three-pronged jet. At a yet higher frequency the three-pronged jet split into a multiple-pronged jet in some cases. Different branches of the multiple-pronged jet of more than two branches merged and split erratically without a discernible periodicity in time. A structurally stable two-pronged jet and a structurally unstable three-pronged jet are shown in Figures 12.12 and 12.13 respectively.

The phenomenon discovered by Lin and Woods is the inverse of that observed by Plateau (1873) and Tyndall (1867). They observed that when a liquid jet was made to ascend obliquely, it scattered into a kind of sheaf. The sheaf was sensitive to sound vibration in the air and could be made to gather into three, two, or even no branches under the appropriate frequency of sound vibration. Plateau reported that this phenomenon could be observed only when the angle made by the jet with the horizontal was between 20° and 50°. Andrade (1941) discovered that a water jet emanating horizontally from an orifice into a water tank can be made to split into two forks by introducing acoustic vibration in the tank. In Andrade's experiments the branching of water jets was not caused by interfacial force. Tyndall (1867) and Brown (1935) showed that the forking of flames from gas burners, which Laconte (1858) observed to be sensitive to the beats and tunes of music in a concert hall, can be controlled by sound waves. Lee and Reynolds (1985) observed that an air jet could be made to bifurcate, trifurcate, or "bloom" by introducing simultaneously a periodic lateral disturbance along the circumference of the orifice and a periodic axial disturbance into air before allowing the jet to emanate at a high speed into the ambient air. In the experiments of an air jet as well as the forking flames, surface tension plays little role. Lee and Reynolds also found that the air emanating from the nozzle did not all end up in various

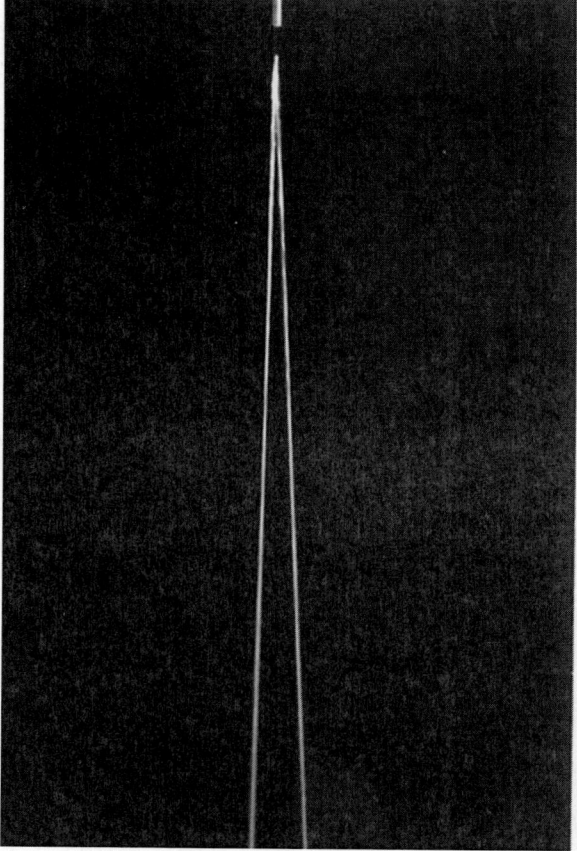

Figure 12.12. A structurally stable two-pronged jet. (From Lin and Woods, 1991).

branches of the jet. Some of the air actually was found between the branches.
This is one of the basic differences between the branching of liquid and gas
jets including the flaming jet. The precise mechanism of the forking of air
jets and flames seems to remain largely unknown.

The nozzle used by Lin and Woods was made of piezoelectric ceramic. It
was erected vertically, clamped at the upper end, and suspended freely at the
lower end. Thus the possibility of lateral vibration of the nozzle cannot be
excluded from their experiments although only the oscillatory displacement
was introduced axially. Both lateral and axial displacements of the nozzle used
by Lin and Woods were measured by Lin and Webb (1994) with a Dantec laser
vibrometer. They discovered that the amplitude of the lateral vibration was
more than 10 times that of the maximum axial displacement, which was of
order 1 μm in their experiments. Hence the liquid jet branching phenomenon

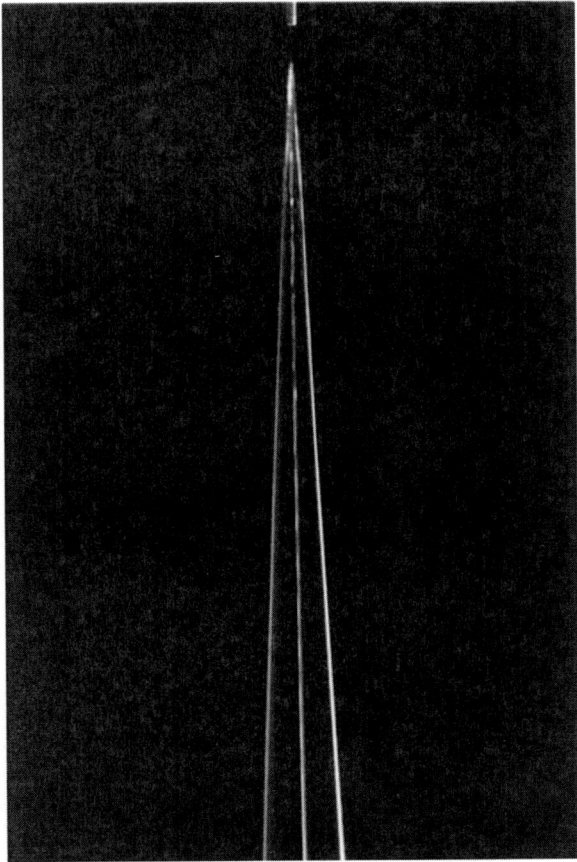

Figure 12.13. A structurally unstable three-pronged jet. (From Lin and Woods, 1991).

observed by Lin and Woods was actually caused by the lateral oscillation of the nozzle tip. The jet branching induced by precisely controlled lateral planer vibration of the nozzle has been reported by Lin and Webb (1994). A fairly sophisticated video system was used to observe the branching phenomenon. When viewed by the human eye with a video camera under white light, the jet appeared to split into multiple-pronged simple jets as in Figures 12.12 and 12.13. However when the branching jets were observed with a stroke light driven at exactly the same frequency as the piezoelectric transducer by splitting the signal from the function generator, the jets turned out to be streams of drops. An example is shown in Figure 12.14.

The experimental results for the two test fluids obtained from the four nozzles exhibit qualitatively the same phenomena. Not all of the liquid jets

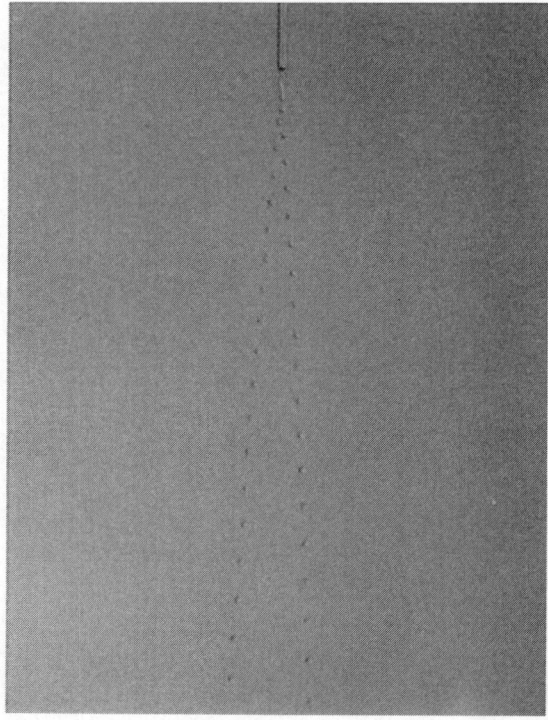

Figure 12.14. Bifurcating streams of drops. (From Lin and Webb, 1994).

are excited at the resonance frequencies of the nozzle because the natural frequencies of the nozzle and the jet do not necessarily overlap. The spatial growth rates k_i of small axisymmetric disturbances in a viscous liquid jet in an ambient gas are plotted in Figure 12.15 against the disturbance frequencies for three sets of flow parameters relevant to the experiment. These results were obtained from the theoretical dispersion equation of Lin and Lian (1990). In their theory, inertial force is assumed to be much greater than the gravitational force. For each curve, there exists a frequency ω_m at which the amplification rate is the maximum, and a cut-off frequency ω_c beyond which the disturbance decays ($k_r < 0$). The values of ω_m and ω_c for relevant flow parameters are given in the last two columns of Table 12.2. The amplification in Figure 12.15 implies that if the jet is left alone without regulated external excitation, the observed droplets in a single stream will correspond to a band of frequencies centering around ω_m for each set of given flow parameters. Their videotape records showed that this is indeed the case. No single stroboscope frequency around ω_m can be found to freeze the drops in space unless a small amplitude

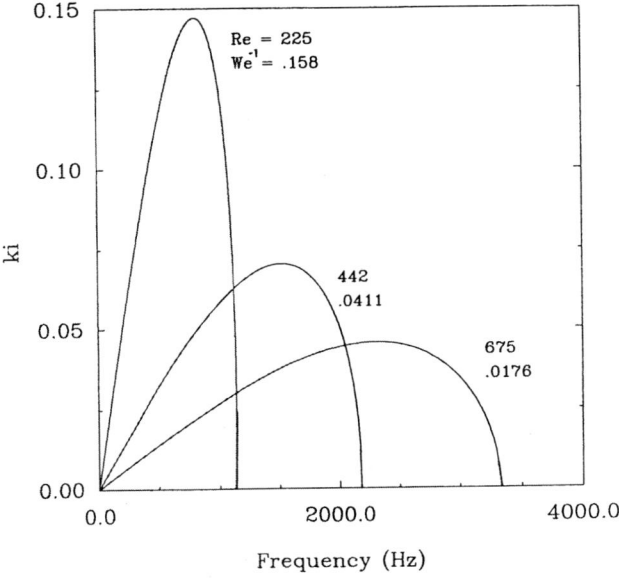

Figure 12.15. Spatial amplification rates. (From Lin and Webb, 1994).

external excitation of the same frequency as that of the stroboscope is applied. Then the drop size and the spacing between drops become very uniform and the image can be frozen in space. The information given in Figure 12.15 alone does not indicate the cause of the jet bifurcation. The jet bifurcation may be due to the first nonaxisymmetric mode, which tends to split the jet into two along the nozzle axis. However, we saw in Section 7.2 that the first nonaxisymmetric mode in a viscous liquid jet in the presence of an inviscid gas is stable in the present flow parameters. This is probably true even when the gas viscosity is taken into account as supported by the experimental evidence in Figure 12.16. The first drop is detached from a thin thread above rather than pinched into two from the side. This picutre also suggests that the vertical distances between successive drops may contain the most unambiguous message on the mechanism of jet bifurcation or even trifurcation.

The results of five experiments are summarized in Table 12.2. The first column identifies each set of experiments. The second column gives the number of measurements used in obtaining the average vertical distance y between two successive drops listed in column 3. The standard deviation of the vertical distance is given in parentheses in column 3. The corresponding average jet velocity and its standard deviation are given in column 4. The fifth column contains the number of branches n observed in each case. The theoretically

Table 12.2. *Branching Alcohol Jets*

Test Case	No. of Measure	y (Std) (mm)	U (Std) (m/s)	n	λ	$(\lambda - y)/\lambda$ (%)	We^{-1}	Re	Fr	ω	ω_m (Hz)	ω_c
1	6	3.10(.47)	1.08(.16)	2	3.26	5.1	0.158	225	763	165	775	1150
2	16	1.25(.58)	1.08(.16)	1	1.32	5.4	0.158	225	763	818	775	1150
3	14	2.40(.38)	1.08(.16)	3	2.12	12.7	0.158	225	763	172	775	1150
4	30	0.96(.03)	2.12(.10)	2	0.96	0.2	0.041	442	2935	1105	1150	2240
5	9	0.81(.38)	2.12(.10)	1	0.98	6.0	0.041	442	2935	2159	1550	2240

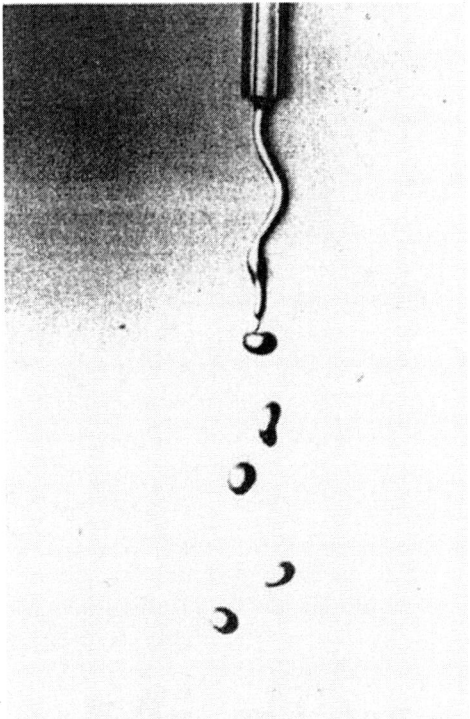

Figure 12.16. The origin of jet bifurcation. (From Lin and Webb, 1994).

predicted distances between drops λ, calculated from the information given in columns 3 to 5, are found in column 6. The seventh column gives $(\lambda - y)/\lambda$, the relative difference between the theoretically and experimentally determined distances. The next three columns give the relevant flow parameters. The last three columns contain the frequencies ω, ω_m, and ω_c.

The theoretical distance between two successive drops is predicted with the assumption that the disturbances leading to the formation of drops are the higher harmonics of the disturbances introduced. The wavelength λ of the n-th harmonic, and the distance between two successive drops in an n-pronged jet, is then given by

$$\lambda = c/(n\omega),$$

where c and ω are respectively the wave speed and frequency of the fundamental disturbance. According to theory (Lin and Lian 1990) and confirmed by experiments $c \doteq U$. The relative difference between the theoretically and experimentally determined vertical distance between two successive drops is quite small for most cases. However, close agreement is not sufficient to confirm the above assumption. This assumption has been checked further with the following additional evidence.

The liquid jet in case 1 is excited at 165 Hz, which is slightly below the first resonance frequency of the nozzle, 172 Hz. The frequency of the second harmonic of the introduced disturbance falls below the maximum frequency ω_c of the growing disturbance in the jet. This permits the growth of the second harmonic of the first axisymmetric mode of disturbances. If the jet is excited at a higher frequency, the second harmonic of which may have a frequency larger than ω_c, then the jet cannot bifurcate because the second harmonic is damped and the jet will remain a single stream of drops. This is indeed the case, as observed in case 2. The excitation frequency is 818 Hz in this case and larger than one half of ω_c. In case 3 the flow parameters remain the same as in the previous two cases, but the jet is excited at the first resonance frequency of 172 Hz to produce a much larger amplitude of oscillation. The frequency of the third harmonic of this introduced frequency is still smaller than ω_c. It appears that while the excitation amplitude in case 1 is not sufficiently large to generate the third harmonic, that in case 3 is large enough to generate the third harmonic. The critical disturbance amplitude above which the jet will bifurcate or trifurcate cannot be predicted on the basis of the present linear theory. Experiments 4 and 5 involved flow parameters different from those for cases 1, 2, and 3. The excitation frequency 2159 Hz in case 5 exceeds ω_m. Its second harmonic cannot be excited because $2\omega > \omega_c$ as explained for case 2. The excitation frequency in case 4 is 1105 Hz, which is slightly larger than the second nozzle resonance frequency of 1085 Hz. Because $2\omega < \omega_c$, the second harmonic is generated and two streams of drops are formed as in case 1. When the excitation frequency is adjusted to the second resonance frequency 1085 Hz, the jet remains two-streamed despite the large excitation amplitude. This is probably because the third harmonic frequency 3ω exceeds $\omega_c = 2240$ Hz. Therefore no trifurcated jet similar to that in case 3 is observed for these flow parameters at this resonance frequency. When the nozzle is oscillated at frequencies higher than the cut-off even at a higher amplitude, no multiple streams of droplets can be found.

We should point out that the three-streamed jet in case 3 occasionally becomes four streamed and then flips back to three streamed. The occurrence of the event is erratic without periodicity. This chaotic behavior was also observed by Lin and Woods. It remains to be explained. The chaotic behavior together with the critical amplitudes required for the generation of different harmonics at the same excitation frequency can be explained only with nonlinear theories.

12.6. Shaped-Charge Jets

The fluid particle in a shaped-charge jet accelerates in the axial direction from the rear to the front tip. The axial and radial velocity components of the basic flow are respectively given locally by (Frankel and Weihs, 1985, 1987),

$$W_0 = \frac{Kz}{Kt+1}, \qquad U_0 = -\frac{Kr}{2(Kt+1)}, \qquad (12.7)$$

where K is the initial rate of the axial stretching. As a result of the stretching, the jet radius decreases with time from the initial value a_0 as

$$a(t) = \frac{a_0}{(Kt+1)^{1/2}}. \qquad (12.8)$$

The corresponding pressure distribution is

$$P_0 = \frac{3}{8} \frac{\rho K^2 a^2}{(Kt+1)^2} \left(1 - \frac{r^2}{a}\right) + \frac{S}{a} - \frac{\mu K}{Kt+1}, \qquad (12.9)$$

where ρ and μ are respectively the density and dynamic viscosity of the jet fluid.

The onset of instability can be investigated again by examining the dynamic behavior of small flow perturbation. However, the basic flow is now unsteady, and the introduced perturbation wavelength increases with time as the jet stretches. The governing equation of the perturbation is not amenable to the eigenvalue solution but has to be solved as an initial value problem. The solution for an axisymmetric disturbance shows that initially the disturbance grows algebraically due to the axial stretching, but later the stretching retards and the surface tension promotes exponential growth. The viscosity has the usual role of damping as well as the role of enhancing instability due to stretching. The longer wavelength disturbance becomes dominant at later stages of the development of a shaped-charge jet.

If the axial velocity is allowed to vary parabolically in the radial direction, the velocity variation reduces the amplification rate, but it cannot suppress completely the onset of instability (Miyazaki and Kubo, 1995).

12.7. Intermittent Sprays

Sprays are produced intermittently in many applications. For example, fuel injection in an internal combustion engine is an intermittent spray. The liquid droplet size distribution even at the same location in a spray varies with the form of the liquid pressure wave applied at the nozzle inlet (Cook and Lin, 1998).

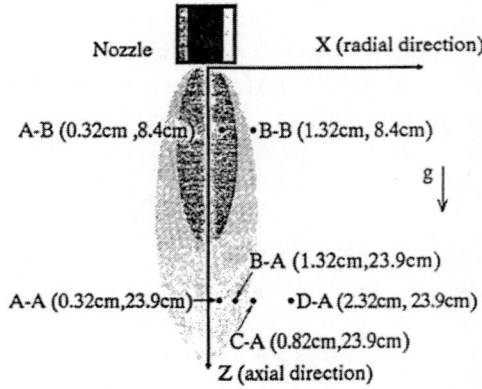

Figure 12.17. Spray sampling locations. (From Cook and Lin, 1998).

The size distribution at various locations in the spray designated in Figure 12.17 was measured with a phase Doppler particle analyzer. The Doppler shift of the scattered laser light by a moving droplet yields the velocity, and the phase shift yields the diameter of the droplet. It has been found that spatial variation in the drop size distribution in an intermittent spray can be manipulated by varying the pressure input as a function of time. For example, spray formation may be suppressed near the nozzle and promoted in the region farther downstream by a gradual pressure increase followed by a sharp drop in pressure during the liquid injection process. The fluid dynamic theory of this phenomenon is still lacking. However, it is of considerable practical significance. Gliken (1985) and Dolenc (1990) pointed out how injection rate variation within one pressure pulse might influence such characteristics of engine performance as specific fuel consumption, noise, nitrous oxide emissions, and smoke emissions. General design criteria show the trade-offs of having different injection rates at different times in the injection cycle. Some works related to intermittent sprays can be found in the Chapter 12 references.

12.8. Viscous Beads

Kliakhandler, Davis, and Bankoff (2000) reported an interesting phenomenon of bead formation in a jet with a fiber core along the jet axis. Figure 12.18 displays three qualitatively different bead train structures in three different flow regimes. The flow regime is characterized by the fiber radius normalized with the unperturbed film thickness H_0, the Reynolds number $Re = gH_0^3/\nu^2$, and the Bond number $B_0 = \rho g H_0^2/S$. Regimes (a), (b), and (c) are respectively

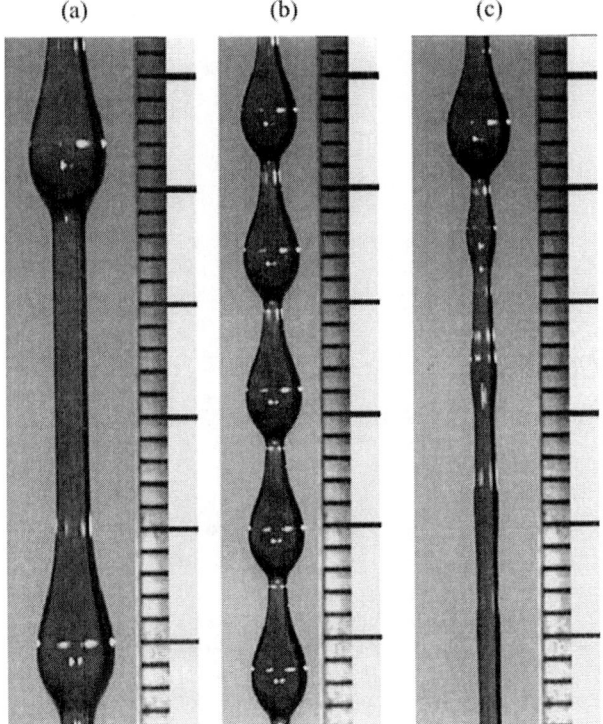

Figure 12.18. The typical structures observed in experiments. Three wave profiles are measured at decreasing flow rates. The marks on ruler are 1 mm apart. (From Kliakhandler et al., 2001).

characterized by $(r_0, B_0^{-1}, Re) = (0.34, 6.2, 0.02), (0.4, 8.6, 0.012)$, and $(0.5, 12.5, 0.07)$, r_0 being the dimensionless fiber radius. They also carried out a numerical simulation based on a lubrication-type model equation

$$h_\tau + \frac{1}{(r_0 + h) B_0} \left[Q(h) \left(\frac{h_{zz}}{(1 + h_z^2)} - \frac{1}{(r_0 + h)(1 + h_z^2)^{1/2}} \right) \right]_{,z} = 0,$$
(12.10)

where τ is the time normalized with H_0/W, W is $g H_0^2/\nu$, and

$$Q = \frac{1}{16} \left[4(h + r_0)^4 \ln\frac{h + r_0}{r_0} - h\left(3h^3 + 12h^2 r_0 + 14h r_0^2 + 4r_0^2\right) \right].$$

Equation (12.10) is asymptotically valid for thin film thickness but was used for thick film thickness to simulate bead formation. A periodic boundary condition is imposed in the axial direction up to 40 λ, where $\lambda = 2\pi/k_m$, k_m being the most amplified wave number of the corresponding linear theory (Lin and

Liu, 1975; Atherton and Homsy, 1976). The results of numerical simulation agree quite well with the experiments in regime (b), but not completely in regime (c).

References

Ambravaneswaran, B., Phillips, S. D., and Basaran, O. A. 2000. *Phys. Rev. Lett.* **85**, 5332.

Andrade, E. N. da C. 1941. The sensitive flame. *Proc. Phys. Soc. Lond.* **53**, 299.

Arnold, V. I. 1989. *Mathematical Method of Classical Mechanics*. Springer-Verlag.

Atherton, R. W., and Homsy, G. M. 1976. *Chem. Eng. Comm.* **2**, 57.

Bachalo, W. D., and Houser, M. J. 1984. *Optical Eng.* **23**, 583.

Brenner, M. P., Eggers, J., Joseph, K., Nagel, S. R., and Shi, X. D. 1997. *Phys. Fluids* **9**, 1573.

Brenner, M. P., Shi, X. D., and Nagel, S. R. 1994. *Phys. Rev. Lett.* **73**, 3391.

Brown, G. B. 1935. On vortex motion in gaseous jets and the origin of their sensitivity to sound. *Proc. Phys. Soc. Lond.* **47**, 703.

Chen, Y. J., and Steen, P. H. 1997. *J. Fluid Mech.* **341**, 245.

Clanet, C., and Lasheras, J. C. 1999. *J. Fluid Mech.* **383**, 307.

Cohen, I., Brenner, M. F., Eggers, J., and Nagel, S. R. 1999. *Phys. Rev. Lett.* **83**, 1147.

Cook, V. F., and Lin, S. P. 1998. *Atomiz. Sprays* **8**, 393.

Day, R. F., Hinch, E. J., and Lister, J. R. 1998. *Phys. Rev. Lett.* **80**, 704.

Dolenc, A. 1990 The injection equipment of future high-speed DI diesel engines with respect to power and pollution requirements, *Proc. Inst. Mech. Eng.* **199**, No. D3, 49–58.

Eggers, J. 1995. *Phys. Fluids* **7**, 941.

Eggers, J., and Dupont, T. F. 1994. *J. Fluid Mech.* **262**, 205.

Frankel, I. F., and Weihs, D. 1985. *J. Fluid Mech.* **155**, 289.

Frankel, I. F., and Weihs, D. 1987. *J. Fluid Mech.* **185**, 361.

Ganan-Calvo, A. M. 1998. *Phys. Rev Lett.* **80**, 285.

Glikin, P. 1985. Fuel injection in diesel engines, *Proc. Inst. Mech. Eng.* **199**, No. D3, 161.

Kliakhandler, I. L., Davis, S. H., and Bankoff, S. G. 2001. *J. Fluid Mech.* **429**, 381.

Laconte, J. 1858. On the influence of musical sounds on the flame of a jet of coal-gas, *Philos. Mag.* **15**, 235.

Ladyzhenskaya, O. 1969. *The Mathematical Theory of Viscous Incompressible Flow.* 2nd ed., Gordon and Breach.

Lee, M., and Reynolds, W. C. 1985. Department of Mechanics Eng. Rep. No. TF-22, Standard University.

Lin, S. P., and Lian, Z. W. 1990. Mechanisms of the breakup of liquid jets. *AIAA J.* **28**, 120.

Lin, S. P., and Liu, W. C. 1975. *AIChE J.* **21**, 775.

Lin, S. P., and Webb, R. D. 1994. *Phys. Fluids* **6**, 2671.

Lin, S. P., and Woods, D. R. 1991. A branching liquid jet. *Phys. Fluids* A **1**, 771.

Lin, S. P., and Zhou, Z. W. 1996. *Appl. Math. Mech.* **17**, 597.

Lister, J. R., and Stone, H. A. 1998. *Phys. Fluids* **10**, 2759.

Miyazaki, T., and Kubo, T. 1995. *J. Phys. Soc. Japan* **64**, 3275–3283.

Nayfeh, A. H. 1970. *Phys. Fluids* **14**, 841.

Nayfeh, A. H., and Mook, D. T. 1979. *Nonlinear Oscillations.* John Wiley & Sons.

Papageorgiou, D. T. 1995. *Phys. Fluids* **7**, 1529.

Peregrine, D. H., Shoker, G., and Symon, A. 1990. *J. Fluid Mech.* **212**, 25.

Plateau, M. J. 1873. *Statique Experimentals Et Theorique Des Liquids Soumic Aue Seules Forces Molecularies*, Canthier-Vallars, Paris.

Pozrikidis, C. 1992. *Boundary Integral and Singularity Methods for Linearized Viscous Flows.* Cambridge University Press.

Rallison, J. M., and Acrivos, A. 1978. *J. Fluid Mech.* **89**, 191.

Rao, S. S. 1990. *Mechanical Vibration.* Edison–Wesley, pp. 394–398.

Shi, X. D., Brenner, M. P., and Nagel, S. R. 1994. *Science* **265**, 219.

Shyy, W., Kan, H. C., Udaykumar, and Tan-Son-Tay, R. 1999. In *Fluid Dynamics at Interfaces* (Eds., W. Shyy and R. Narayanan), pp. 263–277. Cambridge University Press.

Schulkes, R. M. S. M. 1994. *J. Fluid Mech.* **278**, 83.

Stone, H. A., Bentley, B. J., and Leal, L. G. 1986. *J. Fluid Mech.* **173**, 131.

Taylor, G. I. 1959. The dynamics of thin sheets of fluid. III. Disintegration of fluid sheets. *Proc. R. Soc. Lond.* A **253**, 313–321.

Tjahjadi, M., Stone, H. A., and Ottino, J. M. 1992. *J. Fluid Mech.* **243**, 297.

Tomotika, S. 1935. *Proc. R. Soc. Lond.* A **150**, 322.

Tyndall, J. 1867. On the action of sonorous vibration on gaseous and liquid jets. *Philos. Mag.* **33**, 92, 375.

Webb, R. D. 1992. *Theoretical and Experimental Studies of the Stability of a Viscous Liquid Jet with Respect to Non-axisymmetric Disturbances.* M.S. Thesis, Clarkson University, Potsdam, NY.

Woods, D. R. 1990. *Production of Submicron Monodispersed Droplets from a Liquid Jet.* M.S. Thesis, Clarkson University, Potsdam, NY.

Additional References Related to Boundary Integral Method

Stone, H. A., and Leal, L. G. 1989. *J. Fluid Mech.* **198**, 399.

Stone, H. A., and Leal, L. G. 1989. *J. Fluid Mech.* **206**, 223.

Tjahjadi, M., and Ottino, J. M. 1991. *J. Fluid Mech.* **232**, 191.

Youngren, G. K., and Acrivos, A. 1975. *J. Fluid Mech.* **69**, 377.

Additional References Related to Dripping Faucets

Chesters, A. K. 1977. An analytic solution for the profile and volume of a small drop or bubble symmetrical about a vertical axis. *J. Fluid Mech.* **81**, 609–624.

Harkins, W. D., and Brown, F. E. 1919. The determinational of surface tension (free surface energy), and the weight of falling drops: The surface tension of water and benzene by the capillary height method. *J. Am. Chem. Soc.* **41**, 499–524.

Lin, S. P., and Zhou, Z. W. 1996. Atomization of a liquid drop by pulsation. *J. Appl. Math. Mech.* **17**, No. 7, 597–606.

Maxwell, J. C. 1875. In *Encyclopedia Britanica*, 9th ed. Reprinted 1965 in *The Scientific Papers of James Clark Maxwell* (Ed. W. D. Niven) Vol II, p. 541, Dover Publications.

Michael, D. H. 1981. Meniscus stability. *Ann. Rev. Fluid Mech.* **13**, 189.

Rayleigh, L. 1899. Investigation in capillary: The size of drops. The liberation of gas from supersaturated solutions. Colliding jets. The tension of contaminated water-surfaces. A curious observation. *Philos. Mag.* **48**, 321–337.

Scheele, G. F., and Meister, B. J. 1968a. Drop formation at low velocities in liquid–liquid systems. *AIChE J.* **14**, 9–15.

Scheele, G. F., and Meister, B. J. 1968b. Prediction of jetting velocity. *AIChE J.* **14**, 15–19.

Tate, T. 1864. On the magnitude of a drop of liquid formed under different circumstances. *Philos. Mag.* **27**, 176–180.

Wilkinson, M. C. 1972. Extended use of, and comments on, the drop-weight (drop-volume) technique for the determination of surface and interfacial tensions. *J. Colloid Interface Sci.* **40**, 14–26.

Wilson, S. D. R. 1988. The slow dripping of a viscous fluid. *J. Fluid Mech.* **190**, 561–570.

Additonal References Related to Pinch Point Singularity

Brenner, M. P., Shi, X. D., and Nagel, S. R. 1994. *Phys. Rev. Lett.* **73**, 3391.

Keller, J. B., and Miksis, M. J. 1983. SIAM *J. Appl. Math.* **43**, 268.

Keller, J. B., and Miksis, M. J. 1991. *J. Fluid Mech.* **232**, 191.

Additional References Related to Intermittent Sprays

Arai, M., Tabata, M., Hiroyasu, H., and Shimizu, M. 1984. *Disintegrating Process and Spray Characterization of Fuel Jet Injected by a Diesel Nozzle*. SAE Paper 840275.

Chehroudi, B., Chen, S., Bracco, F. V., and Onuma, Y. 1985. *On the Intact Core of Full-Cone Sprays*. SAE Paper 850126.

Cook, V. F. 1993. *Experimental Study of an Intermittent Spray Using Several Nozzle Pressure Waveforms*. M.S. Thesis, Clarkson University, Potsdam, NY.

Ficarella, A., Laforgia, D., and Cipolla, G. July, 1990. Investigation and computer simulation of diesel injection system with rotative pump, *J. Eng. Gas Turbines Power* **112**, 317.

Freudenschub, O., Herdin, G., Schreiner, E., and Schmidt, H. 1987. Fuel injection nozzle for combustion engines, U.S. Patent 4,715,541.

Koo, J., and Martin, J. 1990. *Droplet Sizes and Velocities in a Transient Diesel Fuel Spray*. SAE Paper 900397.

Kuniyoshi, H., Tanabe, H., Sato, G., and Fujimoto, H. 1980. *Investigation on the Characteristics of Diesel Fuel Spray*. SAE Paper 800968.

Ruiz, F., and Chigier, N. 1987. *The Effects of Design and Operating Conditions of Fuel Injectors on Flow and Atomization*. SAE Paper 870100.

Sangeorzan, B., Uyehara, O., and Myers, P. 1984. *Time-Resolved Drop Size Measurements in an Intermittent High Pressure Fuel Spray*. SAE Paper 841361.

Shaoxi, S., Yunyi, G., and Hulmin, S. 1988. *Experimental Study of the Atomization and Evaporation Processes of Diesel Fuel Sprays*. SAE Paper 881253.

Sinnamon, J., Lancaster, D., and Steiner, J. 1980. *An Experimental and Analytical Study of Engine Fuel Spray Trajectories*. SAE Paper 800135.

Takeuchi, K., Senda, J., and Shikuya, M. 1983. *Transient Characteristics of Fuel Atomization and Droplet Size Distribution in Diesel Fuel Spray*. SAE Paper 830449.

Wright, J. A., and Drallmeier, J. A. 1997. Cycle to cycle variation in pulsed fuel sprays, *ILASS 97*, Ottawa, Canada.

Yamaguchi, J. 1998. Two-stage injection reduces DI diesel noise. *Automotive Eng.* **93**, 82.

Review Articles

Eggers, J. 1997. *Rev. Mod. Phys.* **69**, 865.

Rallison, J. M. 1984. *Ann. Rev. Fluid Mech.* **16**, 45.

Stone, H. A. 1994. *Ann. Rev. Fluid Mech.* **26**, 65.

Appendix A

The Orr–Somerfeld System in the Chebyshev Space

The derivatives in the Orr–Sommerfeld equation (9.14) can be transformed into the Chebyshev space defined by (10.23) using the relations

$$\frac{d}{dr} = \frac{dy}{dr}\frac{d}{dy} = a_j \frac{d}{dy} \quad \text{and} \quad \frac{d^p}{dr^p} = a_j^p \frac{d^p}{dy^p} \quad (j = 1, 2, 3).$$

(A1)

Hence the transformed Orr–Sommerfeld equation in y remains the same except that the p-th derivative in r in (9.14) must be replaced by (A1). The same modification must be made in the boundary conditions. Upon substitution of (10.25) into (9.14), we must evaluate the derivatives of the cardinal function $h_{jn}(y)$ at the collocation points. Let $\mathbf{D}_j^p$ represent the matrix whose elements $d_{j,mn}^p$ are defined by

$$d_{j,mn}^p = \frac{d^p h_{jn}}{dy^p}\bigg|_{y=y_{j,m}}.$$

(A2)

The matrix $\mathbf{D}_j^0$ is a unit matrix and the elements of matrix $\mathbf{D}_j^1$ are given by (Boyd, 1989)

$$d_{j,00}^1 = \frac{2M_j^2 + 1}{6}, \qquad d_{j,M_jM_j} = -\frac{2M_j^2 + 1}{6},$$

$$d_{j,mn}^1 = \frac{-y_{jm}}{(1 - y_{jm}^2)} \qquad \text{for} \quad m = n, \quad 1 \leq m \leq M_j$$

$$d_{j,mn}^1 = \frac{C_m(-1)^{m+n}}{C_n(y_{jm} - y_{jn})} \qquad \text{for} \quad m \neq n, \quad 0 \leq m, \quad n \leq M_j.$$

Matrices $\mathbf{D}_j^2$, $\mathbf{D}_j^3$, and $\mathbf{D}_j^4$ are easily obtained by simple matrix multiplication

$$\mathbf{D}_j^2 = \mathbf{D}_j^1 \mathbf{D}_j^1, \qquad \mathbf{D}_j^3 = \mathbf{D}_j^1 \mathbf{D}_j^2, \qquad \mathbf{D}_j^4 = \mathbf{D}_j^2 \mathbf{D}_j^2.$$

(A3)

The Orr–Sommerfeld equation evaluated at collocation points are

$$\left\{ a_j^4 d_{j,mn}^4 - \frac{2}{r_{j,m}} a_j^3 d_{j,mn}^3 + \left[\left(\frac{3}{r_{j,m}^2} - 2k^2 \right) \right. \right.$$

$$\left. - \left(\frac{v_j}{v_2 Re} \right)^{-1} ik W_{j,m} \right] a_j^2 d_{j,mn}^2 + \left[\frac{1}{r_{j,m}} \left(-\frac{3}{r_{j,m}^2} + 2k^2 \right) \right.$$

$$\left. + \left(\frac{v_j}{v_2 Re} \right)^{-1} ik W_{j,m} \right] a_j d_{j,mn}^1 + \left[k^4 + ik^3 W_{j,m} + \left(\frac{v_j}{v_2 Re} \right)^{-1} \right.$$

$$\left. \left. \times ik \left(\frac{W'_{j,m}}{r_{j,m}} - W''_{j,m} \right) \right] d_{j,mn}^0 \right\} \phi_{j,n} = \omega \left(\frac{v_j}{v_2 Re} \right)^{-1}$$

$$\times \left\{ a_j^2 d_{j,mn}^2 - \frac{1}{r_{j,m}} a_j d_{j,mn}^1 - k^2 d_{j,mn}^0 \right\} \phi_{j,n}, \tag{A4}$$

$$(m = 0, 1, 2, \ldots, M_j, \quad j = 1, 2, 3),$$

where

$$\phi_{j,n} = \phi_j(y_{j,n}), \qquad r_{j,m} = a_j y_{j,m} + b_j,$$

$$W_{j,m} = W_j(r_{j,m}), \quad W'_{j,m} = \left. \frac{dw_j(r)}{dr} \right|_{r=r_{j,m}}, \quad W''_{j,m} = \left. \frac{d^2 w_j(r)}{dr^2} \right|_{r=r_{j,m}}.$$

The boundary condition (10.15)-(10.22) can be written as

$$d_{3,0n}^0 \phi_{3,n} = 0, \tag{A5}$$

$$a_3 d_{3,0n}^1 \phi_{3,n} = 0, \tag{A6}$$

$$d_{1,0n}^0 \phi_{1,n} - d_{2,0n}^0 \phi_{2,n} = 0, \tag{A7}$$

$$d_{3,M_3 n}^0 \phi_{1,n} - d_{2,M_1 n}^0 \phi_{2,n} = 0, \tag{A8}$$

$$\frac{1}{r_{1,0}} \left(a_1 d_{1,0n}^0 \phi_{1,n} - a_2 d_{2,0n}^0 \phi_{2,n} \right) - (W'_{1,0} - W'_{2,0}) \xi_i = 0, \tag{A9}$$

$$\frac{1}{r_{3,M_3}} \left(a_3 d_{3,M_3 n}^1 \phi_{3,n} - a_2 d_{2,M_2 n}^1 \phi_{2,n} \right) - (W'_{3,M_3} - W'_{2,M_2}) \xi_0 = 0, \tag{A10}$$

$$\frac{ik}{r_{1,0}}\left(d^0_{1,0n}\,\phi_{1,n}-r_{1,0}W_{1,0}\right)\xi_0=\omega\xi_0, \tag{A11}$$

$$\frac{ik}{r_{3,M_3}}\left(d^0_{3,M_3n}\,\phi_{3,n}-r_{3,M_3}W_{3,M_3}\right)\xi_0=\omega\xi_0, \tag{A12}$$

$$\bar{N}_1\left(a_1^2 d^2_{1,0n}-\frac{1}{r_{1,0}}a_1^1 d^1_{1,0n}+k^2 d^2_{1,0n}\right)\phi_{1,n}$$

$$-\left(a_2^2 d^2_{2,0n}-\frac{1}{r_{1,0}}a_2^1 d^1_{2,0n}+k^2 d^0_{2,0n}\right)\phi_{2,n}$$

$$-\left(\bar{N}_1 W''_{1,0}-\bar{N}_2 W''_{2,0}\right)\xi_i=0, \tag{A13}$$

$$\bar{N}_3\left(a_3^2 d^2_{3,M_3n}-\frac{1}{r_{3,M_3}}a_3^1 d^1_{3,M_3n}+k^2 d^0_{3,M_3n}\right)\phi_{3,n}$$

$$-\left(a_2^2 d^2_{2,0n}-\frac{1}{r_{3,M_3}}a_2^1 d^1_{2,M_2n}+k^2 d^0_{2,M_2n}\right)\phi_{2,n}$$

$$-\left(N_3 W''_{3,M_3}-N_2 W''_{2,M_2}\right)\xi_0=0, \tag{A14}$$

$$\left\{\frac{\bar{N}_1}{Re}\left(a_1^3 d^3_{1,0n}-\frac{1}{r_{1,0}}a_1^2 d^2_{1,0n}\right)-\left[\frac{\bar{N}_1}{Re}\left(\frac{1}{r_{1,0}^2}-3k^2\right)+ikQ_1 W_{1,0}\right]\right.$$

$$\times a_1^1 d^1_{1,0n}+\left(\frac{\bar{N}_1}{Re}\frac{2k^2}{r_{1,0}}+ikW'_{1,0}\right)d^0_{1,0n}\Bigg\}\phi_{1,n}-\left\{\frac{1}{Re}\left(a_2^3 d^3_{2,0n}\right.\right.$$

$$\left.-\frac{1}{r_{1,0}}a_2^2 d^2_{2,0n}\right)-\left[\frac{1}{Re}\left(\frac{1}{r_{1,0}^2}-3k^2\right)+ikW_{2,0}\right]a_2^1 d^1_{2,0n}$$

$$+\left(\frac{1}{Re}\frac{2k^2}{r_{1,0}}+ikW'_{2,0}\right)d^0_{2,0n}\Bigg\}\phi_{2,n}+ikWe_i^{-1}$$

$$\times\left(\frac{1}{r_{1,0}}-r_{1,0}k^2\right)\xi_i=\omega\left(Q_1 a_1^1 d^1_{1,0n}\phi_{1,n}-a_2^1 d^1_{2,0n}\phi_{2,n}\right). \tag{A15}$$

$$\left\{\frac{1}{Re}\left(a_2^3 d_2^3,_{M_2 n} - \frac{1}{r_{3,0}} a_2^2 d_2^2,_{M_2 n}\right) - \left[\frac{1}{Re}\left(\frac{1}{r_{3,M_3}^2} - 3k^2\right)\right.\right.$$

$$\left. + ikQ_2 W_{2,M_2}\right] a_2^1 d_2^1,_{M_2 n} + \frac{1}{Re}\frac{2k^2}{r_{3,M_3}} + ikW_{2,M_2}\, d_2^2,_{M_2 n}\Bigg\}\phi_{2,n}$$

$$- \left\{\frac{\bar{N}_3}{Re}\left(a_3^3 d_3^3,_{M_3 n} - \frac{1}{r_{3,M_3}} a_3^2 d_3^2,_{M_3 n}\right) - \left[\frac{\bar{N}_3}{Re}\left(\frac{1}{r_{3,M_3}^2} - 3k^2\right)\right.\right.$$

$$\left. + ikQ_3 W_{3,M_3}\right] a_3^1 d_3^1,_{M_3 n} + \left(\frac{\bar{N}_3}{Re}\frac{2k^2}{r_{3,M_3}} + ik_3 W'_{3,M_3}\right) d_3^0,_{M_2 n}\Bigg\}\phi_{3,n}$$

$$+ ikWe_i^{-1}\left(\frac{1}{r_{3,M_3}} - r_{3,M_3}\, k^2\right)\xi_0 = \omega\left(a_2^1 d_2^1,_{M_2 n}\,\phi_{2,n}\right.$$

$$\left. - Q_3 a_3^1 d_3^1,_{M_3 n}\,\phi_{3,n}\right),\tag{A16}$$

$$d_1^0,_{M_1 n}\,\phi_{1,n} = 0,\tag{A17}$$

$$a_1 d_1 1,_{M_1 n}\,\phi_{1,n} = 0,\tag{A18}$$

Equations (A4) to (A18) constitute a system of $(M_1 + M_2 + M_3 + 17)$ algebraic equations. The $(M_1 + M_2 + M_3 + 5)$ unknown vector components are $\phi_{1,0}\cdots\phi_{1,M_1}$, $\phi_{2,0}\phi_{2,1}\cdots\phi_{2,M_2}$, $\phi_{3,0}\phi_{3,1}\cdots\phi_{3,M_3}$, ξ_i, and ξ_0. Equation (A4) comes from three fourth-order ordinary differential equations. Each of these equations results in $M_j + 1$ equations; only $M_j - 3$ of which are independent. Removing four equations from each set of governing equations, and adding the 14 boundary conditions, we obtain our final system. We remove equations with $m = 0, 1, M_j - 1, M_j$ $(j = 1, 2, 3)$ in (A4). This is the so-called Lanczos method (1956). The elements of the **B** and **A** matrix in (10.26) can be read off from the coefficients of the unknown vector components. The **B** matrix involves only terms with ω.

The references cited in this appendix can be found in Chapter 10.

Appendix B

Eigenvalue Solution Algorithm

For purely temporal disturbances k is real, and if it is given together with the complete set of the flow parameters, then we can solve the complex wave frequency ω from (10.26) as the eigenvalue. The IMSL library routine GVLCG has been used to obtain ω. We characterize the spatial-temporal disturbances of a given wave number k_r for a given set of flow parameters with the spatial amplification curves $\omega_r = 0$. There are at least two such curves for a given set of flow parameters in the case of convective instability. One corresponds to the sinuous mode and the other to the varicose mode. For each mode, we start with an initial guess of k_i for a given k_r. Then solve for ω_r and ω_i using the IMSL routine GVCCG. If $\omega_r = 0$ the guess was perfect, if not we find k_i by using the Newton integration method with a reduced value of $|\omega_r|$. With the new k_i and the original k_r we update (ω_r, ω_i) by means of the IMSL routine GVCCG. We repeat this procedure until the IMSL routine gives $\omega_r = 0$.

This iterative scheme requires a very close initial guess of k_i for convergence to the amplification curve $\omega_r = 0$. The close initial guess of k_i is achieved as follows. First, we seek the downstream propagating convectively unstable branches. We start our search with a negative k_i in the $k_r - k_i$ plane. The image of k with negative k_i should be an ω with positive ω_r. Then the causality condition that the disturbance vanishes as $\tau \Rightarrow -\infty$ in $z < 0$ is satisfied. With the given k, we obtain ω with subroutine GVLCG in the IMSL library. Two sets of (ω_r, ω_i) with $\omega_r > 0$ are found. We then increase k_i step by step until ω_r changes sign. The value of k_i that causes ω_r to change sign near $\omega_r = 0$ is used as the initial guess for each mode. Once an accurate point on the amplification curve $\omega_r = 0$ is obtained with the iteration method described earlier, a sufficiently close point on the same curve can be obtained with k_i on the neighboring point without having to obtain the first guess of k_i as described. Similarly the upstream branch $\omega_r = 0$ can be obtained by starting the initial guess of k_i with $k_i > 0$. When the downstream and upstream branches touch each other in the upper k-plane a saddle point of $\omega(k)$ is formed. The

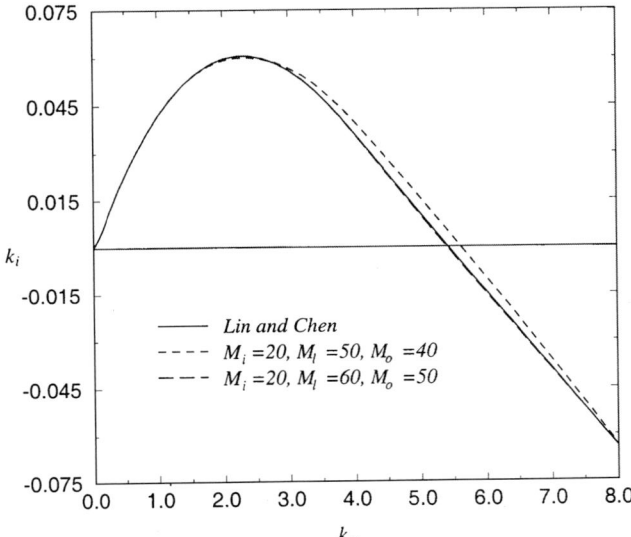

Figure B.1. Amplification curve for a simple jet. $Re = 1000$, $1/Fr = 0.0001$, $We_i = 4761.9$, $We_0 = 0.0$, $Q_1^{-1} = 0.013$, $Q_3 = 1.0$, $\bar{N}_1^{-1} = 0.019$, $\bar{N}_3 = 1.0$, $r_i = 1.0$, $r_0 = 5$, $r_w = 10$. Scaling I is used in this calculation.

image of this saddle point in the upper ω-plane is a pinch point with $\omega_r > 0$. When a pinch point singularity appears, absolute instability results.

The numbers of terms M_j retained in the series are systematically increased until the obtained eigenvalues remain the same up to the third decimal point. To test the possible syntax and computer program errors, the results for the special cases included in the present problem are checked against the known results of axisymmetric Poiseuille flow (Davey and Drazin, 1969) and the cylindrical liquid jet in a concentric pipe (Lin and Chen, 1998).

The known results for a cylindrical liquid jet in a coaxial pipe can be recovered by changing the core gas to a liquid and changing the liquid shell to the same gas as the outer gas. For example, to recover the results in Figure 10.3 of Lin and Chen (1998), we put $Q_3 = 1$, $Q_1^{-1} = 0.0013$, $\bar{N}_3 = 1$, $We_0^{-1} = 0$, $We_1^{-1} = 4761.9$, $\bar{N}_1^{-1} = 0.019$, $r_i = 1$, $r_0 = 5$, $r_w = 10$, $Re = 1000$, and $Fr^{-1} = 0.0001$. Note that the definition of the Weber number here is the inverse of that in Lin and Chen. The solid curve in Figure B.1 shows the results of Lin and Chen; the other two curves are from the present program with two different sets of M_j. The figure also shows how the terms of the series expansion can be increased to improve the numerical accuracy. The known results for the axisymmetric Poiseuille flow can also be recovered by changing all three fluids to the same fluid. This can be achieved by putting

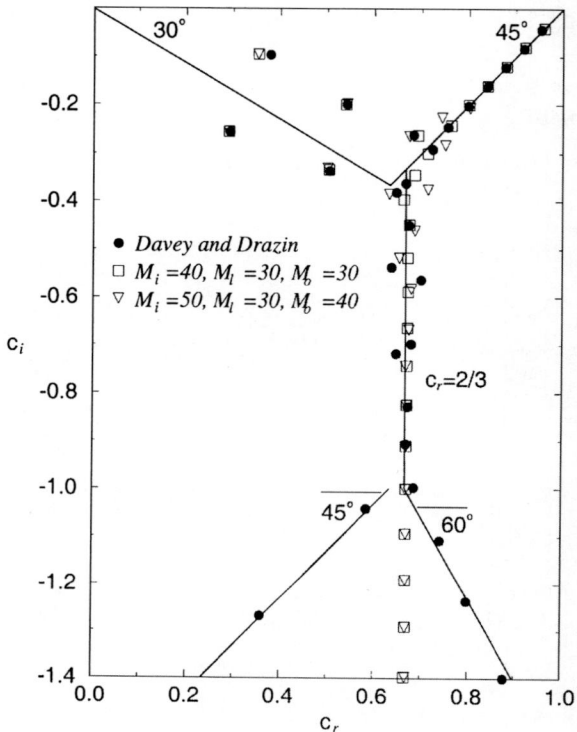

Figure B.2. Eigenvalues of axisymmetric Poiseuille flow. Values of c for $k_r = 0.5$, $k_i = 0$, $Re = 2500$, $1/Fr = 0.0$, $We_i^{-1} = 0.0$, $We_0^{-1} = 0.0$, $Q_1 = Q_3 = 1.0$, $\bar{N}_1 = \bar{N}_3 = 1.0$, $r_i = 1.0$, $r_0 = 1.5$, $r_w = 2.0$. Scaling I is used in this calculation.

$Fr^{-1} = 0$, $Q_1 = Q_3 = 1$, $\bar{N}_1 = \bar{N}_3 = 1$, and $We_i^{-1} = We_0^{-1} = 0$. To compare with the results of Davey and Drazin (1969) for the circular pipe flow, we also put $r_i = 1$, $r_0 = 1.5$, $r_w = 2$, $k_r = 0.5$, $k_i = 0$, and $Re = 2500$. Note that whereas Davey and Drazin used the pipe radius, we use r_i to normalize the distance. Thus, $Re = 2500$ and $k_r = 0.5$ in Figure B.2 correspond to their value of 5000 and 1 respectively. Davey and Drazin used the Galerkin method to obtain the complex wave speed $c = c_r + ic_i = i\omega/k_r$. The comparison is excellent for the three upper branches. However, we have not found the lower two branches of Davey and Drazin for which a theoretical explanation is still lacking (Drazin and Reid 1985). Instead, we find that the central branch $c_r = 2/3$ extends beyond their lower branches. Their two lower branches are most probably due to numerical inaccuracy associated with the Galerkin method. Figure 10.4 also shows how the number of terms M_j may be increased to improve numerical accuracy.

The references cited in this appendix can be found in Chapter 10.

Author Index

Subject Index

PROGRESS IN
INFRARED
SPECTROSCOPY
Volume 2